Good Laboratory Practice
Standards

Good Laboratory Practice

Standards

**Applications for
Field and Laboratory
Studies**

Edited by

Willa Y. Garner
Garndal Associates

Maureen S. Barge
FMC Corporation

James P. Ussary
Ussary Scientific Services

ACS Professional Reference Book

American Chemical Society, Washington, DC 1992

Library of Congress Cataloging-in-Publication Data

Garner, Willa Y., 1936–
    Good laboratory practice standards: applications for field and laboratory
studies / Willa Y. Garner, Maureen S. Barge, James P. Ussary

    p.   cm. -- (ACS professional reference book)

    Includes bibliographical references and index.

    ISBN 0–8412–2192–8

    1. Toxicology laboratories—Standards.

    I. Barge, Maureen S., 1947–   . II. Ussary, James P.
III. Title. IV. Series.

RA1199.G37 1991
615.9′.07—dc20   91–39042

1992 ACS Books Advisory Board

About the Editors

WILLA Y. GARNER received her B.S. degree in chemistry and her M.S. and Ph.D. degrees in pesticide toxicology from the University of Maryland, College Park. She is currently president of Garndal Associates, for which she performs facility inspections and data audits. She also writes and reviews protocols and standard operating procedures to ensure compliance with the Good Laboratory Practice (GLP) Standards. She conducts training courses on the GLP regulations, investigates cases of suspected fraud, and testifies as an expert witness in agricultural chemistry and GLPs.

Before founding Garndal Associates, Garner was a senior consultant for Quality Associates, Inc., in Ellicott City, Maryland. From 1985 until 1989, she worked as a chemist for the National Laboratory Audit Program and the Laboratory Data Integrity Assurance Division of the U.S. Environmental Protection Agency (EPA). From 1976 until 1985, she worked in the Office of Pesticide Programs of the EPA, first as an environmental chemist and environmental scientist and later as section chief and branch chief of the Health Effects Branch and as a science policy analyst.

Garner has published articles on agricultural chemicals, quality assurance, and GLPs. She is a member of a number of professional societies, including the Society for Environmental Toxicology and Chemistry, the American Council on Science and Health, the American Chemical Society, and the Society for Quality Assurance.

MAUREEN S. BARGE is the quality assurance supervisor for research and development in the Agricultural Chemicals Group of FMC Corporation in Princeton, New Jersey. She has been with FMC for eight years, initially with the Residue Chemistry Department and for the past five years with the Quality Assurance Unit. Barge is responsible for monitoring product chemistry, residue chemistry, metabolism, and environmental fate studies, and for the analytical support of toxicology studies.

Barge attended Rutgers University in New Brunswick, New Jersey, where she also received her initial training in pesticide residue analysis. During her eight years in the Department of Entomology's residue chemistry laboratories, Barge worked on crop residue studies for the IR–4 National Pesticide Clearance Research Program, and on worker exposure studies and water analysis studies.

As an active member of the Society of Quality Assurance (SQA), Barge has been part of the program committee and the basic training program. She is also the SQA's liaison to the American Chemical Society, and she founded and continues to work with the SQA Agrochemical Round Table Discussion Group. Barge has been instrumental in bringing the concepts of Good Laboratory Practice Standards to the agrochemical scientific arena by coorganizing two symposia for the ACS Division of Agrochemicals and a symposium for the Society of Environmental Toxicology and Chemistry, and she continues to provide a yearly basic training course for ASTM Committee E–19.

JAMES P. USSARY of Ussary Scientific Services is a private contractor working with the agrochemical industry on compliance with Good Laboratory Practice (GLP) Standards and on environmental chemistry issues. He was manager of quality assurance for ICI Americas, Inc., Agricultural Products Division in Goldsboro, North Carolina, from 1986 until 1990. During the 11 years at ICI before 1986, he was manager of residue chemistry. In 1968, he cofounded ABC Laboratories, Inc., in Columbia, Missouri, and was its president until 1975. Before founding ABC Laboratories, he was an instructor in agricultural chemistry for the University of Missouri in Columbia. He has B.S. and M.S. degrees in Agricultural Chemistry from the University of Missouri. He is the author of 20 scientific publications on analytical chemistry and quality assurance.

Ussary's experience with GLP compliance began in 1979 when the principles of the GLP regulations were integrated into the residue chemistry program at ICI's Agricultural Products Division. The program evolved from laboratory compliance to also include the field portions of residue chemistry and environmental fate studies.

Ussary served on the National Agricultural Chemicals Association GLP Subcommittee that was established to consider industrywide problems associated with applying the GLP regulations to field studies, and he served on the steering committee of the 1990 EPA–industry workshop. In May 1990, he was an industry observer at a training course for EPA inspectors of field studies. He has also conducted numerous international industry workshops on GLP compliance procedures.

Contents

List of Figures and Tables.. xv

Preface.. xix

OVERVIEW

1. Responding to Regulatory Changes in Agrochemical
   Research 3
   Joseph B. Townsend

GOOD LABORATORY PRACTICE REQUIREMENTS

2. Management Commitment to the Good Laboratory
   Practices Process.. 13
   Maureen S. Barge

3. Preparing Standard Operating Procedures
   for Field Studies... 21
   James P. Ussary

4. Preparing Standard Operating Procedures
   for the Laboratory 27
   M. Jean Hornshuh

5. Designing Protocols for Field Studies 43
   Maureen S. Barge

6. Practical Protocols for the Laboratory.. 53
   Gene Burnett

7. Test Substance and Specimen Chain of Custody
   in Field Studies... 85
   Markus M. Jensen

8. Practical Applications for Chain of Custody
   Within an Analytical Laboratory 95
   Robert J. Pollock

9. Test Substance and Analytical Reference
   Standard Characterization and Accountability........................... 113
   S. Rand Fuller and Willa Y. Garner

10. Reporting Study Results 127
    Judy H. Hochman and Willa Y. Garner

QUALITY ASSURANCE RESPONSIBILITIES

11. The Quality Assurance Unit: The Master Schedule
    and the Archives.. 143
    Patricia D. Royal

12. Facility Inspection Conduct: Quality Assurance
    Perspective for Field Studies 155
    David Johnson and Jesse Burton

13. Analytical Phase Inspection of Residue Chemistry
    and Environmental Studies.. 163
    Frances A. Dillon and Jeffery L. Harris

14. Data Audits for Field Studies.................................. 175
    Del W. Huntsinger

15. Laboratory Data Auditing................................... 181
    Harry L. Hyndman

COMPUTER VALIDATION

16. Working Models of Computer System Validation
    and Verification... 203
    Phillip M. Buckler

17. Validation of Laboratory Computer Software........................ 217
    J. Drew Watson

SPECIALTY STUDIES

18. **Product Chemistry from the Formulation Perspective** 227
    John F. Wright

19. **Compliance for Routine and Nonroutine Field Studies** 235
    James P. Ussary

20. **Nonbiased Field Sampling** 241
    Duane D. Ewing

21. **Processing Studies** 249
    Malcolm F. Gerngross

22. **Livestock Studies** 257
    Gino J. Marco, Roger A. Novak,
    and Judy H. Hochman

23. **Specialized Field Testing: Worker Protection** 279
    D. Larry Merricks, Donna H. Merricks,
    and William C. Spare

24. **Mesocosm Studies and Other Aquatic Field Studies
    with Pesticides** 297
    Jeffrey M. Giddings

25. **Terrestrial Field Studies** 309
    Mary E. Johnson and Mark Jaber

26. **Problems with Ecotoxicological Field Studies** 317
    John A. McCann

27. **Agrochemical Groundwater Studies** 333
    Sandra C. Cooper and James M. DeMartinis

28. **Runoff Studies** 343
    Peter N. Coody

REGULATORY IMPACT

29. **Auditing Field Studies: Government Perspective** 361
    Dean F. Hill

30. Is the Federal Insecticide, Fungicide, and Rodenticide Act
Good Laboratory Practices Program at a Crossroads? 375
David L. Dull and Francisca E. Liem

31. Good Laboratory Practice Standards Policies
and Interpretations.. 387
Phyllis E. Flaherty and Stephen J. Howie

32. Economic Impact of Regulations on Field Contractors
and an Agrochemical Company.. 399
James L. Platt, Jr.

33. Harmonization and Prospects for the Future............................ 419
Frederick G. Snyder

APPENDIXES

A. Federal Insecticide, Fungicide, and Rodenticide Act;
GLP Standards 433

B. U.S. Environmental Protection Agency FIFRA Advisories 445
Willa Y. Garner

C. The EPA Enforcement Response Policy for the FIFRA
Good Laboratory Practice Standards... 475

D. Representative Forms Used by Companies for Compliance
with GLP Standards 501

INDEXES

Author Index 557

Affiliation Index 557

Subject Index.. 558

xiv

List of Figures and Tables

Table 6.I.	Pesticide assessment guidelines	57
Table 7.I.	Problems with test substance shipments	87
Figure 8.1.	Master log	98
Figure 8.2.	Analytical reference standard receipt log	99
Figure 8.3.	Analytical reference standard storage log	101
Figure 8.4.	Analytical balance log	102
Figure 8.5.	Field sample transmittal form	104
Figure 8.6.	Freezer log	105
Figure 8.7.	Instrument log for a GC	108
Figure 8.8.	Instrument log for an HPLC	110
Figure 9.1.	Chemical receipt and use log	117
Figure 9.2.	Label formats for reagents and solutions	124
Figure 10.1.	Format for listing study personnel	129
Figure 10.2.	Suggested format for reporting information on test substances and analytical reference standards	129
Figure 10.3.	Format for the quality assurance statement	130
Figure 10.4.	Sample title page	134
Figure 10.5.	Format for "Statement of (No) Data Confidentiality Claims"	135
Figure 11.1.	The responsibilities of the QAU extend to many different people	145
Table 11.I.	Examples of generic phases of studies	148
Figure 16.1.	Data collection system for a scintillation counter	206
Figure 16.2.	Sample data from a scintillation data-base system	208
Figure 16.3.	Data collection system for a scintillation counter with a verification printer	209
Figure 16.4.	The computer acquisition system	210
Figure 16.5.	Sample chromatogram from the peak generator	212
Figure 17.1.	Software development cycle	218
Figure 17.2.	Validation of software	220

Figure 17.3. During the performance audit step of the validation process, the "black box" must check program outputs for given inputs and must reject invalid input 220

Figure 20.1. Randomized sampling design for terrestrial dissipation studies 243

Figure 20.2. Terrestrial dissipation study design when samples are collected randomly within three subplots 244

Figure 20.3. A terrestrial dissipation study design for tractor-mounted soil probes 245

Figure 20.4. Sampling design for a dislodgable residue study on citrus trees using eight trees and offset sampling points 247

Figure 20.5. Sampling design for a dislodgable residue study on citrus trees using five wholly sampled trees 247

Table 22.I. Raw agricultural commodities and feeds derived from field crops 260

Table 22.II. Typical animal sample amounts 264

Figure 22.1. Detailed quality assurance unit checklist 266

Figure 22.2. Decision-making scheme for metabolite identification 268

Figure 22.3. Fractionation scheme for plant and animal material 269

Table 22.III. Feeding trial dosing 271

Figure 22.4. Outline of a livestock study 274

Figure 22.5. Quality assurance protocol-evaluation form 276

Figure 23.1. Sprayer specification sheet 282

Figure 23.2. Informed consent form 288

Figure 23.3. Pump calibration form 290

Figure 23.4. QA inspection checklist for worker-exposure field studies 291

Figure 23.5. Sample or specimen transfer form 293

Figure 23.6. Chain-of-custody form 294

Figure 28.1. Example chain-of-custody form completed to verify shipment of a single field sample 349

Figure 28.2. Example flow meter calibration form for use in runoff studies 352

Figure 28.3. Example sampling grid and random number sequence used to locate random soil sampling locations 354

Figure 32.1.   Distribution of respondents to the 1990
               contract field laboratory survey 402
Figure 32.2.   Number of employees at facilities 402
Figure 32.3.   Number of studies conducted annually 403
Figure 32.4.   Portion of studies that are regulated
               by GLP Standards 403
Table 32.I.   Question 4 responses: Distribution
               of GLP-regulated program 404
Figure 32.5.   Distribution of studies regulated by GLP
               Standards 404
Figure 32.6.   Distribution of cost increases due to GLP
               implementation 405
Figure 32.7.   Staff time spent on GLP issues 406
Figure 32.8.   Improvements due to GLP implementation 407
Figure 32.9.   Activities ranked as the Number 1 staff
               time consumer 408
Figure 32.10. Suggested GLP training sources 410
Table 33.I.   Summation of agreements 426
Table 33.II.   Japanese Good Laboratory Practice
               Standards 428

Preface

THE REVISIONS TO the Federal Insecticide, Fungicide, and Rodenticide Act Good Laboratory Practice (GLP) Standards became effective on October 16, 1989. The objective of the revisions was to expand the scope of the GLPs to encompass all studies submitted to the Environmental Protection Agency in support of pesticide registration. The agrochemical industry struggled with the proposed regulations for several years but was generally in compliance when the regulations were finally enacted. However, GLP compliance is a dynamic process that must be diligently cultivated. The chapters in this book were written by experienced quality assurance (QA) professionals and field and laboratory researchers with the intent of providing concrete ideas for establishing a compliance program and refining the compliance process. This book is more than just a philosophical treatise; it was designed to present practical applications for compliance with GLPs. We hope that this text will be useful to QA professionals, field and laboratory researchers, and managers alike.

The contributors and reviewers gave generously of their time and expertise to make this valuable reference possible. As editors of this publication, we are indeed grateful. We give special thanks to the editorial staff of ACS Books for their support of this effort.

WILLA Y. GARNER
Garndal Associates
Monument, CO 80132

MAUREEN S. BARGE
FMC Corporation
Princeton, NJ 08543

JAMES P. USSARY
Ussary Scientific Services
Goldsboro, NC 27534

December 1, 1991

Overview

Responding to Regulatory Changes in Agrochemical Research

Joseph B. Townsend

Bio/dynamics Inc., Mettlers Road, East Millstone, NJ 08875

The application of the good laboratory practice (GLP) regulations to agrochemical research in 1989 by the Environmental Protection Agency is destined to require profound changes in the manner in which this research is conducted. Although the response in other disciplines should have guided the regulated as well as the regulatory agency in efficiently implementing these new rules, apparently, this has not routinely been the case. The benefits to be obtained in agrochemical research over the next decade are described on the basis of experience gained with GLPs in other disciplines. Several suggestions are made that should assist those attempting to bring their operations into compliance with the GLPs for the first time.

A TIME OF GREAT CHANGE IS OCCURRING in the agrochemical industry. This change is in the form of the new Good Laboratory Practice (GLP) Standards that have been recently applied to agrochemical research. Although new standards have recently been instituted for other areas of chemical research as well, the focus of this book is on agrochemicals. These regulations are not new; are not necessarily good; do not apply just to laboratories; and in the strict sense of the word, are not standards; but they now have the effect of law. But take heart, there is good news: These same regulations have previously been applied to other research disciplines, and those disciplines survive to this day.

This chapter describes some of the regulatory events that have occurred since the Environmental Protection Agency (EPA) finalized the

2192-8/92/0003$06.00/0

GLP regulations in October 1989, draws some parallels with past events, and then provides suggestions on coming into compliance with the regulations.

Regulatory History

Until recently there has been a pattern to governmental regulation of applied research. Historically, restrictions on research were most often instituted after a clear and predictable chain of events.

1. A scandal or catastrophe, or sometimes simply the perception of a catastrophe promulgated by the press, occurred.

2. Demagogues, many of whom were politicians concerned mainly with reelection, blamed the industry and the government agency with oversight responsibility for allowing this catastrophe to happen.

3. The responsible agency, often with good reason, announced that the problem would not have occurred if the agency had more and better resources, and the agency promised to crack down on industry.

4. Congress and the administration felt compelled to respond and enacted legislation to correct the transgressions, usually without allocating sufficient funds for the legislation to be enforced effectively.

5. The affected agency dutifully initiated new regulations.

6. The regulated community began its struggle to understand and comply with these constraints until a new crisis arose and the cycle began anew.

Although the promulgation by the EPA of the GLP Standards that will govern research on agrochemicals followed this pattern to some extent, there were for the first time distinct differences. No new massive catastrophic event or scandal occurred to captivate the attention of Congress or the public and to propel the EPA to action. The scandals that are now surfacing in the area of analytical chemistry occurred after the fact and did not cause the EPA to apply the GLPs to agrochemical research. The GLP regulations were instituted essentially between scandals. Politicians have not made a separate issue out of agrochemical research. No new legislation forced the EPA to act.

A number of reasons caused the EPA to apply these constraints to the agrochemical industry. The most compelling reason was the result of a new and subtle type of public activism that has been growing for years—a demand for action that differs from the activism of the 1960s and

1970s. Environmental activism was evident in those years, but today the difference is a public demand for accountability. The GLP Standards are just one of many requirements that pesticide researchers will be obliged to address in the next decade. This new activism is characterized by the proliferation of sophisticated pressure groups, legal actions, and political activity, all carefully planned and orchestrated on a scale not seen even 20 years ago when the EPA was first established as an independent agency. The result is that the public is setting the agenda; citizens are no longer content simply to allow government officials, and the attendant government bureaucracy, to act for them, particularly when the environment is at stake. When the environment is involved, this new activism demands a response from the government. As never before, the government is following rather than leading. Members of the agrochemical industry must recognize that the EPA was compelled to apply the GLPs, whether they clearly recognized it or not, and that public pressure will compel the EPA to apply further restrictions in the future. Compliance with the GLP Standards is a public demand, and the EPA is simply the messenger. In other words, the GLP Standards were inevitable.

The response of the EPA and the regulated community to this new public activism was, and continues to be, disappointing. The promulgation of the GLP Standards by the EPA and the response of the regulated community was not very different from the response to regulations enacted in the past. Apparently, little was learned from past experience. The response of the agrochemical industry to these new regulations seemed to ignore the sense of urgency placed on them by the public.

For example, as in the past, apparently many in the industry are slow in coming into compliance with the regulations. Although many companies began to institute GLP programs when the regulations were proposed, a number of companies erroneously assumed that compliance could be accomplished overnight, and these companies waited for finalization of the regulations before instituting compliance programs. These companies essentially wasted the 2 years allowed for coming into compliance by not seriously addressing the issues. The changes to the Federal Insecticide, Fungicide, and Rodenticide Act (FIFRA) regulations were published for comment in December 1987, and the changes became final in October 1989, a period of a little less than 2 years, yet many in the industry appeared to be shocked when the regulations became final and compliance became mandatory.

I have had occasion over the past several months to review facilities and to observe the reactions of some members of the agrochemical community to the advent of these regulations. When organizations are found not to be in compliance, the questions are why did they not react in 1987 instead of in 1989 and why did they not use the available 2 years to come into compliance? Why, for example, did the issue of one study director versus multiple study directors become so important in 1989? By the time the question was raised, the official comment period had expired. This issue could have been discussed with the agency, and perhaps even settled, or at least made more palatable, in the 22 months allotted for discussion. With the realization that compliance is neither easily nor quickly attained comes the desire to have those two years back.

Many individuals still do not accept these new regulations and continue to argue their merit. Although most industry groups are genuinely trying to understand and to come into compliance with the GLPs, there are some individuals who do not understand that the time for questioning the wisdom of the EPA for even proposing these regulations is over and that the time for acceptance and compliance is at hand.

In 1976 and 1977, and again in 1982 and 1983, when the Food and Drug Administration (FDA) and EPA GLP regulations were applied to toxicological research, the reaction was the same as that of the agrochemical industry today. Some toxicological researchers refused to comply. Many of the comments heard today were heard then: "These GLPs will make the cost of research prohibitive", "I will spend all my time doing paperwork, when will I do research?", or "I will not be a study director because I cannot be everywhere at once and I cannot be a master of all the disciplines you are asking me to supervise." "Management"[1] did not move quickly to enlighten the dissenters then, and in many instances, management is not doing it now.

Anyone who has been involved in implementing the GLPs may have heard individuals castigate the EPA for instituting these regulations. This reaction also occurred when the FDA first implemented the GLPs and in 1962 when major changes were made in drug research. Dissenters must be made aware that the time for questioning the wisdom of the regulations is over and that the time for compliance has begun. Compliance is not difficult once the process is understood, and there are benefits to compliance. If you recognize yourself in my characterization, you may want to take some time and reorder your priorities, not only because your opinions can be costly to your company, but also because they can threaten your career. A winnowing process will take place during the coming decade as it did in toxicological research in the 1980s. The greatest success will go to those who are the first to learn to use the GLPs. The agrochemical industry will find that it cannot afford anything less than full acceptance of the GLP Standards by its employees both in substance and in spirit.

The EPA also has reacted in essentially the same manner as it did when it instituted new regulations in other disciplines. Instead of profit-

---

[1]Management is defined here as the person or persons responsible for setting policy at a facility. For a description of the duties of testing facility management, see § 160.31 of the GLP Standards.

ing from previous experience and being innovative in assisting the industry in complying with the GLPs, evidently little forethought was given to the impact of these regulations, and even less effort was expended in attempting to help the regulated community understand and comply with the regulations. In essence, the EPA appeared simply to publish the regulations with heavy reliance on the FDA regulations and interpretations and then waited for the comment period to expire without giving much thought to what interpretations would be required of it and with the intention of using fines to enforce compliance. During recent public sessions dedicated to discussions of the GLPs, some EPA spokespersons, who should be conversant with the regulations, were apparently caught off guard by some very obvious questions.

Although there are indications that the EPA is now moving to rectify these oversights and is setting up mechanisms to help industry comply with the regulations, I believe that the EPA had an implied responsibility to educate industry, or at least to describe the interpretations they intended to make, before the regulations became final. The EPA is providing oversight to an industry that needs the EPA's support and assistance possibly more than any of the other industries for which the EPA has responsibility. The EPA should work more to assist the agrochemical research community and, except for fraud or refusal, should reject the use of punitive measures as compliance tools.

Meaning of Compliance

Compliance with GLP Standards requires an understanding of the regulations. The GLP Standards are a new set of rules that will permeate and change the way that work will be performed in the laboratory and in the field. These rules, which take up only seven pages in the *Federal Register*, will call into question almost all of the professional work habits and procedures that have been learned and developed over the length of a career. Other people, many of whom are not even trained in specific fields of research, will be required to review and pass judgment on that work. New techniques must be learned, and the real meaning of terms such as documentation, audit trail, raw data, test system, and "experimental start date"[2] must be understood. Protocols must be planned in greater detail, and must be carefully documented before proceeding with the research. Puzzling requirements, such as saving empty storage containers or including expiration dates and storage conditions on innocuous reagents, must be followed. Resumes alone will no longer be sufficient proof of ability.

[2]The EPA defines "experimental start date" as the first date the test substance is applied to the test system.

Agrochemical research personnel will be required to demonstrate their proficiency in the tasks that they are expected to perform.

All of this and more is involved in compliance with GLPs, but the benefits will be remarkable. Scientific work will be much more precise, and findings will be clearly supportable. Experiments will be easily understandable and reproducible. Once the GLPs have been understood and applied and once a laboratory has been brought into compliance with the GLPs, research costs will be no greater than they are now and may possibly be lower.

Recently I reviewed data from an analytical chemistry study that was conducted in the mid-1970s before the advent of the GLPs and therefore before our laboratory was in compliance. I was shocked at what we had accepted as a study only 15 years ago, and I came to the realization that only in retrospect can the changes that the GLPs have made be clearly seen.

To understand the benefits of GLP compliance, those of you who are in compliance with the GLPs or who are well on the way to compliance should compare a pre-GLP study to a recent comparable study conducted in compliance with the regulations. The comparison can be a convincing argument for the benefits of GLP compliance.

Although the GLP Standards are not perfect, not only will they markedly improve the quality of scientific work industry-wide and raise public confidence in agrochemical research, but also they will assist in weeding out the negligent, the incompetent, and the fraudulent. Agrochemicals are not viewed favorably in all segments of our society, even in the face of serious dislocations in the worldwide food supply. Some of the predictions for population growth and food supply in the next century are bleak, and the question of whether the public approves of agrochemicals is less important than the fact that agrochemicals provide one of the few hopes we have that we may be able to postpone the predictions of Malthus. In trying to convey this message to the public, incidences of negligence and fraud do not help the image of the agrochemical industry. The GLPs do not and cannot totally eliminate sloth or malfeasance, but they do raise the price and they reduce temptation dramatically. In the GLP Standards, safeguard provisions, such as requirements for accountability and third-party oversight, make dishonest and slothful research difficult and protect those who are trying to work in a challenging discipline.

Practical Methods for Complying with the GLP Standards

In the first part of this chapter, compliance with GLP regulations was discussed; the remainder of the chapter deals with attaining compliance.

Several general suggestions that would facilitate implementation of the GLPs in any laboratory are contained in the following list.

1. If it has not already been done, as soon as possible a program should be initiated in your laboratory or research organization, possibly in conjunction with your GLP training sessions, to encourage everyone to understand and accept the GLPs. Your organization must demonstrate that compliance is a serious matter and that noncompliance is unacceptable. Because silence or half-hearted efforts at compliance are often interpreted as disinterest or even resistance, your organization or company must emphatically show where it stands, and it must do so early. The price paid for a misunderstanding could be catastrophic. Full compliance is a long-term task that, even with everyone's cooperation, is difficult. Compliance is possible only when everyone is convinced that it is to their advantage to be in compliance.

Your company must do its best to convince its employees that conformance with the GLP Standards is not simply a matter of abiding by the law, but that significant benefits will result for your organization and for each individual within your organization. If an organization does not immediately make a specific, clear statement as to its expectations in regard to the GLP Standards, it in essence has made a statement that will eventually be regretted.

2. Make sure that your quality assurance unit (QAU) is staffed with competent, experienced, interested people, and then use them. Have the QAU personnel point the way to compliance with the GLPs. One of the implicit jobs of the QAU is to train and assist all laboratory personnel in GLP matters and to provide interpretations of the regulations. The quality of the QAU is crucial to the success of an organization and may provide a measure of how well the organization is complying with the law. During an inspection of a laboratory, for example, I find that an assessment of the competence of the QAU can often reveal the management's commitment to GLP compliance. The QAU is no place to park untrained, disinterested people who do not have a clear commitment to quality assurance (QA), nor should the responsibility be assigned as an additional duty to a busy scientist who has little interest in the task. At the start, it is important that your QA officer be trained and experienced in the profession and completely conversant with the GLPs.

Before leaving the subject of quality assurance, some precautions should be mentioned. Do not expect your QAU alone to be responsible for GLP compliance. Do not expect QA personnel to be your safety net; they are human, and they make errors. The primary responsibility for GLP compliance still remains with the scientist.

3. Make certain that management at all levels are at least conversant with the GLPs. Several sections of the GLP Standards describe the responsibilities of management. Although there are exceptions, management is probably the most poorly informed group about the regulations in

the laboratory. Although no one realistically expects all management people to be thoroughly conversant with every aspect of the GLPs, all persons responsible for operation of an organization must know that there are new rules governing the conduct of their business. There are clear rules for management, and there are penalties for noncompliance.

4. Establish procedures and assign responsibilities to provide for the regulatory interface between your laboratory and the EPA. Establish, for example, what your inspection policies will be, who will host regulatory inspections, who will be authorized to speak for the company, and what in your company is considered proprietary. Conduct mock inspections and make sure that the person assigned the task of hosting inspections has an in-depth knowledge of not only the GLPs but also the interpretations made by the EPA. Consider your QA officer for this job.

Because the resources provided to the EPA are extremely limited, the knowledge and understanding of the EPA inspectors often varies; this lack of consistency requires that you closely examine the inspector's findings, contest them if there is a misunderstanding, and provide explanations where needed. Your objective, however, should truly be one of assistance and cooperation, not antagonism and confrontation.

Since 1977, my company has been inspected with monotonous regularity by the EPA. After some initial nervousness, we have found that give and take with the agency is possible and that the EPA's inspections can indeed be of benefit to us. Although some of the findings made by the inspectors have been inappropriate or incorrect, many findings have been quite appropriate and have pointed out areas in the laboratory that clearly needed attention. Inspections provide us with a measuring device by which we can gauge our progress toward full compliance with the GLPs, and they can do the same for other organizations. As confidence in your laboratory's level of GLP compliance grows, your fear of inspections will decline.

Conclusion

Agrochemical research in all respects will never again be the same. The agonies we will experience in the next decade, if the effects of the GLPs on toxicological research in the 1980s can be used as a guide, will be excruciating, but the results of our efforts, the acceptance of the GLP Standards as a necessary and functional part of our jobs, will pay extraordinary dividends for ourselves, for our companies, and for our ultimate boss, the public.

RECEIVED for review September 17, 1990. ACCEPTED revised manuscript August 17, 1991.

Good Laboratory Practice Requirements

Chapter

2

Management Commitment to the Good Laboratory Practices Process

Maureen S. Barge

FMC Corporation, P.O. Box 8, Princeton, NJ 08543

The primary necessities for a successful Good Laboratory Practice (GLP) Standards program are the understanding and commitment of all levels of "management"[1]. Just as business direction and ethical climate are dictated by upper-level management, the objectives of the company's GLP program must receive management direction. Only managers can allocate the resources necessary to maintain a GLP program and can instill a sense of importance to the program. This direction and commitment instilled by upper-level managers must then be implemented with the same conviction by middle managers or department managers. Companies cannot hire a quality assurance staff and then absolve themselves of the entire program. The quality assurance unit can be responsible for monitoring the GLP program and can be a consultant on GLP issues, but it cannot control and manage the entire program. To dis-

---

[1]Management is defined here as the person or persons responsible for setting policy at a facility. For a description of the duties of testing facility management, see § 160.31 of the GLP Standards.

2192-8/92/0013$06.00/0 © 1992 American Chemical Society

cuss attaining success it is necessary also to discuss failure. Where have we failed in the past? How can we build in success and how can we continue to keep management's interest to ensure success in the future? These topics are discussed in this chapter.

The History of Good Laboratory Practices Programs

Most Good Laboratory Practices (GLP) programs began as a result of an Environmental Protection Agency (EPA) or Food and Drug Administration (FDA) mandate, and unsurprisingly, these programs were resisted and resented by industry. The force-feeding of the GLP Standards created major compliance and management problems. The frustration with compliance that laboratories experienced when the GLP Standards originated for health-effects studies mushroomed when the regulations were expanded to include analytical, residue, metabolism, and environmental studies but were not modified or clarified accordingly.

During the developmental years of GLP programs, many companies and laboratories hired or created positions for quality assurance (QA) specialists or auditors. The new QA professionals were charged with developing and implementing GLP programs, although in many cases the programs were called QA programs. The QA professionals rarely received adequate formal training or guidelines. Even worse, the QA professionals were rarely given the proper authority, credibility, or support.

Not unexpectedly, the classic time and money arguments surfaced. Often, priorities were set by managers who may not have been fully committed to the GLP process, and projects that affected profits, like meeting EPA registration deadlines, were given priority over developing and maintaining a costly, time-consuming GLP program.

Despite these handicaps, QA units (QAU) attempted to develop GLP programs; standard operating procedures (SOPs), inspection and auditing procedures, and methods for archiving were developed. In many cases, the early programs were not allowed to mature, and the result was either rubber-stamp programs or programs that fueled antagonism between the QAU and laboratory personnel. Unfortunately, such conditions still exist today in many companies and laboratories.

Transition and the People Factor

As with many major changes, the people factor often made the transition to GLP programs an emotionally charged issue, thus making success difficult to achieve. People who were accustomed to working without the restrictions of GLP Standards (*1*) seemed to go through a period of mourning for the old ways. These people seemed to experience all four emotional stages of mourning:

1. Shock: First, people quit thinking and did little work. They were in a state of disbelief and seemed to band together for reassurance.

2. Denial: A period of anger and denial arose, coupled with resistance. People tried to hold on to the old ways and to dwell in the past.

3. Acknowledgment: The denial period was followed by a sense of sadness over the loss of the past and a period of "letting go". During this period, people began to look for ways to make the required changes work.

4. Adaptation: Finally came the time when people became ready to establish new routines and to do what was called for, that is, to comply with the GLP Standards.

To add to the transition problems, not everyone progressed through the four stages of mourning at the same pace. The transition period was a frustrating time for everyone involved. First-line managers, many of whom were also going through the mourning period were especially frustrated; they were often caught between a QAU expecting cooperation and a scientific staff demanding support from their manager.

Quality Assurance Programs vs. Good Laboratory Practices Programs

One of the major flaws in the development of GLP programs was the lack of any differentiation between GLP and QA. The QA requirements in the GLP Standards comprise less than one page of text; the remaining pages address laboratory compliance with the GLP Standards and the responsibilities of management. Briefly, the GLP Standards require testing facilities to establish a QAU or function whose major responsibilities are monitoring studies for compliance with the GLP Standards and notifying managers in writing of study status with respect to the GLP Standards. The GLP Standards detail some specific requirements for accomplishing these responsibilities, such as maintaining inspection schedules, carrying out periodic inspections of studies, providing written documentation and records of such inspections, and submitting written status reports to the managers and the study director.

An implied QA responsibility is the development, implementation, and documentation of QA procedures dealing with the day-to-day operations of the QAU. However, the GLP Standards neither require nor imply that a QAU is, or should be, responsible for the day-to-day management of laboratory studies, laboratory practices, or laboratory personnel. Who, then, is responsible for developing and implementing the GLP program and managing day-to-day laboratory operations?

Managers of the Good Laboratory Practices Program

Like any other important company policy (safety, for example), a company's GLP program must be manager-driven to succeed. If "managers" are defined as those within a company who control money, people, and resources, then clearly the responsibility for developing and enforcing a successful GLP program properly belongs to the department or laboratory managers who directly control the money, people, and resources involved in day-to-day laboratory operations.

The benefits of changing to a manager-driven program are clear. The resulting GLP programs can be tailored to the specific nature of the studies conducted by that department. Also, the manager would not be caught between laboratory personnel and QA because the manager's objective would be to achieve compliance.

Because the new programs could be developed with input from each department's laboratory scientists, the scientists would more readily accept the GLP program. The scientists would feel assured that they are considered the experts in their particular field, and because no one likes to be associated with mediocrity, they would feel a greater responsibility for the quality of the program.

The QA staff should act as consultants, auditors, and inspectors in conjunction with the laboratory compliance effort.

Engineering the Change

Commitment of management to the GLP program must be clear. Managers must accept the fact that the GLP Standards are the building blocks toward quality science and that compliance with these standards will enhance the company's chances of attaining regulatory success. Although compliance with the GLP Standards does not guarantee good science, a good GLP program will make the good science, or the lack thereof, more visible. Management could issue a philosophy statement regarding compliance with GLP Standards, and this statement could clarify the roles of managers, scientists, and QA professionals. In addition, managers should emphasize that success with GLP compliance requires teamwork.

Education is one of the keys to successfully engineering the change to a manager-driven GLP program. All levels of management should attend courses on the requirements of compliance with the GLP Standards. Managers must develop a working understanding of the concepts of the GLP Standards if a manager-driven system is to work. All levels of laboratory personnel, particularly new employees, must be similarly educated by managers and QA professionals if employee input and cooperation are expected. Everyone involved must clearly understand the rules that must be followed when submitting a study to the EPA. Everyone, regardless of job function, is critical to obtaining and maintaining the EPA registrations that are required for agrochemicals to reach the marketplace.

In order to assist in the development, refinement, and maintenance of the GLP program and to allow realistic and functional cooperation between QA staff and laboratory staff, QA staff members should attend scientific staff meetings regularly. This interaction should enable laboratory personnel to view the QA staff as a part of the team and in addition give QA staff some scientific perspective.

Continual refinement of the GLP process is necessary. The GLP Standards and related compliance programs should be a frequent agenda item at the staff meetings of upper-level managers. The cost of a GLP program is significant, but the cost of noncompliance is far greater. The possibility of fines for noncompliance is a threat, but the cost of redoing a study affects the time line for marketing and sales, and ultimately, company profits.

Unfortunately, there are always people who do not understand the initial message no matter how clearly it is given. For individuals who do not or cannot accept the fact that compliance with the GLP Standards is a fact of life, managers have the option of giving those individuals the opportunity of hearing the message again at yearly performance appraisals. This type of leverage is best used as a last resort.

Focus on the Benefits of the Process

Managers should foster an environment of team effort with a focus on the benefits of the GLP Standards. It is hard to deny that a good GLP program increases the quality and reliability of laboratory studies.

Protocols and SOPs provide valuable benefits for planning and executing studies. Protocols can be considered contracts between the manager and the study director. The protocol is also the initial communication tool for each study. SOPs can eliminate needless supervision, aid the training process, and add consistency and uniformity to data generation.

In the concept of team effort, QA personnel often can be allies. Often QA personnel will argue against attempts to shorten study time lines or report-writing schedules. Managers and study directors can be reminded that the process of organizing data and writing, typing, auditing, and reviewing a report is not a simple task and that in some cases, changing existing deadlines may be unreasonable. QA reports are in the best interests of a department manager. The QAU can help a department improve the final product, improve the department's image, and perhaps ultimately accelerate the time to product registration.

Contributions of the Quality Assurance Unit

How a QAU can help the GLP compliance process by being educators, consultants, and monitors of the program has already been discussed. In the development process, the most difficult task facing the QAU is the enormous challenge of trying to influence others. Handling people is a vital part of a QA professional's job. Learning this skill is critical; courses on managing people, on interpersonal relationships, and on negotiation skills are extremely helpful. However, the QAU should not attempt to usurp a manager's authority.

Some positive QA strategies for assisting managers with the development and maintenance of a successful GLP program are contained in the following list.

- The QAU should become a resource. Because the QAU has the opportunity to observe the organization of data by other laboratories when visiting contract laboratories or dealing with sponsors, it can share information on how others are handling data or SOPs.

- The QA staff must become technically competent and must be viewed by others as the resident resource on compliance. QA personnel should work hard to maintain a current awareness of GLP issues.

- The QA staff should maintain an open-door policy regarding discussion of GLP issues with the laboratory and field staff. Above all, QA personnel should not dictate. Instead, the QAU should make suggestions for achieving compliance and provide advice on how to make the GLP program work to the company's advantage.

- The role of the QA professional is that of an observer and reporter. The QAU must avoid being viewed as a police officer, judge, or jury. However, the QAU must make it clear that when necessary QA personnel can and will reserve the right to discuss compliance problems with higher level of management.

- QA personnel must also attempt to create an atmosphere in which people do not perform tasks merely because the QAU said so, without understanding or caring why an action is suggested.

- QA personnel can make subtle changes that may ease the emotional strain of auditing on laboratory or field staff, for example, using a green pen for auditing rather than a red pen, which is demonstrative and authoritative.

Conclusion

One of the key elements in implementing a successful GLP program is the commitment of management. Management must exhibit the leadership skills needed to guide the program to success. Otherwise, the program will either drift aimlessly or progress to a tug of war between the scientific staff and the QAU. Time and resources must be provided for planning, building, and maintaining a quality program. If managers take responsibility for GLP compliance and advocate GLP Standards as a minimum set of standards for producing quality studies, any company can build and sustain a successful GLP program. Managers must shift the emphasis from quality for the sake of compliance to total quality in producing scientific data.

Managers who take the time to become GLP literate should be applauded. As with many things, managers can be responsible for change and success but cannot create that change or success alone. Managers must cultivate the expertise and establish the atmosphere of team cooperation to institute change. The leadership, commitment, and involvement of managers is the forerunner to this change and the key to success.

Reference

1. "Final Rule for Good Laboratory Practice Standards Under the Federal Insecticide, Fungicide, and Rodenticide Act", *Code of Federal Regulations* Title 40, Pt. 160; *Federal Register* 54:158 (August 17, 1989), pp 34067–34074.

RECEIVED for review April 22, 1991. ACCEPTED revised manuscript September 17, 1991.

Chapter
3

Preparing Standard Operating Procedures for Field Studies

James P. Ussary[1]

ICI Americas, Inc., Agricultural Products Group,
Goldsboro, NC 27533–0208

Standard operating procedures (SOPs) are required by GLP regulations for all routine activities critical to the successful outcome of studies conducted at a test site. SOPs for field studies that are conducted at multiple sites must be written and maintained to obtain acceptable and consistent quality and to allow for variations in procedures and equipment normally found among test sites. This chapter describes a system for writing, reviewing, distributing, and managing SOPs for field trials conducted for agricultural products development and describes what information those SOPs should contain.

STANDARD OPERATING PROCEDURES (SOPs) ARE REQUIRED by all good laboratory practices (GLP) regulations in existence today. SOPs are written procedures for the various routine activities involved in conducting a study. Management-approved instructions necessary for the scientist to perform the activity and to take corrective action if something goes wrong are contained in the SOPs. The purpose of this chapter is to provide a guide for writing and managing SOPs used in agricultural product field studies.

The experimental phase of an agricultural product study often begins in the field with the test chemical application and ends in the analytical

[1]Current address: Ussary Scientific Services, 1614 Boyette Drive, Goldsboro, NC 27534

chemistry laboratory. These studies involve magnitude of residue, crop rotation, soil dissipation, and plant and soil metabolism. In the field, the scientist is responsible for establishing and maintaining test plots; obtaining, storing, and applying test chemicals; collecting, maintaining, and shipping samples for analysis; and recording field data. Much of this work is routine and can be described in general SOPs.

For complex field trials that are not normally considered routine, detailed SOPs are required for the various specialized functions. For example, in a study to measure potential for a chemical to contaminate groundwater, an SOP that gives specific instructions about collecting and handling water samples may be needed to ensure against inadvertent contamination of the samples. A mesocosm study (small-pond aquatic effects) will need SOPs that address the various types of sampling. Radiochemical studies conducted in the field may require an SOP that gives very detailed instructions for test chemical application, but instructions for applying commercial products with a tractor-mounted sprayer may be somewhat general.

Contents of Field SOPs

Requirements for the contents of SOPs for field studies are basically the same as those for laboratory studies; however, SOPs for equipment and scientific practices for studies that are to be done at multiple sites by various scientists should be written in a form that is general enough to permit, as much as good science will allow, the use of familiar techniques by each scientist and to allow for variations in similar items of equipment. Among field scientists a common variation occurs in the technique used to calibrate chemical sprayers. The technique used is not important if the result is an accurate sprayer calibration and if the data clearly indicate the technique that was used. SOPs should instruct scientists to record the details of the calibration technique. Although the SOP should specify that each nozzle must be checked and that the ground speed of the rig must be timed, exactly how that should be done does not need to be specified. On the other hand, the SOP should be specific about the frequency of sprayer calibration and the acceptable variation of the outputs of the spray nozzles.

For SOPs to be effective, they should contain the following elements:

- a descriptive title

- an identifying number, including a version number to show if the SOP has been revised

- a statement of purpose and scope

- the date the SOP goes into effect
- the schedule for review (for most field SOPs an annual review is sufficient)
- the author's name
- the dated signature of the appropriate manager who authorized the SOP
- a brief description of equipment if the SOP is for technically important equipment
- instructions for achieving the purpose of the SOP, as general as possible but specific enough to ensure that the intent of the SOP is achieved
- maintenance instructions for equipment, possibly including a checklist to be followed prior to using the equipment
- malfunction instructions (If an item of equipment malfunctions while being used in a study, the event should be recorded in the study records, the study director should be notified, and an opinion about the effect on the study should be written.)
- specification of who (often by job title) has overall responsibility for equipment

Subjects for Field SOPs

If a function that is technically related to a study is not a simple, self-evident procedure, that function needs an SOP. SOPs should be written for all technical functions and for all other functions required by the GLP regulations. Examples of SOPs that should be written for agricultural-chemical field studies are given in Appendix 3.I of this chapter. SOPs should give instructions for management of test chemicals; operation and management of all technically important equipment; establishment, labeling, and management of test plots; sample collection and maintenance of sample identity and integrity; and recording and maintenance of data. Also necessary are SOPs describing maintenance of employee qualification records, safety procedures, and writing and management of the SOPs.

Field SOP Authors

An SOP should be written by someone who is experienced with the type of studies to be described in the SOP, and often the field scientists are the individuals most familiar with the equipment and techniques used in field studies. For multisite studies that involve different personnel at each site, writing SOPs at each site is usually not satisfactory. Following this procedure results in a large number of similar SOPs that are difficult to manage and may result in unacceptable variations in scientific or GLP compliance standards within a study. The most satisfactory procedure is to assign one person to write each SOP and then obtain group acceptance.

Responsibility

Responsibility for field studies is the same as that for laboratory studies. Management is required by the GLP regulations to approve all SOPs and to be satisfied that these procedures ensure the quality and integrity of the data produced in a study. The study director must ensure that the SOPs are followed or that deviations from the SOP are properly documented. The SOPs that are pertinent to the study in progress must be available to the scientist at the work site, and the scientist must understand and follow the SOPs and document any deviations in the raw data. Organization of this responsibility is somewhat more difficult for SOPs used at several sites. The manager responsible for the field portions of the studies can gain assurance from experience and from the advice of others that the procedures are satisfactory, but the study director often cannot personally guarantee that the approved SOPs were actually followed. Ensuring that SOPs were followed must often be accomplished through discussions with field managers and project coordinators, telephone conversations with the field scientist, and inspections by the quality assurance unit (QAU).

Distribution and Retention

Distribution of field SOPs should be controlled from a central point. Following this procedure ensures that each field scientist has the latest edition of each SOP and that a copy of each SOP and all revisions are submitted to the archive. A copy number should be affixed to each copy of an SOP. Different organizations use red ink, punch numbers, embossed numbers, or imprinted numbers so that unauthorized copies can be easily identified.

Field scientists may want individuals who are not on the distribution list to have a copy of one or more of the SOPs. These copies should be requested from the central source, and the copy recipients may need to be added to the distribution list to receive updated copies during only the experimental period of a particular study.

A historical file of all SOPs must be kept. Keeping field SOPs in the archive at the point of distribution is often more convenient than retaining them in a field office.

Appendix 3.I: Subjects of SOPs for Field Studies

1. Test chemicals: the procedure for obtaining, storing, and maintaining a bulk chemical inventory and for dispensing test chemicals.

2. Test chemical dilutions: the procedures that will be acceptable for measuring the volumes of test chemicals and for making dilutions for application to test plots.

3. Test chemical application equipment: for description, purpose, calibration, operation, and maintenance of each sprayer, granular applicator, wick applicator, or other equipment used to apply test chemicals to the test systems.

4. Balances: for operation, calibration, and maintenance of chemical balances.

5. Freezers: for operation, temperature calibration and monitoring, and security of freezers used to store samples.

6. Weather stations: for operation, calibration, and maintenance of any electronic or mechanical equipment used to obtain climatic information.

7. Sample collection: for sample collection techniques; sample sizes; and labeling, storage, and transport of samples.

8. Test plots: for labeling of plots, security of plots, and buffer zones between plots.

9. Personnel: for policy and procedures for maintaining curricula vitae and training records of study personnel.

10. Data: for management and security of field data. If data forms are used for recording field data, an SOP giving instructions for com-

pleting the forms may be helpful. The procedure for transferring field data to the laboratory or office and the custody of records to be kept should be described.

11. Standard operating procedures: for policy and procedures for writing and maintaining SOPs. This should describe the required format, numbering system, review policy, and distribution procedure.

12. Safety: any specialized safety procedures such as use of respirators or protective clothing.

RECEIVED for review September 17, 1990. ACCEPTED revised manuscript August 17, 1991.

Chapter
4

Preparing Standard Operating
Procedures for the Laboratory

M. Jean Hornshuh

**Agricultural Products Department, Experimental Station,
E. I. du Pont de Nemours and Company, Wilmington, DE 19880–0402**

*Standard operating procedures (SOPs) are a required com-
ponent of a testing facility's compliance program, according
to good laboratory practice (GLP) regulations and international GLP guidelines. Specifically, SOPs are the standard
procedures that trained personnel must follow to ensure the
quality and integrity of the work performed during a study.
If properly developed and followed, SOPs ensure consistency
and good definition for a laboratory's research program,
regardless of who conducts the research. This chapter provides some practical suggestions and general guidelines for
writing SOPs and for managing an effective SOP system.
Examples are given of typical SOP topics within the realm
of laboratory operations.*

Good LABORATORY PRACTICE (GLP) STANDARDS ARE THE BASIC PRINCI-
PLES that were developed to ensure that studies are conducted with good
planning, appropriate execution, and complete documentation. To that
end, standard operating procedures (SOPs) are a critical component

in the foundation of each testing facility's GLP compliance program. The GLP Standards apply to studies submitted under the Federal Insecticide, Fungicide, and Rodenticide Act (FIFRA) and the Toxic Substances Control Act (TSCA) (1, 2). SOPs are also a component of the *Guidelines on Good Laboratory Principles* prepared by the Organization for Economic Cooperation and Development (OECD) to be used internationally (3). Field studies in the agrochemical industry are another area of testing that is heavily governed by SOPs, and field study SOPs are covered in Chapter 3. A GLP compliance program for health effects studies and pharmaceutical studies submitted to the Food and Drug Administration also includes SOPs (4).

Definition, Purpose, and Value of SOPs

The GLP Standard for both FIFRA and TSCA (160.81 and 792.81, respectively) (1, 2) states "A testing facility shall have standard operating procedures in writing setting forth study methods that management is satisfied are adequate to ensure the quality and integrity of the data generated in the course of a study." The points to emphasize in this definition are contained in the following list.

1. SOPs must be documented in writing.

2. Managers must approve SOPs.

3. Adherence to SOPs ensures the quality and integrity of the study data.

The basic purpose of an SOP is to provide a "how-to" document that study personnel, whether in-house or at a contract laboratory, can use while they perform various routine study operations. Thus, developing clearly written, unambiguous SOPs that can be understood and followed by both experienced and inexperienced laboratory personnel is critical. Another intent of a sound SOP system is to minimize the introduction of systematic error into a study by ensuring that all personnel use the same procedure for a specific operation in the study. Additional information can be obtained from an article on SOPs by A. E. Parks (5). This paper (5) outlines the regulatory requirements and some of the fine points of SOPs.

The benefits of an effective SOP system are numerous. SOPs provide a basis for uniformity, consistency, and accountability. They can also be used as a training vehicle and guide for new personnel. SOPs eliminate the repetitive documentation of each day's work and describe actions or events that affect the integrity of the study. From a historical perspective,

SOPs are one aspect in the study records that provides an outline of how a study was conducted; they can be included in the study records or referenced and archived in a different location.

SOPs can present challenges as well. SOPs can lock personnel into a procedure that is difficult to change, and SOPs require a great deal of time to write, update, and distribute. Moreover, some laboratory procedures cannot be standardized. To minimize these negative aspects, SOP deviations must be clearly documented, and an SOP system should incorporate a scheduled review-and-revision process.

Writing Standard Operating Procedures

When an SOP Is Needed. Three concepts define the need for an SOP:

1. SOPs concern data that result from measurements, calculations, recorded observations, and procedures.

2. SOPs ensure the quality and integrity of data.

3. SOPs focus on activities that have some impact on the study results.

Therefore, SOPs are routine, standardized procedures and should be required for methods and procedures that have an impact on data generated during the course of a study. General or study-specific procedures can be handled in the protocol or in a laboratory notebook.

SOP Preparation. Anyone who sees the need can initiate an SOP. The author should be someone who is willing to write the SOP, who has the time to devote to the effort, and who needs the SOP to conduct his or her work. The author must have a thorough understanding of the technical work that is being performed and should be able to present the procedure in a logical and orderly format. If the SOP affects people who work in different functional areas, the author must consider the impact of the SOP on the other workers. This process requires patience. The author must be capable of and willing to exchange ideas with colleagues and to fine tune the wording of an SOP during the development process. A common obstacle is that other projects are often given higher priority, forcing SOP development to take a back seat. Management must actively support this documentation function and allow authors adequate preparation time.

Laboratory Input. The author should circulate a draft of the SOP to obtain input from other technical personnel. If a technical person writes the SOP for personal use, the main input should come from that person or from other technical personnel who have a knowledge of the work being performed. If the SOP will be used by a number of people who work in different functional areas, qualified and interested personnel from each functional group should review and critique the SOP. In the latter case, it is important to obtain as much input as possible before the SOP is finalized so it will be accepted by those who will be expected to use it.

SOP Approval. When an SOP is complete, the author signs and dates the document and sends it to "management"[1] for approval. Some testing facilities also have the quality assurance unit (QAU) review, approve, and sign the SOP before sending the SOP to management for approval. The appropriate person in management, usually a research supervisor or higher level manager, should review and approve the document by dating and signing it. Management who approves SOPs should understand the procedure described in the SOP and how it pertains to the study conduct. Management should also recognize how the procedure fits into the total laboratory operation.

SOP Format. Certain general elements and a consistent format should be standard to all SOPs within an organization. The scope of the SOP should describe to whom and what the SOP applies, and the purpose should describe the SOP objectives. Another prerequisite of an effective SOP is flexibility. That is, it must meet the needs of the user and yet be general enough that other personnel may use it if it applies to work in their area.

To achieve these goals, the author must focus on the key points of the SOP, which outline the critical aspects of the procedure and provide sufficient, yet not restrictive, detail. The text of the document should describe the sequence of events involved in the procedure. The presentation should be straightforward and easy to follow.

Next, the author must ensure that the SOP contains the following general elements:

- a meaningful title with key descriptive words
- identification and version numbers
- date the SOP goes into effect
- author's printed name and dated signature
- approving manager's printed name and dated signature
- scope and purpose of the SOP
- procedural text

---

[1]Management is defined here as the person or persons responsible for setting policy at a facility. For a description of the duties of testing facility management, see § 160.31 of the GLP Standards.

- parts revised (if a revision was made)
- references (if applicable)
- referenced supplements to the SOP (if applicable)
- all pages numbered, with first page giving total number of pages

In addition to these general elements, the following sections should be common to equipment SOPs:

- calibration instructions
- maintenance instructions
- malfunction and repair instructions
- reporting documentation requirements
- frequency of use
- responsible personnel
- log books and records

Managing Standard Operating Procedures for the Laboratory

SOP Review and Revision. SOPs should be reviewed periodically to determine if the written procedure still reflects present laboratory practices. If no changes have occurred, the SOP can stand until the next scheduled review. The SOP custodian should keep a record of all SOP reviews by documenting the date and the name of the reviewer. If the SOP must be revised, the revised draft should be circulated to potential users and personnel who have knowledge of the work being done. Comments should be returned to the author within a short, but defined, period of time. Ideally, the author of an existing SOP should revise the document. If the author has been reassigned to other duties, the author's replacement or someone designated by management should assume the responsibility for SOP review and revision.

If the SOP affects a number of laboratory personnel, the revised draft can be submitted to a committee that comprises personnel from the different functional areas and from management. The committee should review the changes and agree to them before the revised procedure is finalized.

The need for a revision is paramount if a deviation from the SOP has been documented in the study records of several different studies. QAU audits are also excellent tools for recognizing procedural changes that have evolved. Such audits focus attention on the need for SOP reviews and on new procedures that must be documented.

SOP Distribution. The appointment of an SOP custodian will contribute immensely to the success of any SOP system. For example, each testing facility must maintain an active file of all current SOPs used at the facility as well as a historical file of SOPs that have been revised and removed from active use. Custodial responsibilities, in addition to SOP maintenance and distribution, should include assigning SOP numbers, cataloging SOPs, and maintaining the SOP master list.

There are various approaches to SOP distribution. Some laboratories distribute SOPs at technical meetings. Some facilities require that staff read the SOP in the presence of the distributor. One laboratory has the user sign for an SOP when it is distributed so the facility has a record of those who received the SOP. Another facility uses electronic mail to notify its staff that a new SOP is ready for use and has been placed in the active SOP file. One laboratory posts new and revised SOPs on a central bulletin board so that each employee can read and be apprised of the facility's active SOPs. When distributing revised SOPs, one laboratory requires personnel to return the old procedure to the SOP custodian before the new SOP is handed out. Some laboratories prepare SOP notebooks containing all current SOPs used by the staff; the maintenance of these manuals is typically the responsibility of the SOP custodian. Each approach to distribution is valid, but all SOP programs must have a mechanism for informing potential users that a new SOP has been written or that an existing SOP has been revised or retired.

Types of SOPs

SOPs for laboratory operations can be organized in several ways. The various categories, numbers, and types of SOPs used by test facilities are numerous. For the purpose of simplicity, SOPs pertaining to laboratory operations can be separated into three general functional categories: procedures, equipment, and test methods. A fourth category, product-chemistry SOPs, will also be discussed.

The number and types of SOPs may vary among testing facilities and should be based on the site's needs and the work performed. Quality assurance (QA) professionals from several agrochemical companies and contract facilities have contributed their input to the following lists of SOPs. The following sections provide an overview of these categories.

Procedural SOPs. Procedural SOPs can be subdivided into several categories. Suggested SOP categories may include, but are not limited to, the categories in the following list (suggested titles related to each category can be found in the Appendix 6.I, "Suggested Titles for SOPs"):

- general
- test systems
- animal test systems

- laboratory operations
- facility and personnel management
- study directorship
- quality assurance unit
- test, control, and reference substances
- computer operations
- archiving

Equipment SOPs. In general, laboratory equipment can be sorted into two categories: equipment that produces actual raw data and equipment that maintains the proper laboratory conditions for the study. Examples of equipment that produce actual raw data for the study and require equipment maintenance and calibration SOPs include balances and scales, pH meters, high-performance liquid chromatographs (HPLC), and gas chromatographs (GC). An expanded list of equipment SOPs can be found in Appendix 6.I.

Many people do not interpret the GLP Standard to include equipment that does not provide actual raw data. However, others believe that SOPs should be prepared for these types of equipment because proper maintenance of the equipment is necessary to create the laboratory conditions needed for a study. Examples of laboratory equipment for which standard operational and maintenance procedures might be appropriate are water baths, vacuum pumps, autoclaves, and drying ovens. An expanded list of these types of equipment can be found in Appendix 6.I to this chapter.

Equipment SOPs should detail the methods, materials, and schedules used in routine inspection, cleaning, maintenance, testing, calibration, and standardization of equipment. The GLP regulations also specify that actions taken to correct equipment failure or malfunction should be documented in the equipment SOP (*1, 2*).

Testing facilities may maintain a large and unmanageable number of active equipment SOPs. One way to reduce the size of the equipment SOP file is to develop one general SOP for each type of instrument commonly used in the laboratory, such as balances, pH meters, HPLCs, or GCs. The general portion of the SOP should address the purpose, scope, general maintenance, and calibration procedures, and a reference to the instrument logbook also would be included for management-approved specifics concerning maintenance and calibration for a particular instrument. A separate standardized information sheet would cross-reference specific information in the manufacturer's equipment manual and list other pertinent criteria for the specific instrument, such as make, model, serial number, location, and custodian. A copy of the general SOP and the specific instrument sheet can be maintained with the instrument logbook. Similarly, the active equipment SOP file would contain the general

SOP for each type of commonly used instrument and a copy of the specific instrument sheet for each instrument associated with the general SOP. This approach reduces the size of the active equipment SOP file and the number of equipment SOPs. Moreover, if changes are required, the specific instrument sheet is the only part of the equipment SOP that requires revision and archiving.

Test Methods. Test methods are procedures and can be considered a type of SOP. Many test methods are used to process and analyze different matrices in the laboratory (e.g., analysis of pesticide residues on crop samples or analysis of pesticide impurities in the formulated product). These methods can be established as specific to a test substance or the type of matrix being analyzed. Test methods can be held in an active method archive separate from the active and historical SOP file. Again, test methods must be straightforward and sufficiently detailed without being restrictive. If the method is too restrictive, the user may spend excessive time documenting SOP deviations for the study records. Examples of general test methods may include a combination of the following:

- preparing standard solutions for a specific compound
- processing specific matrices for a specific compound type
- analyzing specific matrices for a specific compound type
- processing samples for a family of compounds
- analyzing samples for a family of compounds

Product-Chemistry SOPs. Most testing facilities will not be concerned with product-chemistry studies in which the physical and chemical characteristics for the technical and formulated product are determined. However, the GLP Standard states that "testing facilities that do perform product-chemistry analysis and characterization functions will have written SOPs for study methods that management believes adequately ensure the quality and integrity of the data generated in the course of the study that will be submitted to the EPA" (*1, 2*). In product-chemistry studies, the test substance in most cases is also the test system. Suggested examples of product-chemistry SOPs have been cited in the GLP Standard (*1, 2*) and can be found in the Appendix 6.I.

Many SOPs for product-chemistry studies could be categorized as a type of test method SOP. Test requirements for physical and chemical properties can be found in the *Pesticide Assessment Guidelines*, Subdivision D, Series 61, 62, and 63 (*6*). Test method SOPs may include color, odor, dissociation constant, solubility, storage stability, and vapor pressure. An expanded list of physical property test method SOPs can be found in Appendix 6.I.

Some physical property tests, such as color and odor, can be clearly described in the protocol, but other tests, ambiguously described in the EPA *Product Chemistry Guidelines* (6), may benefit from an SOP. Standard methods have been developed by such organizations as Collaborative International Pesticide Analytical Council (CIPAC), American Society for Testing and Materials (ASTM), American National Standards Institute (ANSI), and the Organization for Economic Cooperation and Development (OECD) for a number of physical property tests such as storage stability, viscosity, flash point, or dielectric breakdown voltage, and these standard methods have been cited in the EPA *Product Chemistry Guidelines* (6). The testing facility could use a standard method from one of the above organizations as a guide in developing a test method SOP. If this approach is taken, the testing facility should clearly reference the standard method that was used as a guide.

The Inherent Challenges of SOPs

SOPs can present many challenges. One of these challenges is the preparation of a flexible SOP that meets staff needs. Such flexibility, which requires creativity and adequate preparation time, reduces efforts associated with documenting deviations in the study records.

Another challenge is to obtain sufficient, knowledgeable input to generate an SOP. Dedication on the part of the author and on the part of contributors and reviewers is required. The development of a realistic time line is especially critical to SOP completion. To this end, management must support SOP authors, allowing them sufficient time to write, review, and revise SOPs.

Some large facilities have found it difficult to develop a numbering system that meets the needs of an expanding GLP compliance program. One approach is to exchange ideas with QA personnel from other companies and testing facilities and then use those ideas that will best work for your testing facility.

The efforts of a dedicated, organized SOP custodian will facilitate SOP system operations. Among the custodian's job responsibilities are updating SOPs in an efficient manner and maintaining an accurate, historical SOP file that can be easily retrieved.

Finally, instilling in laboratory personnel the importance of knowing and following SOPs and documenting any changes that occur is difficult. Management support of the GLP compliance program must be readily apparent to all facility staff. One suggestion is for management to issue a directive that knowing and following SOPs is a job requirement.

Conclusion

Overall, three requirements stand out as essential to the existence of any SOP program. First, an SOP custodian should be assigned to oversee the SOP system and to maintain an updated list of all active and historical SOPs so that potential users know when new or revised ones are issued. Second, there must be management support for the compliance program. Third, there must be a directive from the managers to all facility staff that knowing and adhering to SOPs is a job requirement. Fulfillment of these three requirements will facilitate the process of developing a sound SOP system.

References

1. "Federal Insecticide, Fungicide, and Rodenticide Act (FIFRA); Good Laboratory Practice Standards, Final Rules", *Code of Federal Regulations Title 40, Pt. 160; Federal Register* 54:34052 (August 17, 1989).

2. "EPA Toxic Substances Control Act (TSCA) Good Laboratory Practice Standard, Final Rules", *Code of Federal Regulations* Title 40, Pt. 792; *Federal Register*, 54:34034 (August 17, 1989).

3. "OECD Principles of Good Laboratory Practice. Good Laboratory Practice in the Testing of Chemicals", Organization for Economic Cooperation and Development, Washington, DC, 1982.

4. "FDA Good Laboratory Practice Regulations, Final Rules" *Code of Federal Regulations* Title 21, Pt. 58; *Federal Register"*, 52:33768 (September 4, 1987).

5. Parks, A.E. in *Good Laboratory Practices: An Agrochemical Perspective.* Garner, W; Barge, M., Eds.; ACS Symposium Series 369, American Chemical Society: Washington, DC, 1988.

6. Beush, G. J., et al., "Pesticide Assessment Guidelines Subdivision D: Product Chemistry", U.S. Environmental Protection Agency: Washington, DC, 1982.

RECEIVED for review September 17, 1990. ACCEPTED revised manuscript August 17, 1991.

Appendix 6.1: Suggested Titles for SOPs

For simplicity, SOPs pertaining to laboratory operations can be separated into three functional categories: procedures, equipment, and test methods (including product chemistry).

Procedures. Procedural SOPs can be subdivided into several categories. Suggested SOP categories and suggested related procedures include the following:

• general

— SOP preparation
— SOP numbering
— SOP file system maintenance
— laboratory notebook data entry
— error correction and documentation
— telephone conversation documentation
— significant figures and number rounding
— protocol preparation, review, and distribution
— report preparation, review, and distribution

• test systems

— identification, care, handling, transfer, and proper placement
— observation methods
— sampling methods
— room preparation
— room maintenance
— test-system disposal

• animal test systems

— preparation of animal rooms
— room maintenance and environmental monitoring
— feed, bedding, supply, and equipment storage
— project number and personnel assignment
— animal identification and randomization
— animal quarantine
— water sampling
— diet preparation and storage

— animal housing, care, handling, transfer, and observation
— handling moribund or dead animals
— euthanasia
— necropsy or postmortem examination
— histopathology
— specimen collection and identification
— cage washing and sanitation
— animal waste collection, storage, and disposal

• laboratory operations

— receipt and distribution of test, control, and reference substances
— storage of test, control, and reference substances
— sample collection
— sample randomization procedures
— identification and labeling of specimens and samples
— sample tracking, receiving, storage, and processing
— sample shipping
— data collection, handling, storage, and retrieval
— soil testing procedures and methods of classification
— laboratory glassware cleaning
— reagent and solution labeling
— solvent inventory maintenance
— water-quality monitoring
— preparation and documentation of standard solutions
— environmental chamber record keeping and maintenance
— walk-in freezer and cold room maintenance and security
— study documentation, definitions, and numbering

• facility and personnel management

— preparation and maintenance of current resumes
— appropriate attire for performing study
— employee training
— documentation and maintenance of employee training record
— management of the SOP file
— site safety
— chemical hazard handling
— laboratory security
— visitors on-site
— insect and rodent control
— project number and personnel assignment
— procurement of supplies

— procurement of animals
— emergency backup procedures

- study director

  — duties and responsibilities
  — inspections and audits
  — recording and reporting study progress
  — relationship with upper managers
  — relationship with contract facility
  — relationship with sponsor facility
  — training

- quality assurance unit (QAU)

  — master schedule
  — final report reviews
  — study records reviews
  — protocol reviews
  — in-life audits
  — facility inspections
  — contract facility inspections
  — archiving QAU reports and records
  — QAU file maintenance
  — QAU personnel training
  — interaction with technical personnel
  — role in training laboratory personnel
  — conducting an EPA inspection visit
  — conducting sponsor inspection visits (applicable to contract facilities)

- test, control, and reference substances

  — identification, characterization, handling, mixing, and sampling
  — shipping, receipt, use, storage, archiving, and disposal
  — handling test substances labeled with radioactive isotopes
  — disposal of radioactive waste

- computer operations

  — system operation
  — system security

- — system crash procedures
- — hard disk management
- — validation procedure for computer programs

- archiving

  - — submission and retrieval
  - — security and limited access
  - — maintenance
  - — environmental conditions
  - — filing index and tracking system
  - — archiving final report and supporting study records
  - — specimen and sample storage
  - — storage of magnetic and other media
  - — procedures in case of fire or other accidents

Equipment. In general, laboratory equipment can be sorted into two categories: equipment that produces actual raw data and equipment that maintains the proper laboratory conditions for the study. Examples of equipment that is used to produce actual raw data for the study and that require equipment maintenance and calibration SOPs include the following:

- balances and scales

- pH meters

- liquid scintillation counters

- HPLC, GC, MSD, NMR spectrometers, and other complex analytical equipment

- analytical detectors

- autosamplers

- thermometers

- water-purification systems

- refrigerators and freezers

- combustion instrumentation

- moisture-balance equipment

- data station and software

- TLC radioactivity analyzers

The following are examples of laboratory equipment that is used to maintain the proper laboratory conditions and for which standard operational and maintenance procedures might be appropriate:

- compressed gas cylinder use, handling, and storage
- rotary evaporators
- water baths
- homogenizers
- fume hoods
- floor-model centrifuges
- microcentrifuges
- autoclaves
- vacuum pumps
- Soxhlet extractors
- drying ovens
- freeze dryers

Test Methods. *General Test Methods.* Many test methods are used to process and analyze different matrices in the laboratory. These methods can be established as specific to a test substance or to the type of matrix being analyzed. Test methods are procedures and are a type of SOP. Examples of general test methods include the following:

- preparing standard solutions for a specific compound
- processing specific matrices for a specific compound type
- analyzing specific matrices for a specific compound type
- processing samples for a family of compounds
- analyzing samples for a family of compounds

Product Chemistry. The following are suggested examples of product-chemistry SOPs:

- receipt, identification, storage, handling, and mixing
- method of sampling the test, control, and reference substances

- test-system observation
- data handling, storage, and retrieval
- equipment maintenance and calibration
- labeling reagents and solutions

Many SOPs for product-chemistry studies can be categorized as a type of test method SOP. Test method SOPs include the following:

- color
- physical state
- odor
- melting or boiling point
- oxidation–reduction properties
- density, bulk density, or specific gravity
- pH
- dissociation constant
- solubility (aqueous and solvent)
- octanol–water partition coefficient
- vapor pressure
- viscosity
- storage stability
- flammability or flash point
- corrosion characteristics
- explodability
- dielectric breakdown voltage

Chapter
5

Designing Protocols for Field Studies

Maureen S. Barge

FMC Corporation, P.O. Box 8, Princeton, NJ 08543

The study protocol can be considered both a planning and a communication tool. Although a protocol is a requirement of the Good Laboratory Practice (GLP) Standards, it should never be prepared merely to comply with the regulations. No documentation is more crucial to a well-designed and well-executed study than the protocol. Although the GLP Standards provide certain key requirements for every protocol, scientists must incorporate sufficient flexibility in the protocol design to provide for the proper balance between exact definitions of what must be done and what is feasible in the real world. This flexibility is especially needed in the area of field testing where many conditions, such as the weather and the growing season, are not under the control of the scientist. What is the balance between rigidity and loss of control? What parameters can be used to guide the protocol preparation and the GLP process to guarantee that the study design ensures compliance and results in the desired objectives of the study? These questions are addressed in this chapter.

As BOTH A PLANNING AND A COMMUNICATION TOOL, the study protocol is crucial to a well-designed and well-executed study. A protocol should

2192-8/92/0043$06.00/0 © 1992 American Chemical Society

never be prepared merely to comply with the Good Laboratory Practice (GLP) Standards. Although the GLP Standards provide certain key requirements for every protocol, scientists must incorporate sufficient flexibility in the protocol's study design to provide for the proper balance between exact definitions for what must be done and what is feasible in the real world. This flexibility is especially needed in the area of field testing where many conditions, such as the weather and the growing season, are not under the control of the scientist. A fine line can be drawn in attempting to strike a balance between rigidity and loss of control. A standard operating procedure (SOP) should be developed to set a format that ensures compliance. However, to guarantee that the study design ensures compliance and results in the desired objectives of the study, the writer must keep in mind the purpose of the study.

General Concepts

As a generic set of standards, the GLP Standards attempt to cover all facets of a study from people and tools to records and retention. The GLP Standards do not set standards of performance and in many areas give the reader sufficient latitude to interpret the standards to suit the nature of the science. The protocol is an example of such an area. The protocol should be written on the basis of good science and should describe what will be done in the study. Then the study should be conducted according to the protocol.

GLP Standards force everyone to stop, think, and plan studies more effectively. The philosophy should be that all work is done once and is done correctly. Once a protocol is written, it should be used, not placed in a file and forgotten. In the early years of developing GLP Standards, many people wrote protocols because they were required, then forgot about them until a quality assurance (QA) inspection discovered that the protocol was not being followed. This abuse is not the intent of the protocol.

Purpose of the Protocol

Whether it be a toxicology, analytical chemistry, environmental fate, nature of the residue, metabolism, or ecotoxicology study, the general principles of GLP Standards still apply. Writing and reviewing protocols for field- or laboratory-mandated studies should not be an arduous task. All protocols serve a common purpose.

The protocol may be considered

- an outline or study design for a given experiment
- a planning tool that clearly indicates the objectives and methods of the study
- a contract between the study director and the manager and sponsor
- a vehicle for communicating information to all personnel involved in the study

Much like a newspaper article, the protocol should provide the answers to the questions who, what, where, when, how, and why.

Uses of the Protocol

As a tool, the protocol has a variety of uses. The study director uses the protocol to plan and design the study. What needs to be done? How will it be accomplished? What tools are needed? What is the work schedule? Study personnel use the protocol for understanding what needs to be done and how the study director wants it done. The QA unit (QAU) needs the protocol to schedule and prepare for inspections and audits. Managers and sponsors may use the protocol to ensure the proper assignment of the resources necessary to complete the study and to ensure that the study design will achieve the study objectives. Lastly, regulatory agencies use the protocol to assess planning and execution of the study and whether the study fulfills regulatory requirements. What did the scientist say would be done and was it done?

Above all, protocols should be developed to design scientifically valid studies. Resources available for designing regulated studies include the EPA Pesticide Assessment Guidelines (PAGs) (*1*), which provide the scientific guidelines; the standard evaluation procedures, which explain what the EPA reviewer will be looking for during the agency review of a report; and the data reporting guidelines, which explain how the report should be formatted and what should be reported. In addition to the PAGs, the Federal Insecticide, Fungicide, and Rodenticide Act (FIFRA) Phase III Acceptance Criteria (2) are an excellent check for study design.

Preprotocol Meeting

For nonroutine studies or the study director's first GLP study, a preprotocol meeting can help in the planning stages. When attended by the study director, critical study personnel, managers, and QA personnel, this meeting should reduce the time needed for review and should produce a protocol that everyone has agreed upon. In addition, this personal interaction will facilitate better understanding of all the issues, both science and compliance.

Who Reviews the Protocol

Protocols should be reviewed and signed by the study director, the department manager, the sponsor, and a member of the QA staff. The only signature required on the protocol by GLP Standards is the study director's; however the protocol must be sponsor approved. The best way to acknowledge sponsor approval is with a signature on the protocol. The QAU is not required to review or sign the protocol, but early QA monitoring of a study results in fewer and less severe problems in the later stages of the study. The protocol may be more easily changed before it is signed than after, when documentation of amendments is required. Any other managerial signatures that are deemed necessary for accountability within the framework of the individual facility should also be obtained.

Review Issues

The key issues to be addressed in reviewing the protocol include the following questions:

- Is the objective of the study clearly stated?
- Is the objective reasonable and attainable?
- Is the action plan delineated (who, what, where, when, how, and why)?
- Is the protocol readily understandable?
- Does the protocol comply with the GLP Standards and SOPs?

SOPs and Checklists

SOPs should exist to assist the author on the format and content of the protocol, providing details on not only what is required by GLP Standards but also what is required by the laboratory "management"[1]. Protocol audit checklists can be developed to assist in writing and auditing protocols. However, although checklists are excellent training tools and good reminders, using checklists must not be a rigid exercise. Each type of study may have different study requirements. For example, the protocol for a study with radiolabeled material will contain more information regarding the test substance than one for a study being conducted without radiolabeled material, and the protocol for a study with an established method of analysis will contain details of that method, but a protocol for a study being done for metabolite identification may not be able to provide a detailed method of analysis but merely a general description of what will be attempted in the process. When writing or reviewing protocols, always be cognizant of the purpose of the study.

Protocol Revisions

According to GLP Standards, changes to the protocol must be documented by the study director. However, a good and widely accepted practice is for all those who signed the protocol also to sign any changes to the protocol. The sponsor's acknowledgment of any changes is particularly important because without sponsor concurrence, the protocol is no longer sponsor approved. Historically, changes to the protocol have been divided into deviations and amendments. *Deviations* are unplanned, usually noted after the fact, one-time mistakes or oversights; for example, a temperature not recorded or a data point omitted. These usually cannot be corrected, and they should have minimal impact on the study. *Amendments*, on the other hand, are planned additions, deletions, or changes to the study design (e.g., a change in study director, additional analyses, additional harvest intervals, or the addition or deletion of a trial). To eliminate confusion, one form can be devised to document any change to the protocol. Merely refer to the changes as modifications, revisions, or alterations rather than the historic nomenclature of amendments and deviations. Keep in mind, however, that although deviations are documented after the fact, changes to the study design should be documented prior to the change actually taking place.

---

[1]Management is defined here as the person or persons responsible for setting policy at a facility. For a description of the duties of testing facility management, see § 160.31 of the GLP Standards.

Protocol Development for Field Studies

The following discussion will focus on the field portion of a study; the laboratory portions will be discussed in Chapter 6. The protocol should be formatted with general information followed by the study design in a chronological order of events. The protocol should begin with a descriptive title. For consistency with the final report, the title of the protocol should be the same as the title to be used on the final report. Keep in mind that according to Pesticide Registration Notice 86–5, the final report title must use the same language as the corresponding PAGs subdivision.

Each study should be assigned a unique identifier or study number, and this number should be displayed prominently on the protocol. In addition, putting this study number on the top of each page will eliminate confusion if pages should become detached. Some companies also assign a protocol number, but this may just add to the complexity of the tracking system without any real benefit to the study.

Statement of Purpose. The purpose of a study should be more than to generate samples or determine residue levels. The study's true purpose can be to support a registration, reregistration, label change, or experimental use permit.

Study Responsibility. Include the name of the study director and the name and address of the sponsor and the testing facility. Although not required, the protocol is also a good place to put telephone numbers as well as addresses. This section should also include any other principle parties involved in the study, such as any scientist responsible for a major portion of the study, a contract laboratory that will analyze the samples, or a processing facility. With the complex studies required for registration of a product today, the study director must have help in those areas outside his or her area of expertise. Rarely is an entire study conducted on one site. Provide as much detail as possible during this planning stage of the study to ensure proper execution as the study progresses.

Study Dates. Study dates can be given as a month and year. These are proposed dates for when the experiment will begin and end. The study initiation date is the date the study director signs the protocol. The protocol is required to contain the proposed "experimental start date"[2] and "experimental termination date"[3] for the study. For multifaceted studies, providing experimental start and termination dates for specific phases of a study (i.e., field, processing, and laboratory phases of the study) is helpful. Breaking the study dates into phases will help the QAU plan for inspections and audits more effectively and will assist the study director in keeping the entire study on schedule. Although not required by the regulations, providing the study completion date is extremely important as an element in communication. Completion of the last analysis does not mean that the final report is written. Clear understanding on the length of time from study termination to report completion will help prevent misunderstandings with those waiting for the report and will emphasize to those conducting the study the importance of keeping to the schedule that was agreed upon.

Test, Control, and Reference Substances. Identification of the test substance should include the trade name and formulation, the common name, and the chemical name. The test substance is defined as the chemical that is applied to the test system, but this definition cannot be taken too literally. In some cases, the chemical will be applied to the soil (preplant incorporated), and the test system to be analyzed might be corn or soybeans. Include a list of the reference substances (analytical standards) that are known at the time the protocol is written; the option always exists to modify the protocol to add new metabolites that may be identified after the study begins.

The justification of the test system should include the type of crop and crop grouping along with the reason for selecting that particular crop. The justification may be stated in the purpose of the study.

Study Design. Study design is where the true study planning begins. Here is where the quality of the science is evident. For field studies, start with the trial locations; identify each trial, trial location, and the specific person responsible for the trial. For studies that are conducted at various locations across the United States, a scientist is usually assigned to conduct the trial. Each location should have a designated person responsible for that location and a unique identification number for tracking purposes.

The treatment program is one of the most important aspects of any protocol. Provide explicit directions for application or dosing, including

[2]The EPA defines "experimental start date" as the first date the test substance is applied to the test system.
[3]The EPA defines "experimental termination date" as the last date on which data are collected directly from the study.

the rate and the formulation to be used. Due to the nature of field studies, give schedules as approximate dates if possible. Rain can easily destroy a schedule, and requiring specific dates will only lead to numerous
unwanted protocol deviations. When necessary, name specific types of
equipment and the plot size to be used.

Sampling and Shipment. Specifying the required samples is another
critical aspect of the protocol. Provide explicit sampling intervals, growth
stages, and required sample size. If duplicate or replicate samples are
required, this need should be stated. For example, the samples required
to obtain a representative sampling in an orchard study might be a certain
number of trees, and the minimum number of trees should be stated in
the protocol. Anything unique about the study should be explained; otherwise the protocol can direct that SOPs are to be followed.

Additional questions to be considered regarding sampling and shipment include:

- How are the samples to be handled?

- Where should the samples be shipped after harvest (e.g., laboratory or
  processing facility)?

- What are the requirements for storage and shipment (i.e., ambient or
  frozen) and should they be shipped overnight delivery or by refrigerated truck?

Records To Be Maintained. Provide a list of examples of records
that should be maintained as a reminder to field personnel. Records to be
maintained include the protocol and all amendments, notebooks and calculations, storage temperature records, maintenance and calibration
records for equipment, test substance inventory and usage log, and receiving and shipping records.

In addition, specify where all the original raw data and records are to
be archived at the close of the study. Specification of an archive location
is especially important for sponsors to communicate when dealing with
many different cooperators or contractors in various geographic locations.
Specify where the data should be sent and when.

Protocol Modifications. A statement regarding the handling of protocol amendments is not required, but reminding everyone that changes to
the protocol must be made according to SOPs is a good idea. A simple
statement regarding the procedure for handling amendments and deviations should be sufficient.

Statement of Compliance. Another suggestion that is not a requirement is a GLP compliance statement and a statement on the QAU's commitment to the study. From the beginning, the study director and all study personnel should understand that the study will be run in accordance with GLP Standards. GLP studies require the involvement of the QAU. As part of the contract of compliance, the study director is declaring what will be done; why, when, how, where, and by whom it will be done; and that the QAU will monitor the study.

Signatures. The protocol should be signed and dated by the study director and the sponsor. To follow the contract concept and the team effort needed to complete the study, the QA officer should also sign the protocol. QA staff should audit the protocol for compliance to GLP Standards and SOPs. QA personnel do not approve the scientific design of the study.

Distribution of the Protocol

The protocol should be distributed to all those who sign it and to all study personnel. When mailing protocols to various locations throughout the United States, a color-coded envelope will help the recipient to recognize readily the protocol or amendment, thereby ensuring prompt attention.

Conclusion

The design of the field portion of the study is only part of the protocol; for completeness, the processing and laboratory portions of the protocol must be added. During the protocol development process, it will become apparent that the protocol is much more than a document needed for compliance. Flexibility where possible and rigidity when necessary are important factors to remember when developing the protocol. However, the protocol is of little value if it is not read and used during the course of the study. Design the document for easy reading. Too much text will discourage the reader. Tables and charts are good visual aids that attract the reader's attention. As with any other document, there is no point in writing it if no one is going to read it.

References

1. "Pesticide Assessment Guidelines," U.S. Environmental Protection Agency, Office of Pesticide Programs, Washington, DC, October 1982.

2. "FIFRA Accelerated Reregistration Phase III Technical Guidance," U.S. Environmental Protection Agency, Pesticides and Toxic Substances (H-750C), Washington, DC, December 24, 1989.

RECEIVED for review April 22, 1991. ACCEPTED revised manuscript September 17, 1991.

Chapter

6

Practical Protocols for the Laboratory

Gene Burnett

Agricultural Division, Ciba–Geigy Corporation, P.O. Box 18300, Greensboro, NC 27419

Protocols are the driving force behind every study that is performed under the Federal Insecticide, Fungicide, and Rodenticide Act Good Laboratory Practice (GLP) Standards. The GLP regulations require that each study have an approved written protocol and specify the information that must be in the protocol to meet minimum standards. For some studies, the GLP-required information may be neither necessary nor applicable. A literal interpretation of the GLP protocol requirements may result in a redundant, unyielding protocol. Protocols should be written first to address the study objectives and the users' needs and then to conform to the GLP requirements. Once the study director signs the protocol, the study is initiated and is subject to GLP Standards.

THE PROMULGATION OF THE GOOD LABORATORY PRACTICE (GLP) STANDARDS containing the requirement of approved written protocols for all nonclinical studies is the result of the deficiencies found by the Food and Drug Administration (FDA) in some health-effects studies in the mid-

1970s. The FDA found that some studies did not have adequate study plans and that studies were planned while being performed rather than being planned before the study was started. The FDA testimony before Congress in 1975 resulted in the industry-wide standards known today as GLP regulations.

The Federal Insecticide, Fungicide, and Rodenticide Act (FIFRA) GLP Standards, 40 *Code of Federal Regulations* (CFR) Part 160 (*1*), which became effective on October 16, 1989, expanded the scope of the GLP regulations to include GLP compliance for those studies routinely performed in a laboratory environment (e.g., residue-chemistry, metabolism, environmental-fate, and product-chemistry studies). The FIFRA GLP Standard 40 CFR § 160.120(a) states that each study must have an approved written protocol that clearly states the objectives of the study and all methods for the conduct of the study. This chapter will describe some practical approaches to developing protocols for laboratory studies and will also provide examples from actual laboratory protocols.

Purpose of Protocols

Protocols are the driving force behind every study. The study protocol is an outline that answers the questions what, when, how, who, where and even why. Protocols are vital documents for the reconstruction of a given study. Reconstruction of the study is necessary if the study needs to be repeated for any reason or for the purposes of retrospective auditing. The study protocols are also a communication tool for "management"[1] that describes exactly how the study is to be conducted. Management uses protocols for estimating costs, scheduling, and allocating the needed resources to perform the study. The allocation of adequate resources is necessary to conduct the study in a timely manner and to help ensure that the study will be performed in a scientifically sound manner and in compliance with GLP regulations.

The sponsor has two "customers" at the EPA who need to be satisfied: a scientific customer (the data reviewer) and a compliance customer (the compliance auditor). The clients also have a set of requirements that need to be considered when developing a study protocol. The "two customers" concept is inferred in 40 CFR § 158.80(a) (*2*), which states in part:

> The agency will determine whether the data submitted to fulfill the data requirements specified in this part are accept-

[1] "Management" is defined here as the person or persons responsible for setting policy at a facility. For a description of the duties of testing facility management, see § 160.31 of the GLP Standards.

able. The agency will evaluate the conduct of each experiment in terms of whether the study was conducted in conformance with the design, good laboratory practices were observed, and results were reproducible.

Definition of Protocols

The FIFRA GLP Standards state in § 160.120(a) that each study must have an approved written protocol. Thus, to adequately define a protocol, the definition of a study must be considered. The FIFRA GLP Standard § 160.3, "Definitions", provides the following definition for the term *study*:

> Study means any experiment at one or more test sites, in which a test substance is studied in a test system under laboratory conditions or in the environment to determine or help predict its effects, metabolism, product performance (efficacy studies only as required by 40 CFR 158.640), environmental and chemical fate, persistence and residue, or other characteristics in humans, other living organisms, or media. The term "study" does not include basic exploratory studies carried out to determine whether a test substance or a test method has any potential utility.

The use of such key words as "its effects" and "other characteristics" infers that all scientific endeavors, except the "basic exploratory studies", are included in the definition of study in § 160.3. The FIFRA GLP Standard § 160.1(a), "Scope", provides the following additional clarification to the definition of a study:

> This part prescribes good laboratory practices for conducting studies that support or are intended to support applications for research or marketing permits for pesticide products regulated by the EPA.

40 CFR § 160.1(a) basically states that studies are defined by the data that is submitted to the EPA and that all submitted data must be in compliance with GLPs. Further clarifications as to what constitutes a study can be found in the EPA documents "Data Requirements for Registration", 40 CFR Part 158 (*2*) and Pesticide Registration Notice 86–5 (PR 86–5) (*3*). 40 CFR § 158.32(b) (*2*) notes that:

> All data must be submitted in the form of individual studies. Unless otherwise specified by the agency, each study should

address a single data requirement and be listed separately in the bibliography.

The Data Requirements for Registration and the relationship of the requirements to the Pesticide Assessment Guidelines (PAG) are summarized in Table 6.I. The cited sections of 40 CFR Part 158 provide detailed information on the types of data and the minimum amount of data and information that need to be generated and submitted to support a marketing or research permit. PR 86–5 adds further clarification of the definition of a study on page 4, section C, which states:

Studies generally correspond in scope to a single Guideline requirement for supporting data, with some exceptions discussed in section C.1.

Section C.1 discusses some exceptions to a single data requirement resulting from the fact that some studies address data requirements that are "broader than normal scope or for other reasons." Included in this category are safety studies, such as health-effects (toxicological) studies, product-chemistry studies, and residue-chemistry studies. The exception for product-chemistry studies is that all product-chemistry data should be submitted as a single study for data that supports an end-use product produced from registered manufactured products. Residue-chemistry studies referred to in section C.1 of PR 86–5 include data generated according to §§ 158.690–158.740, "Biorational Pesticides Data Requirements" (Guidelines M).

In essence, 40 CFR Part 160, 40 CFR Part 158, and PR 86–5 support the concept that a study is defined by a protocol and represents, for the most part, a single data requirement of 40 CFR Part 158. The regulations state that all data submitted to the EPA in support of or intending to support a marketing or research permit, minus the exceptions provided in the study definition of § 160.3, must be performed in compliance with GLP Standards, and one of the requirements of GLP Standards is to have an approved written protocol. Otherwise, the exception must be addressed in the GLP compliance statement (§ 160.12). The final report should also be kept in mind when developing protocols. One study requires one protocol and results in one final report.

Development of Laboratory Protocols

Protocol Requirements. 40 CFR § 160.120(a) (*see* Appendix A at the end of this book) lists the minimum information required by the GLP regulations to be contained in a study protocol. A strict and literal

Table 6.I. Pesticide Assessment Guidelines

Data Requirement 40 CFR Part 158	Title	Subdivision	Example Study
§§ 158.150–158.190	Product Chemistry	D	Stability, solubility, octanol–water partition coefficient, volatility
§ 158.490	Hazard Evaluation: wildlife and aquatic organisms	E	Avian and aquatic toxicology
§ 158.340	Hazard Evaluation: humans and domestic animals	F	Toxicology
§ 158.640	Product Performance	G	Efficacy
§§ 158.20–158.740	Experimental Use Permits	I	All studies, limited data
§ 158.540	Hazard Evaluation: nontarget plants	J	Phytotoxicity, seed germination
§ 158.390	Reentry Protection	K	Worker exposure, dissipation
§ 158.590	Hazard Evaluation: nontarget insect	L	Honeybee
§§ 158.690–158.740	Biorational Pesticides	M	Growth regulators, pheromones and hormones
§ 158.290	Environmental Fate	N	Hydrolysis, photolysis, soil–water metabolism
§ 158.240	Residue Chemistry	O	Magnitude and nature of the residues
§ 158.440	Spray-Drift Evaluation	R	Droplet size, drift-field evaluation

interpretation of the 15 sections listed in § 160.120(a) may result in a redundant and unyielding protocol, which, as a result, may be difficult to follow.

An exception to the GLP protocol content requirements is provided in 40 CFR § 160.135(b) (*see* Appendix A at the end of this book) for product-chemistry studies. According to § 160.135, §160.120(a)(5)–(12) and (15) are applicable to the following studies but not applicable to any other product-chemistry studies:

1. Complete GLP Compliance [§ 160.135(a)]

   - stability

   - solubility

   - volatility

   - octanol–water partition coefficient

   - persistence: bio-, photo-, and chemical degradation

2. Partial GLP Compliance [§ 160.135(b)]

   - all physical and chemical characterization studies not listed in § 160.135(a)

Protocol Organization. The elements of information of a protocol need to be organized in a logical and useful manner. The data should be arranged in the protocol in much the same sequence as the information will be needed and used during the study. A protocol written to flow with the conduct of the study will usually help to guide the study participants through the study. Protocols organized simply to mimic § 160.120(a) as far as the type and order of information is concerned will usually fall short in providing the needed guidance. Protocols must have the information required in § 160.120(a) but not necessarily in the order presented in that section. The following is a generic protocol outline that shows placement of information in a protocol. Within each section heading is an explanation of the contents of the section and, in most cases, a reference to an attachment that exemplifies the section. The corresponding FIFRA GLP Standard reference number is included.

Study Title [§ 160.120(a)(1)] (Appendix 6.I). The title should be brief but descriptive enough to indicate clearly the nature of the study. The title should also include the name of the test substance and test system.

Unique Study Identification Number (Appendix 6.I). A unique study number needs to be assigned to the study and should be prominently placed on each page of the protocol. Each page of the protocol should be numbered and paginated to show the total number of pages of the protocol.

Approval Signatures [§ 160.120(a)(14)] (Appendix 6.I). The study director must sign and date the protocol, and the protocol must also bear the date of sponsor approval. The latter requirement is easily satisfied by the dated signature of the sponsor. If the study is being conducted at the sponsor's facility, then the testing facility management becomes the sponsor. The assignment of an alternate study director to act in the capacity of the study director is a worthwhile practice in the event that the study director is unavailable for an extended period of time (i.e., vacation) and decisions concerning the study must be made. Protocols for studies that are performed at contract laboratories should also bear the dated signature of the principal investigator and management of the contract laboratory. The protocol can be viewed as a contract, and those who sign it are attesting to their commitment to the study (i.e., protocol). Numerous organizations also require a quality assurance unit (QAU) representative to sign the protocol. Once the study director signs the protocol, the study becomes a GLP-regulated study.

Sponsor and Testing Facility Addresses [§ 160.120(a)(3)] (Appendix 6.I). The name and address of the sponsor and the facility using the test substance or evaluating a test system must be clearly indicated. In some cases there will be two or more testing facilities. A street address should be given; a post office box may also be given.

Study Personnel (Appendix 6.II). The names of a management (testing facility or sponsor); the study director; and, when applicable, the study coordinators, principal investigators, or study monitors who are involved with the study should be listed. Protocols must name a single study director who will be responsible for the overall conduct of the study.

Study Dates and Schedules [§ 160.120(a)(4)] (Appendix 6.III). The proposed "experimental start date"[2] and "experimental termination date"[3] of the study must be provided. The proposed experimental start

---

[2]The EPA defines "experimental start date" as the first date the test substance is applied to the test system.

[3]The EPA defines "experimental termination date" as the last date on which data are collected directly from the study.

date is the anticipated date of application or dosage (biological part) or of the first analytical activity (analytical part). The proposed experimental termination date is the anticipated date of last sample collection (biological part) or the anticipated date of last sample analysis (analytical part). A proposed schedule is not only a valuable planning tool but it also provides a clear timetable from the study director to the study participants as to when critical events in the study are to take place. The date convention, that is, month, day, year or day, month, year, should be defined by the testing facility.

Study Purpose and Objectives [§ 160.120(a)(1)] (Appendix 6.IV). This section clearly states the purpose, objectives, and type of study. The stated objectives should correspond to the protocol title and to the objectives that will be stated in the final report.

Chemical Substances [§ 160.120(a)(2)] (Appendices 6.IV and 6.V). The test, control, and reference substances must be identified in this section. Test, control, and reference substances are administered or added to the test system. The reference substances (analytical standards) are used to establish a basis of comparison between the test system and the test substance with respect to specific chemical measurements. The identification should include the common name or code number, batch or lot number, purity, expiration or reanalysis date, and storage conditions. For radioactive substances, the specific activity and either the chemical purity or the radiochemical purity (or both) should be reported.

Test System [§ 160.120(a) (Refs. 6,7)] (Appendix 6.VI). The test system to which the test substance is administered or added must be identified. The method by which the test system (i.e., animals or plots) will be identified (i.e., ear tags, labels, or location) should also be stated. This section also includes, when applicable, the total number of animals or plants, species, strains (varieties), body-weight range, age, sex, and source of supply. For some studies, the test system may not be easily identifiable or may not exist at all. For example, in the case of some product-chemistry studies, the test substance may be the test system. Containers can become the test system. In computer-validation studies, the hardware and software under evaluation become the test system.

Every effort should be made to list the sample identification numbers (i.e., plant, soil, animal tissue, and stability vials).

Justification of Test System [§ 160.120(a)(5)] (Appendix 6.VII). This section explains why the test system was chosen. If a standard test system is used because it is the referenced system in an EPA guideline, the guideline should be cited. Following are additional examples:

- Soybean samples will be analyzed in order to obtain residue data to support a label expansion of Femiron 3.3 S on soybeans in accordance with the EPA Pesticide Assessment Guidelines, Subdivision O, Residue Chemistry (reference number 171–4).

- The rat was chosen because of its historical use in this type of test and in accordance with the EPA Health Effects Test Guidelines (40 CFR 798.2250).

A detailed discussion is required only when the study design requires special choices to be made when selecting the test system.

Experimental Design and Procedures [§ 160.120(a)(8)–(12)] (Appendix 6.VIII). This section describes how the overall study will be conducted. It should contain a complete overview of the study plan. This section most often addresses chemistry, sample generation (in-life), sample analysis (analytical), and data-evaluation phases. This section may include or refer to diagrams or graphs that clearly explain how the study is set up and may cite standard operating procedures (SOPs) whenever possible in lieu of detailing routine procedures. The methods for the control of bias (including the randomization procedure, untreated controls, and fortified controls) should be stated. Specifics should be given in subsections when applicable; these specifics are discussed in the following sections.

Chemistry Phase. This section should address such topics as dose preparation, recovery fortifications, and when applicable, information concerning stability (pure compound and in a mixture), homogeneity, solubility, and concentration.

Sample-Generation Phase (In-Life Phase). For animal or plant selection and husbandry or culture, selection criteria (i.e., animal health and weight, plant condition at harvest), housing conditions (e.g., cage, pail, or soil type, room temperature and humidity, and light cycle), and diet must be defined. The diet description or identification (including drinking or greenhouse water) should include specifications for acceptable levels of contaminants that are reasonably expected to be present in the dietary materials and are known to be capable of interfering with the purpose or conduct of the study (if present at levels greater than those established by the specifications). Planting parameters such as ranges of acceptable row and seeding rate for seed-treatment studies should also be defined. Temperature and humidity parameters for storage-stability studies should be stated. Provisions are to be made to address what to do in the event of animal death or crop loss.

For test substance application, the protocol should describe how the test substance will be administered to the test system (e.g., injection into stalk of plant, band width, post-over-top, broadcast, oral gavage, or intravenous) and the reason for the choice of a particular method. This section should also describe the amount of test substance that will be administered and the frequency of application of the test substance (e.g., 2.0 mg/kg body weight/day for 10 weeks; 50 mL injected every 2 weeks for 6 weeks; one spray application at 0.14 lbs of active ingredient per acre). In the analytical portion of a residue study, the frequency and use of procedural recoveries should be discussed.

Sample Analysis Phase (Analytical Phase). For tests, analyses, and measurements, the type and frequency of measurements taken or data gathered during the study, such as environmental conditions, soil-moisture determinations, stability in diet or dose, body weights, animal observations, egg production, necropsy specifications, and harvest and sampling intervals, must be defined. The procedures that will determine how much of the test substance is in or on the test system must be provided. Analytical methods or SOPs may be cited. Otherwise, sufficient detail must be given to describe the procedures. Nonapproved (i.e., draft or nonvalidated) procedures or contract laboratory methodologies should be attached to the protocol as proposed methods. All attachments should bear the protocol number, a protocol attachment number, and separate pagination.

Data Evaluation Phase. This section should detail how the numerical data will be handled during the study and included in the final report (i.e., average, standard deviation, or chi-square test), how outliers will be handled (the criteria for rejection or acceptance of data, such as limits of standard deviation), and how the raw data will be used (e.g., if two GC runs are obtained for each sample, the averaged results will be used in the final report).

Records To Be Maintained [§ 160.120(a)(13)] (Appendix 6.IX). The records section should identify completely what records are to be generated and maintained and by whom, where these records are to be kept during the study, and where these records are to be stored or archived once the study is terminated, discontinued, or completed. This section should also address what reports (e.g., interim reports, analytical reports, contract laboratory reports, and final report) will be issued and by whom, and where the copies and originals of these reports will be archived.

References or Attachments (Appendix 6.X). When applicable, list all cited material in the protocol. Provide sufficient information to ensure that the documents can be secured by the study participants.

Optional Information (Appendix 6.XI). Optional information may be included at the study director's prerogative and inserted in the protocol where applicable. Examples of optional information are

- radiation safety

- study alterations (details on how to handle alterations)

- statement concerning GLPs

Protocol Details. 40 CFR § 160.120(a) states that the approved written protocol must "clearly indicate the objectives and all methods for the conduct of the study." The protocol must contain or cite all information that is needed to perform the study. If all of this information is not available at the time the protocol is developed, this fact should be noted in the protocol, and the information should be provided by a protocol amendment when it becomes available. The protocol should be supplemented, wherever possible, with details available in applicable SOPs, standard methods, and reference literature. Citing the necessary material will reduce the text of the protocol. Listing the pertinent SOPs, methods, and literature allows the study director to provide exact instructions as to how the study should be performed. If cited material (e.g., a sponsor's draft analytical method) would not be readily available to the study personnel, the document should be included as an attachment to the protocol and should be part of the overall pagination (*see* Appendix 6.XII). Specific SOPs should be cited in the text of the protocol, and the SOP number, revision or edition number, and title should be included on the reference pages of the protocol (*see* Appendix 6.X). In some cases, more than one SOP may describe a procedure. In such cases, the protocol should specify the appropriate SOP to minimize guesswork on the part of the study participants, interruptions of the study director, and the possible use of an inappropriate SOP. Including the SOP revision or edition number in the protocol can help eliminate the possibility of using an invalid, obsolete SOP. Listing in the protocol the specific SOPs to be used in a study also aids the QAU in preparing more effectively for in-progress inspections. The amount of detail to be contained in a protocol can be looked at another way. Study protocols, supported by SOPs and cited or included documents, should be documents that can stand alone and that contain enough information to reconstruct or repeat the study if necessary.

Protocols, for the most part, describe how the study will take place rather than how the study will not take place. The protocol should not discuss events that will not take place unless questions will arise if the matter is not addressed. The choice of words in protocols can provide any flexibility the study director may need for the study. For example, instead of stating that "the samples will be collected six hours after application" state that "the samples will be collected five to seven hours after application". Protocols do not necessarily have to be exact as long as the flexibility that is built into the protocol does not adversely affect the conduct of the study.

Protocol Alterations. Alterations to a study must be documented and acknowledged by the study director. There are two types of protocol alterations: a protocol *amendment* and a protocol *deviation*. Protocol amendments are anticipated (foreseen circumstances), planned, permanent, and deliberate systematic changes [*see* 40 CFR §§ 160.33(a), (c), 160.120(b), and 160.185(a)(2)]. Protocol amendments are usually documented on standardized protocol-amendment forms that are signed by the study director and maintained with the protocol. Management must sign protocol amendments to change study directors. Management should also sign protocol amendments that terminates the study. Protocol amendments must clearly document what is being changed (e.g., section, part, page) and how it is being changed (e.g., 40 mL to 400 mL, 3 days to 36 days). In addition, the amendment must also give the reason for the change. Protocol amendments must be issued prior to the planned change. At times, however, protocol amendments are issued after the change because the study cannot be held up to allow a protocol amendment to be generated and signed by the study director. In these cases, the amendment may include an "effective date" indicating when the changes became effective.

Protocol deviations are unanticipated (unforeseen circumstances), unplanned, unintentional, or uncontrolled variations of the study [*see* 40 CFR §§ 160.33(c) and 160.81(a)]. A deviation is any departure, both in a positive sense (more than what is required) and a negative sense (less than what is required), from approved procedures. All deviations must be documented in the raw data, and the study director must be informed. There must be a record that the study director is aware of the deviation; that is, any deviations recorded in a laboratory notebook must bear the study director's dated signature or initials. This requirement includes deviations from SOPs and analytical methods.

Conclusion

Protocols must be written to satisfy the FIFRA GLP Standards and to provide investigators with a clear plan for how a given study is to be performed. The requirements of 40 CFR Part 160 must be critically reviewed and understood before beginning to draft a protocol. After establishing the contents of a protocol, the information must be organized in a useful manner.

Acknowledgment

I thank N. J. Atherton for her contribution to the text of this paper and P. Scarbrough for her editorial expertise. I also thank C. R. Ganz and R. M. Speth for their invaluable assistance in the overall continuity of this chapter and in ensuring accuracy.

References

1. "Good Laboratory Practice Standards, Final Rules", *Code of Federal Regulations* Title 40, Pt. 160; *Federal Register* 54:34052 (August 17, 1989).

2. "Data Requirements For Pesticide Registration, Final Rule", *Code of Federal Regulations Title 40, Pt. 158; Federal Register*, 49142881 (October 24, 1984) and amended *Federal Register*, 53:15991 (May 4, 1988).

3. Akerman, J. W. "PR Notice 86-5, Notice to Producers, Formulators, Distributors, and Registrants", Attachment I: Pesticide Assessment Guidelines, U.S. Environmental Protection Agency, Washington, DC, July 29, 1986.

RECEIVED for review September 17, 1990. ACCEPTED revised manuscript September 13, 1991.

Appendix 6.I. Study Title, Identification, Signatures, and Addresses

AGRICULTURAL DIVISION
CIBA-GEIGY DIVISION

RESIDUE CHEMISTRY DEPARTMENT
PROTOCOL NUMBER 59-88-PART B-1

DETERMINATION OF RESIDUES OF _____ IN FIELD CORN AND ROTA-
TIONAL CROP SAMPLES FOLLOWING APPLICATIONS OF _____ TO FIELD
CORN

Project No.:
Purchase Order No.:

STUDY DIRECTOR:	APPROVAL BY:
TITLE: Project Scientist	TITLE: Manager, Environmental & Contract Studies
SIGNATURE:	SIGNATURE:
DATE: *6/18/90*	DATE: *6/19/90*
STUDY MONITOR:	PROTOCOL AUDIT BY CIBA-GEIGY OAU
TITLE: Senior Group Leader	QA AUDITOR: *[signature]*
SIGNATURE:	DATE: *June 19, 1990*
DATE: *6/14/90*	
CONTRACT LAB STUDY DIRECTOR:	CONTRACT LAB MANAGEMENT APPROVAL BY:
TITLE: Research Associate	TITLE: Group Leader
SIGNATURE.	SIGNATURE.
DATE: *6/27/90*	DATE: *6/27/90*
SPONSOR:	TESTING FACILITY:

CIBA-GEIGY Corporation
Agricultural Division
410 Swing Road
P. O. Box 18300
Greensboro, NC 27419

PROPOSED EXPERIMENTAL START DATE, PART A: 4/6/88

PROPOSED EXPERIMENTAL STARTING DATE: 7/1/90

PROPOSED EXPERIMENTAL TERMINATION DATE: 8/31/90

PROTOCOL

METABOLISM OF   IN GOATS

Study No.:   F-00088
Project No.:   010345
Greensboro Protocol No.:   245-89

Study Director: Date: _12/12/89_

Alternate Study Director: Date: _12 Dec 89_

Section Manager: Date: _12/12/89_

Director, Metabolism   Date: _12/14/89_
Department, Greensboro:

Approved by:
Director, Environmental   Date: _12/12/89_
Health Center

Conducted by: CIBA-GEIGY
                           Environmental Health Center
                           400 Farmington Avenue
                           Farmington, CT 06032

Sponsor: CIBA-GEIGY
                           Agricultural Division

          Headquarters:
                           410 Swing Road
                           P.O. Box 18300
                           Greensboro, NC 27419-8300

Appendix 6.II. Study Personnel

AGRICULTURAL DIVISION PROTOCOL NUMBER 78-90
METABOLISM DEPARTMENT
PAGE 9 OF 26

PERSONNEL

An Alternate Study Director may be specified by protocol amendment if the Study Director is unavailable for an extended period of time.

All personnel listed below will be placed on the distribution list to receive copies of this protocol and its amendments:

1.   Manager, Plant Metabolism and Environmental Chemistry
2.   Senior Group Leader, Plant Metabolism
3.   Senior Chemist, Study Director
4.   Biological Research Specialist I, Biological
   Coordinator
5.   Radiation Safety Officer, Radiation Safety
6.   Senior Group Leader, Chemical Synthesis, Chemistry
   Coordinator
7.   , Associate Chemist, Project Analyst
8.   Shipping and Receiving

Appendix 6.III. Study Dates and Schedules

AGRICULTURAL DIVISION PROTOCOL NUMBER 78-90
METABOLISM DEPARTMENT
PAGE 12 OF 26

TABLE IA - PROPOSED TREATMENT SCHEDULE - BIOLOGICAL PHASE

DATE (1)	SCHEDULED ACTIVITY	
6/21/90	Pre-application Soil Sample	
6/22/90	First spray application of control & cores	soil
6/29/90	Harvest ~1/5 of the plants; soil cores	
7/2/90	Ship samples to Greensboro	
7/10/90	Second spray application of control & cores	soil
7/17/90	Harvest ~1/5 of the plants; soil cores	
7/19/90	Ship samples to Greensboro	
9/3/90	Harvest all of the plants; soil cores	
9/6/90	Ship samples to Greensboro	

TABLE IB - PROPOSED SCHEDULE - CHEMICAL PHASE

DATE (1)	SCHEDULED ACTIVITY
6/19/90	AG Chemical Synthesis Group (CSG) will provide the Study Director with ɔf (ᴵ(2)
6/19/90	The Study Director will formulate, radioassay, and chromatograph the test compound
6/20/90	The Study Director will give the formulated test compound to the AG CSG to ship via overnight delivery to the Biological Coordinator
7/9/90	The Study Director will formulate, radioassay, and chromatograph the test compound
7/9/90	The Study Director will give the formulated test compound to the AG CSG to ship via overnight delivery to the Biological Coordinator

   (1)   Dates are only approximate; actual dates of application will be determined at the descretion of the Biological Coordinator

   (2)   If the second label is not available a protocol amendment will be issued to propose a schedule fro this compound

Note: The Study Director should be notified immediately of any significant deviations from this schedule.

F-00088: METABOLISM OF [IN GOATS

Chemistry Coordinator:
(responsible for Chemical Synthesis   Senior Group Leader
Group support)   Greensboro, NC

Biological Coordinator:
(responsible for biological activities   Senior Group Leader
conducted at the Vero Beach   Vero Beach, FL
Research Center)

Veterinarian:

                                Vero Beach, FL

6. SCHEDULE

Study Initiation:
Proposed Acclimation of Animals:
Proposed In-Life Starting Date:
Proposed Sacrifice Date:
Proposed Experimental
   Termination Date:
Proposed Draft Report Date:
Proposed Final Report Date:

7. TEST SUBSTANCE DATA

The test substance for this study will be a mixture of [
and non-radiolabeled   A protocol amendment will be
issued indicating the estimated dosage of the test substance (as ppm in
the feed), as well as stating the chemical reference number, specific
activity, radiochemical purity, chemical purity, reanalysis date, and
other physicochemical data relevant to the test substance. The Chemical
Synthesis Group will have documented the data regarding characterization
of the test substance prior to the start of dose administration.

a. Nonlabeled Compound

1. Identification:

2. Source: Chemical Synthesis Group
                   CIBA-GEIGY Corporation
                   Greensboro, NC

3. Safety Precautions: See   Safety Information Supplement
(attached) for handling the test substance, accidental exposure,
spill handling and disposal of the nonlabeled compound. For
handling and disposal of the radiolabeled material, follow the
Standard Operating Procedures for the
(No. M15) and Greensboro Metabolism Laboratory (No. 2.4).

b. Radiolabeled Compound

1. Identification: [

Appendix 6.IV. Study Purpose and Objectives and Chemical Substances

AGRICULTURAL DIVISION PROTOCOL NUMBER 78-90
METABOLISM DEPARTMENT
PAGE 2 OF 26

STUDY OBJECTIVES

The   test   compound,   is a herbicide under development by
CIBA-GEIGY Corporation, Agricultural Division.
                                   will be applied to
                   The objectives of this study are 1) to
determine   the   uptake following   applications with
          2) to determine the distribution of the   and its
metabolites   in individual plant parts; and 3) to characterize and identify
the major metabolites of the test compound.

Application of the   label of   will be defined
in a separate protocol if this radiolabel becomes available for test
application. This study will be conducted in accordance with this protocol
and applicable CIBA-GEIGY Standard Operating Procedures.

TEST AND REFERENCE SUBSTANCES

The radioactive test chemical to be applied in this study will be
                                    (Figure 1). The test chemical
will be supplied by the CIBA-GEIGY Chemical Synthesis Group, Greensboro,
North Carolina. The specific activity of the test chemical will be
approximately   micro curies per milligram (µCi/mg). The purity of the
test chemical will be at least   The synthesis reference code,
radiochemical purity, chemical purity, specific activity, and   the
formulation lot number of each of the test chemical will be provided as a
protocol amendment(s) prior to the first application.

The test substance,   test chemical, will be shipped
by overnight carrier to   days prior to the scheduled
application dates. The shipments will contain the following information:
name of the test substance, spray treatment number, amount of test
substance, the radiochemical purity information, and the experiment number.
The test substances will be applied to the plants within   hours of
receipt.

The radiochemical purity, expiration date and lot number of the
chemical will be provided by the Chemical Synthesis group and documented by
the Study Director in an amendment to this protocol. A protocol amendment
will also be used to document this information for any analytical standards
of metabolites (reference chemical) provided by the Chemical Synthesis
group.

Appendix 6.V. Chemical Substances

RESIDUE CHEMISTRY PROTOCOL 20-90 PART B
AGRICULTURY DIVISION
CIBA-GEIGY
PAGE 2 OF 5

STUDY OBJECTIVE

The objective of this part of the study is to determine residues
of   in celery and
celery fractions that have been prepared for consumption.

TEST AND REFERENCE SUBSTANCE

A. Field Test Substances

Refer to the field trial notebooks located in the Residue
Chemistry Department Archives under Protocol 20-90-A, Project
Number   , and Field Test Number

B. Analytical Reference Substances

1.

Analytical Section Inventory Number: S87-1185-1
Residue Chemistry Department Inventory Number: B05900
Purity: 96.7%
Storage Conditions: Freezer
Reassay Date: 2/92

2.

   Analytical Section Inventory Number: S89-1459
   Residue Chemistry Department Inventory Number: B05901
   Purity: 95%
   Storage Conditions: Freezer
   Reassay Date: 6/91

3.

   Analytical Section Inventory Number: 54216
   Residue Chemistry Department Inventory Number: B05902
   Purity: 99%
   Storage Conditions: Freezer
   Reassay Date: 6/91

4.

   Chemical Synthesis Lot Number: DPS-I-59-2
   Residue Chemistry Department Inventory Number: B05903
   Purity: 99.4%
   Storage Conditions: Freezer
   Reassay Date: 6/92

Appendix 6.VI. Test System

                    RESIDUE CHEMISTRY DEPARTMENT PROTOCOL
                    PROTOCOL 8-90
                    AGRICULTURAL DIVISION
                    CIBA-GEIGY CORPORATION
                    PAGE 2 OF 10

TEST SYSTEM

This storage stability study will be conducted using weathered
crop samples. These samples were frozen immediately after
harvesting in the field and shipped directly to the Residue
Chemistry Department Sample Receiving Section for long-term
storage and preparation for analysis. Analysis will be conducted as soon as possible after the samples are prepped and
delivered to the analysis laboratory.

Samples and the project numbers to be considered are listed in
the following table:

Project Number	Crop	Biological Protocol Number
	Potatoes	43-89-A
	Spinach	164-89-A
	Cranberries	147-88-A
	Peppers	23-89-A
	Corn	32-89-A
	Peanuts	46-89-A

All substrates will be prepared according to SOP 7.21 (Revision 0).

Each sample will receive a label with its own unique identification code. Refer to the residue stability report sheets for the lab code and proposed storage intervals for these samples.

After the initial analysis, a Storage Stability Test Number will be assigned to identify the samples used for the remainder of the storage stability study. An amendment to this protocol will be issued to document these new numbers.

PROTOCOL 17-90-C
PAGE 2 OF 6

OBJECTIVE(S)

The objective of this study is to validate the Production Technical Analytical Services' (PTAS) LAS computer system to ensure that it meets the users' needs and that the data it generates is accurate, reliable and secure.

TEST SYSTEM

Hardware: Hewlett-Packard Company HP-1000 model A-900 computer.

Operating System: Hewlett-Packard Company RTE-A 5.0.

Applications: Hewlett-Packard Company HP-3350 LAS D.00.00
Hewlett-Packard Company LOOP Subsystem D.00.00
Hewlett-Packard Company CPLOT C.00.02

SYSTEM OVERVIEW

The Production Technical Analytical Section uses a Hewlett-Packard A-900 computer with a loop-based data acquisition system to receive chromatographic data from either gas chromatographs or liquid chromatographs. LAS (Laboratory Automation System) is a collection of computer programs running on the A-900 that converts this data into analytical results. The system provides a central location where raw chromatographic data is collected, analyzed, reported and archived.

Raw data is defined as the paper report that is printed following an analysis. The report can be traced back to a file containing area slices representing a time sequence of digitized voltages received from the instrument during the run. For peak areas, an integration program calculates a baseline and sums the slice data for peaks it detects. (On occasion, peak heights are also calculated using similar algorithms). The peak areas are proportional to the amount of the material present in a sample. The user supplies peak identities (as retention times) and sample weights. LAS calculates standard factors which are applied to the samples to give the assays. Area distributions are calculations where the individual peak areas are divided by the total peak area in a run, giving the fractional proportion of that component in the sample. System utilities are included that tabulate the data, print the results and plot the chromatograms. The data is backed up on tape and periodically archived for storage and later retrieval.

RESIDUE TEST REPORT

FIELD TEST NUMBER: _____

TEST SUBSTANCE:
TEST SYSTEM: __WILD RICE__

ANALYTICAL SECTION

PROTOCOL NUMBER: _117-89-PART 3_

RESULTS

SAMPLE CODE	GRAMS AI/ACRE	FORMULATION	LAST APPLICATION DATE(S)	SAMPLE DATE(S)	PHI (DAYS)	DATE EXTRACTED	DATE ANALYZED	RESIDUE PPM*
STRAW								
1-1-A	CONTROL	--	--	3/14/89	--			
2-1-A	100		7/10/89	3/14/89	35			
2-1-B	100		7/10/89	3/14/89	35			
4-1-A	75+75		7/17/89	3/14/89	28			
4-1-B	75+75		7/17/89	3/14/89	28			
6-1-A	75+75	(Aerial)	7/17/89	3/14/89	28			
6-1-B	75+75	(Aerial)	7/17/89	3/14/89	28			
GRAIN								
1-2-A	CONTROL	--	--	8/14/89	--			
2-2-A	100		7/10/89	8/14/89	35			
2-2-B	100		7/10/89	3/14/89	35			
4-2-A	75+75		7/17/89	3/14/89	28			
4-2-B	75+75		7/17/89	3/14/89	28			
6-2-A	75+75	(Aerial)	7/17/89	3/14/89	28			
6-2-B	75+75	(Aerial)	7/17/89	3/14/89	28			

RECOVERIES   % RECOVERED

     1-1A + 1.0 PPM
     1-2A + 0.05 PPM

COMMENTS: RESULTS ARE NOT CORRECTED FOR CONTROL VALUES.
          RESULTS ARE CORRECTED FOR PROCEDURAL RECOVERIES <100%.
          *RESIDUES DETECTED AS   AND CONVERTED TO
                      EQUIVALENTS USING THE FACTOR

Appendix 6.VII. Justification of Test System

GENERAL PROCEDURES

Sample Identification

All samples will be labeled with at least the study number, a unique sample designation, and isotope.

Sample Storage

For the   , all samples or subsamples after analysis will be kept in a freezer for possible additional analyses. After extraction, all soil samples will be air dried and stored at room temperature until all analyses are completed. Samples from   once all data have been analyzed.

JUSTIFICATION FOR TEST SYSTEM AND METHOD OF APPLICATION

The test system and method of application were selected to comply with the EPA Pesticide Assessment Guidelines, Subdivision N, Sections 162-1 and 162-2.

EXPERIMENTAL PROCEDURES

Preliminary Investigations

Preliminary investigations will be performed to define the experimental parameters of the definitive study. These parameters will include, but not be limited to, extractability of   and   products from soil, the chromatographic behavior of   reference standard(s), and   products in preliminary study samples, and an estimate of the degradation rate of   in soil.

Separation of the   products will be accomplished by using   Confirmation of the number degradation products will be conducted using

Portions (approximately   of sieved soil (2 mm) that has been air dried appropriately to maintain microbial activity will be used. At   soil samples in amber bottles will be fortified with   and, if necessary, nonradiolabeled   dissolved in   at a concentration of approximately   The volume of test material added to each sample will be less than   The solvent will be evaporated under nitrogen, if necessary, and the bottles will be capped and gently hand tumbled. The moisture content of the soil in each bottle will be adjusted to approximately 75% FMC (measured at 0.33 bar) by addition of water. Seven fortified soil samples will be placed in an aerobic incubation chamber connected to a series of traps for collecting organic volatiles and One sample will be removed at approximately 1, 3, 5, 7, 9, 14, and 30 days after fortification and analyzed as described below.

Appendix 6.VIII. Experimental Design and Procedures

EXPERIMENTAL DESIGN

The Multiresidue Method Testing will be conducted according to
EPA Protocols A, C, D, and E, and "FDA Pesticide Analytical
Manual, Volume I." Replicate analyses are to be conducted at
each level of fortification of the test substances. The
reagent blank will not be run in replicate. The levels of
test substance fortification for each test system are outlined
in Table I. In accordance with Protocols A-E, matrices will
be fortified at 0.05 ppm, at 0.25 ppm, at 0.5 ppm, at 12.0 ppm
according to EPA Protocols and current tolerances for
                      In the cases where 0.05 ppm is the current
tolerance, the higher fortification level will be determined
by the requirements in each Multiresidue Protocol.

ROUTE OF ADMINISTRATION (Procedural Recovery)

The test substances will be administered in solvents specific
to chemicals and analytical tests to crop samples at the
levels described in Table I, each in duplicate. These
fortified samples will be carried through the procedures of

                    RESIDUE CHEMISTRY PROTOCOL 59-88 B-1
                    AGRICULTURAL DIVISION

                    PAGE 3 OF 20

individual crops are provided in Protocol 59-88 Part A. A list
of samples to be analyzed is provided on the attached Report
Sheets. Each sample to be analyzed will have a label bearing,
at least, the field test, sample code, and project number.

Any additional samples which will be analyzed during the course
of the study will be documented in a protocol amendment.

TEST SUBSTANCE APPLICATION

Details are provided in Protocol 59-88-Part A.

EXPERIMENTAL DESIGN

Samples were generated, collected, shipped, and prepared for
analysis under Protocol 59-88-Part A. The samples listed on the
attached Report Sheets will be analyzed using Analytical Method

TEST/ANALYSES/MEASUREMENTS

Method   will be used and a screening level for the analyses will be   ppm. Any modifications to the method will be
documented in the raw data and detailed in the analytical report
to the Sponsor. Residue data will be reported with corrections
for procedural recoveries (<100%) in each analytical set. Residues will not be corrected for control values except for the
recovery samples.

RECORDS TO BE MAINTAINED

Copies of the protocol, protocol amendments if any, and analytical methods will be made available to the Contract Lab Study
Director. The contract laboratory analysts will maintain records in which they will record all procedures, weighings, observations, etc., relevant to the experimental work. Chromatograms, computer printouts, etc., will be clearly labeled and
maintained by the contract laboratory analysts until the conclusion of the study. All the raw data will be approved by the
Contract Lab Study Director and audited by the Quality Assurance
Unit of the contract laboratory before submittal to the Sponsor.
The raw data are subject to audit by the Sponsor and/or Sponsor
QAU.

AGRICULTURAL DIVISION PROTOCOL NUMBER 121-90
RESIDUE CHEMISTRY DEPARTMENT
PAGE 3 OF 25

Sample number G-0:   Control grapes
Sample number G-01A: Control grapes fortified with   ppm
                        replicate A.
Sample number G-01B: Control grapes fortified with   ppm
                        replicate B.
Sample number G-05A: Control grapes fortified with   ppm
                        replicate A.
Sample number G-05A: Control grapes fortified with   ppm
                        replicate B.
Sample number G-0.1:   Control grapes fortified with   ppm
                        no replicate.

JUSTIFICATION OF TEST SYSTEM

Analysis of control and   fortified control apple and
grape samples by "Draft" Analytical Method AG-579 (Appendix I)
will be performed to determine the accuracy and precision of
the method for

EXPERIMENTAL DESIGN

CIBA-GEIGY "Draft" Analytical Method AG-579 (Appendix I) will
be used to determine

Fortified Samples - "Draft" Analytical Method AG-579 (Appendix
I). See also the "TEST SYSTEM" section.

Modifications - Any modifications will be documented with
protocol amendments.

The experiments will consist of the analysis of control and
fortified control apple and grape samples fortified at or
above the screening level of "Draft" Analytical Method AG-579
(Appendix I).

The sets of samples to be analyzed in this study are outlined
in Table I. The accuracy of the method used in this study
will be confirmed by the recovery results from the analyses of
fortified control samples. The precision of the method will
be determined by the reproducibility of the amounts of
                determined by the method.

The control of bias in the study will be accomplished by the
use of control samples for all fortification experiments.
Other experimental design details are to be found in Appendix
I, "Draft" Analytical Method AG-579.

EXPERIMENTAL DESIGN

As the prepped samples from each substrate arrive at the
laboratory, do the following:

1. On day zero weigh, as described in SOP 7.23 (Revision 0),
    three 10.0-gram subsamples from the control and two
    10.0-gram subsamples from the treated sample into 500-ml
    round bottom flasks.

2. Fortify two control samples with   one at the
    screening level of   ppm and the other in the range
    expected in the residue samples for the substrate.

3. Analyze a control sample, the two freshly fortified samples,
    and two weathered residue samples by procedures described
    under TEST, ANALYSES, MEASUREMENTS.

4. Place remaining control and treated sample in freezer
    immediately after subsampling.

After evaluating the residue results from analysis of the
substrates at various PHI's, select two substrates at PHI's
which are best suited for a weathered residue study (i.e., the
most residues at the highest levels) and proceed as follows:

1. On the date chosen for analysis of the first storage
    interval, remove from the freezer the samples selected as
    most suitable for the storage stability study, and weigh
    into 500-ml round bottom flasks control and treated samples
    for analysis as described in Step 1 above.

2. Analyze the first storage interval samples (2-month inter-
    val) as described for the 0-day analysis. Place the remain-
    ing samples in the freezer immediately after subsampling.

3. On each subsequent analysis date, remove appropriate samples
    from freezer storage, subsample, and analyze as described
    for the 0-day analysis. Typically, for each substrate at
    each analysis interval, the sample scheme is as follows:

    (1) Control.
    (2) Control freshly fortified at the screening level with

    (3) Control freshly fortified near the level of expected
         weathered residue levels with
    (4) Weathered residue sample.
    (5) Replicate of (4).

Samples will be analyzed at intervals from 0-day through
24-months unless the study is extended by the Study Director.
The reason for the extension will be documented in an amendment.
The proposed analysis dates are as follows:

Interval	Proposed Analysis Date
2 month	3/30/90
6 month	6/30/90
12 month	1/30/91
18 month	6/30/91
24 month	1/30/92

Appendix 6.IX. Records To Be Maintained

> RESIDUE CHEMISTRY PROTOCOL 59-88 B-1
> AGRICULTURAL DIVISION
> CIBA-GEIGY CORPORATION
> PAGE 4 OF 20

RECORDS TO BE MAINTAINED

Copies of the protocol, protocol amendments if any, and analytical methods will be made available to the Contract Lab Study
Director. The contract laboratory analysts will maintain records in which they will record all procedures, weighings, observations, etc., relevant to the experimental work. Chromatograms, computer printouts, etc., will be clearly labeled and
maintained by the contract laboratory analysts until the conclusion of the study. All the raw data will be approved by the
Contract Lab Study Director and audited by the Quality Assurance
Unit of the contract laboratory before submittal to the Sponsor.
The raw data are subject to audit by the Sponsor and/or Sponsor
QAU.

The contract laboratory will write an analytical study report
following EPA reporting requirements as defined by the Pesticide
Assessment Guidelines (Subdivision O, Guideline 171-4) and the
associated Addenda on Data Reporting, and the EPA FIFRA GLP
Standards, 40 CFR Part 160.185. All raw data, including original chromatograms, will be submitted to the Sponsor. A copy of

the analytical report including copies of the raw data, will be
archived at the contract laboratory. All original study documents submitted to the Sponsor will be archived in the Residue
Chemistry Departments Archives. The Study Director will submit
a summary report (ABR report) that will summarize and evaluate
the biology data generated under Protocol 59-88-A and the data
reported by the contract lab. The ABR report will be archived
in the Residue Chemistry Department Archives.

All samples and sample extracts from this study will be retained
by the contract laboratory until permission for their disposal
is obtained from the Sponsor.

GOOD LABORATORY PRACTICES

All applicable sections of the EPA-FIFRA Good Laboratory Practice Standards, 40 CFR Part 160, will be followed throughout the
conduct of this study.

Appendix 6.X. References or Attachments

AGRICULTURAL DIVISION PROTOCOL NUMBER 78-90
METABOLISM DEPARTMENT
PAGE 10 OF 26

REFERENCES

1.   Handbook for Radioactive Studies Carried Out   On CIBA-GEIGY Research
      Stations, Agricultural Division, CIBA-GEIGY Corporation, 410 Swing
      Rd., Greensboro, NC 27419.

2.   Somody,   C. and Ellis, J.,   SOP 24.85, Rev. #1, "Operation,
      Calibration, and Maintenance of Custom-Built Remote-Controlled
      ^{14}C-Sprayer".

3.   Byers, R., SOP 7.13, Rev. # 0, "Packaging and Shipment of Radioactive
      Plant and Soil Samples."

4.   Blair, S., SOP 7.14, Rev. # 0, "Metabolism Sample Receipt and
      Log-in."

5a.   Simoneaux, B. J. and J. Rogers, AG-223, "Blending of Soils and
      Homogenization of Biological Materials for Radioassay and
      Extraction."

5b.   Nelson, P., SOP 4.60, Rev. #0, "Operation of the Computrac Max 50
      Moisture Analyzer."

6.   Nelson, P., SOP 7.8, Rev. #0, "Homogenization of Soil Samples for
      Metabolism Studies."

7.   Torbett, M., SOP 4.67, Rev. #1, "Operation, Maintenance and
      Calibration of Manual Harvey OX-400 Oxidizers"

8.   Simoneaux, B. J. and T. Capps, SOP 6.15, Rev. #0, "Isolation, Characterization, and Identification of Metabolites in Various Matrices"

9.   CIBA-GEIGY Agricultural Division Protocol 247-89.

10. Pesticide Assessment Guidelines, Subdivision O, Hazard Evaluation, Residue Chemistry Series 171-4 (a)(1)&(2), <u>Nature of the Residue: Plants</u>, Addendum 3 on Data Reporting (12/1986); US NTIS PB87-208541.

Appendix 6.XI. Optional Information

F-00088: METABOLISM OF [IN GOATS

14. FINAL REPORT

A biological report containing pertinent information and tables or forms describing the conditioning, treatment, and necropsy stages of this study will be submitted to the Study Director after completion of the biological phase. A final report will be prepared by the Study Director upon completion of all phases of the study. The biological report will be included as an integral part of this final overall report.

15. DATA RETENTION

The notebooks, forms, a photocopy of the biological report, and all other records pertaining to the biological phase will be archived at the

Data regarding the characterization and purity determination for the test substance and preparation of capsules will be archived in the Chemical Synthesis Group,

All other data and correspondence pertaining to this study, along with the originals of the protocol and all final reports will be archived in the Technical Information Center at the

Study samples will be stored in the   Laboratory for as long as they afford reevaluation.

16. QUALITY ASSURANCE

Quality Assurance Units (QAU) will monitor the study and audit the final report. The Biological and Chemistry Phases and the biological report will be monitored and audited by the Greensboro QAU. The Analytical and Characterization Phases, plus the overall final report, will be monitored and audited by the   QAU. The nature of each of the study phases have been defined previously in this protocol.

17. ALTERATION OF PROTOCOL

All changes or revisions of this protocol and the reasons for the changes will be documented, signed, and dated by the study director and maintained with the protocol. All persons who receive the final protocol will receive copies of any protocol amendments.

18. RADIATION SAFETY

The use and handling of radioactive materials will be in conformity
with the provisions, requirements and conditions of Florida Radioactive
Materials License 261-1 (D84), North Carolina Radioactive Materials
License 41-450-1, and Connecticut Radioactive Materials License 06-
19068-01.

Appendix 6.XII. Referenced Material

AGRICULTURAL DIVISION PROTOCOL NUMBER 121-90
RESIDUE CHEMISTRY DEPARTMENT
APPENDIX I   "DRAFT" AG-579
PAGE 7 OF 25

Analytical Method for the Determination of   in Apples and
Grapes by High Performance Liquid Chromatography With Column-Switching

"DRAFT"
METHOD NO. AG-579
CIBA-GEIGY CORPORATION
RESIDUE CHEMISTRY DEPARTMENT
POST OFFICE BOX 18300
410 SWING ROAD
GREENSBORO, NC 27419

SUBMITTED BY: TITLE: Chemist

SIGNATURE:

APPROVED BY: TITLE: Senior Group Leader
                                     Method Development

SIGNATURE:

DATE:

Appendix 6.XIII. Protocol Alterations

AG DIVISION PROTOCOL AMENDMENT FORM

AMENDMENT NUMBER: **# 2 ᵗ** PROTOCOL NUMBER: 78-90

TITLE: UPTAKE AND METABOLISM OF
TREATED WITH

Study or Project Number: Date Effective   6/22/90

CHANGE(S):   (Amendment #1: Changes 1-5)

6. EXPERIMENTAL DESIGN, Preparation of Spray Solutions, page 5:

The vial containing the test substance will placed in the refrigerator the afternoon before preparation of the second spray solution.

Reason(s): Difficulty was experienced in dissolving the formulation for the first spray application which had been stored in the freezer.

7. EXPERIMENTAL DESIGN, Preparation of Spray Solutions, page 5:

The formulation is dissolved in   of and   aliquots are taken immediately after mixing.   is added to give a final concentration of   water (v/v). The solution is diluted to a final volume of approximately   and the final volume is recorded.   aliquots are taken immediately prior to application. The solution is applied until the spray bottle is completely dry and therefore aliquots are not taken and the completion of the application.

Reason(s): The Biological Coordinator recommended the use of   to ensure even distribution of the spray solution. Aliquots not taken at completion because entire solution can be applied in less than one minute.

8. EXPERIMENTAL DESIGN, Preparation of Spray Solutions, page 5:

Both bottles of control formulation were dissolved for   An additional bottle of control formulation will be sent to the Biological Coordinator.

Reason(s): The bottles contained approximately one-half the amount of formulation that was applied to   that was one-half the size of   are the same size.

STUDY DIRECTOR _____ DATE 6/27/90

Amendments to be distributed per Protocol Distribution List.

Page 1 of 1

Test Substance and Specimen Chain of Custody in Field Studies

Markus M. Jensen

Jensen Agricultural Consultants, Inc., Washington, LA 70589

Good laboratory practices for field residue studies have been implemented to provide accurate and complete data in all phases of field trials. The strict documentation involved in these studies, starting with the receipt and storage of the test substances, through application to the test system and finally specimen collection and shipment, is necessary to ensure a complete and trackable chain of custody. If at any point throughout this process the test substance or sample is altered or misrepresented, the integrity of the data generated will be questionable. In this chapter, the various phases in the chain-of-custody process are examined. Various sample-identification systems, documentation requirements, and industry trends are discussed.

GOOD LABORATORY PRACTICE (GLP) STANDARDS have expanded the scope of the Environmental Protection Agency's (EPA's) regulations to encompass residue chemistry and environmental-fate studies in field research. Strict documentation is necessary to provide accurate and complete data on all phases of field trials as well as on the handling of test, control, and reference substances.

A clear, precise data trail from initial test substance receipt through storage, application, specimen collection, and return of shipment ensures a complete and trackable chain of custody. If at any point throughout this process substances are altered or misrepresented, the integrity of the data generated will be questionable. In this chapter, the various phases of the chain-of-custody process will be examined. Some problems, solutions, and industry trends will also be discussed.

Field research is not a predictable science from one year to the next because of changing environmental conditions and varying test systems, cultural practices, and protocol requirements. Throughout the chain-of-custody process, the area that consistently changes in response to varying conditions is sample generation or collection. Documentation of sample collection is vital because samples are obtained in an unpredictable sampling environment. However, sample collection is not the only chain-of-custody phase where problems exist.

Test Substance Shipment and Receipt

Test and reference substance shipment and receipt is one area of concern. Table 7.I illustrates various problems with initial test substance shipments during a 3-year period beginning in January 1988 and ending July 1990 at Jensen Agricultural Consultants, Inc., a research facility in southern Louisiana. The data represent the number of times various problems arose with shipments from the projects of approximately 20 sponsor companies and management firms. Although some companies had no problems in this regard, the results in Table 7.I are indicative of the extent to which these problems are present throughout industry.

In 1989, problems occurred for the following reasons:

1. GLPs were not finalized until mid-October.

2. Numerous projects were cumbersome for some companies.

3. Many companies were just beginning to implement GLP programs for field research.

The reasons for additional documentation in the chain of custody were

- The wrong chemical arrived or the batch or lot numbers were incorrect (amendments were required).

- Additional chemical was required (mainly in aerial trials).

Table 7.I. Problems with Test Substance Shipments

Problem	1988	1989	1990
Wrong chemical	0	1	0
Wrong batch or lot number	0	2	2
Not enough test substance	1	1	1
Test substance not analyzed	0	2	0
Container arrived empty	1	0	0
Sent to wrong trialist	0	1	0
Test substance used for multiple projects	0	2	1
Arrived damaged	1	2	1
Arrived late	1	3	2
Arrived via U.S. mail	1	2	1
No MSDS form	1	2	1

- Return for new test substances was necessary because the original shipment had not been analyzed.

- Authorization by the study director was required to use the same chemical for different projects (some through the sponsor company and others through a management firm) because the designated test substance was not available for some of the projects.

- Substances were damaged during shipment.

- Late arrival of test substances caused applications to be made after the expected application date.

In 1990, test substance shipments improved in most areas, and this trend should continue through the 1990s as companies and management firms become more accurate and timely with substance shipments. Timely arrival of the shipment is crucial because late arrival causes a delay in application. Delayed applications can result in numerous problems: Target pests may be well past a stage of targeted control, extended preharvest interval (PHI) or treatment-to-sampling interval (T–S) timings may be disrupted, and target crops may be adversely affected (such as cotton with open bolls or rice at heading).

Study directors are usually in charge of test substance shipments, although sponsor company field representatives or analytical laboratory personnel are sometimes given this assignment. Labeling, packaging, handling, and shipping of all test substances must be properly implemented and documented.

When the test substance arrives at the field testing facility, receipt of the test substance must be documented, usually in a logbook, along with all pertinent information. This information includes but is not limited to

- test substance name and identification

- sponsor company

- project number

- transport carrier

- package identification (ID) number

- batch and lot number

- formulation

- date

- amount received

- condition upon arrival

- signature of person receiving material

- storage location and conditions

Other items may include

- container type and number

- Chemical Abstracts Service (CAS) number

- code numbers

- expiration date

Items to check upon receipt are damage to container, Material Safety Data Sheet (MSDS) forms, comparing the contents with the packing list, storage requirements, and comparing the test substance with the requirements of the protocol. The last item is most important because at this point it can be determined if lot numbers, amounts, and formulations are sufficient to carry out the objectives of the study protocol.

Test Substance Use and Tracking

Field research application methods vary widely across the United States and are determined by the nature of the test system, test substance formulation, method of application (e.g., hand sprayer, ground tractor, or aerial), and protocol requirements. Many protocols require collection of verification samples during application and may include samples of the test substance in product form, tank mixture samples, field spikes, Mylar sheets, and Petri dishes that are placed in the field at the time of application. Whenever and for whatever purpose test and reference substances are used, the information accounting for their use must be documented accurately.

Residue manuals or field notebooks should have a test substance use log section; additional pertinent information can also be entered in the daily notes section. This information must be recorded in a facility test substance use logbook and must include the amount of test substance used, the date, the use or purpose, and the signature of the person removing the substance. If deemed necessary, the amount of excess mixture and method of disposal of excess mixture and the rinsing material may be documented in the logbook.

Many test substance log sheets contain other information pertaining to application, such as method of application (e.g., ground or aerial), timing of application (e.g., preplant incorporated, preemergence, or postemergence), type of application (e.g., band or broadcast), amount and pH of carrier, and adjuvants. The test substance use section is also a convenient section for recording such information.

The amount of test substance is documented to account for use throughout the entire field trial. Many such trials may have numerous applications in one season. Because numerous applications are made, both the actual and the theoretical amount of test material remaining may be recorded as the totals may not be the same. For example, if the test substance arrives in 8 1-L containers, each container may have more or less than 1000 mL. The cumulative effect of this disparity along with numerous applications and the small amounts of test substance remaining in each empty bottle or container may result in a significant difference between actual and theoretical values. The same principle applies for test substances in wettable powder or granular form measured in units of weight. Therefore, recording and documenting the amount of test substance used at each application is imperative.

Sample Generation and Identification

Samples generated during field trials can vary in number from as few as two (untreated control and treated) in a raw agricultural commodity (RAC) study to as many as several hundred or more for soil–water dissipation trials. Regardless of the total, each sample must have a unique identification code or number to separate it from all other samples. A mislabeled or unlabeled sample is worthless. This identification system may vary among companies, protocols, and field investigators. Whatever system is used, it must fulfill the requirements of the sampling schedule and be understood by all participating parties throughout the chain-of-custody process.

Numerous types of samples can be generated during field trials, including RAC; soil, water, and sediment samples; field spikes; Mylar sheets; and Petri dishes. Preparation of all sample tubes, bottles, jars, and residue bags before the day of sampling allows the transition period from field collection to freezer to be timely and accurate. Color-coding containers to match field spike solution vials and prelabeling soil tubes and bottles with a permanent marker ensure sample identification integrity because sampling under adverse conditions such as mud, rain, snow, and dust is often unavoidable. With this procedure, samples can be wiped off, dried, and properly labeled before being frozen.

All samples must be recorded on a residue sample record and custody form. These forms vary throughout industry and are a descriptive summary of pertinent sample information that provides the analytical laboratory proper identification upon receipt and evaluation. Items on this form include but are not limited to

- project number
- sample ID numbers
- dates sampled
- treatment rates
- plot or replicate number
- sample timing
- time to freezer
- time entered freezer
- freezer temperature
- sample description
- date shipped
- box number
- shipper signature

This form may be in triplicate for all parties concerned, such as the study director, processor, QAU, company field representative, or analytical laboratory.

Various methods may be used for sample identification. RAC studies may have CK–001 (untreated) or TR–002 (treated) to identify each treatment. Multiyear projects with various rates applied at different intervals (preplant incorporated, preemergence, or postemergence) or encompassing many subplots or rotational crops need a flexible but manageable identification system. An example of this type of study would be one with a 1× rate requiring 10 soil cores to be taken at two crop-sampling intervals (midpoint and harvest) from eight rotational crops planted 30, 90, and 360 days after treatment (DAT). A sample from this project could be identified as TR–14–B–7 where TR represents treated, 14 represents the sample number (carrot roots at final harvest), B represents the 90 DAT plot, and 7 represents the particular rotational crop (carrots). The advantages of such a system are that field personnel can identify the sample in the field and that additional samples may also be included if they are needed because of high residues or if adverse weather conditions necessitate replanting of crops. Once a system is learned, sample recognition is quick and accurate.

In 1989, many companies began using computer-generated bar code labels for sample identification. This worked well for RAC studies, but for complex field studies where hundreds of labels were printed in advance, problems were numerous. Many labels were omitted, duplicated, had treated and check (control) reversed, or had misprinted information. Other labels were smeared or became distorted due to condensation on the tube or bottle as a result of sudden and extreme temperature variations (such as a 98 ° F soil core being placed into a −20 ° F freezer).

Samples need a unique identification system so that the chain-of-custody process is trackable. In addition, each sample should be identified as treated or control, top or bottom (if applicable), project number, and date sampled. This entire process is time consuming and requires preparation and detailed documentation to provide samples that adequately represent the sampling schedule described in the protocol.

Return Shipment and Documentation

Sample shipment to an analytical laboratory must be conducted in such a manner as to maintain sample integrity. Care in packaging and labeling residue boxes is necessary, and samples of similar rates of application must be boxed together from control to lowest to highest rate.

For example, pretreatment (before the first application) samples should be packaged and stored in control conditions. The most common method of shipment is by ground in a freezer truck. Freezer trucks should have a backup generator and a constant thermograph to monitor and record the temperature. Sample shipment from the field research facility to the analytical laboratory must be documented and may take 7–10 days. The second most widely used method of sample shipment is overnight air express. Because of flight delays and carrier performance records, study directors should specify which carriers are considered acceptable. Sufficient dry ice, usually triple the weight of the sample box and contents, is required when shipping by overnight air express.

A verification of sample arrival form should be included along with sample custody forms. This form is to be returned to the shipper upon arrival from the laboratory; the form describes the condition of the arriving samples. This form includes information pertaining to the shipper, receiver, test protocol, project number, number of samples, number of boxes, date shipped, date received, and condition of samples upon arrival, and it must contain the signature of the receiver.

GLPs include test substance retention and storage container requirements for the duration of a study. Standard operating procedures should specify procedures for handling and returning test substances to the sponsor company or other designated party and should state the necessity of maintaining all original shipment papers, MSDS forms, and shipping containers. Study directors use various methods and times for the return of any unused test substance. Greater than 75% of study directors surveyed approved returning the test substance upon field trial completion.

The remainder of the study directors surveyed determined that the analytical portion of the study should be completed before shipping the test substance. The study director is responsible for this decision and should provide the necessary information on how and where to ship the unused test substances and containers. Study directors may have test substances returned directly to them, sent to various laboratories, or picked up by sponsor field representatives, or, with extra documentation, may allow registered products to remain and be used by the field researcher conducting the study in a manner consistent with EPA regulations.

A verification of test substance arrival form completes the chain-of-custody process concerning field trials. Once the designated receiver signs and returns this form verifying the condition, amounts, and lot or batch numbers of the test substance, the field trial is complete, and all chain-of-custody documentation should be maintained and archived as raw data.

Conclusion: Looking Ahead with Confidence

The chain-of-custody process in field studies is constantly being refined. Chain-of-custody forms and SOPs relating to the process have expanded with the nature of GLP field research. The 1990s should provide exciting challenges, and by sharing ideas and experience, the chain-of-custody process will become more accurate, beneficial, and timely for all. Following are the key points to remember:

- Make sure the test substance arrives on time.

- Compare the test substance with the requirements of the protocol.

- Prepare, prelabel, check, and recheck all application, sampling, and chain-of-custody forms in advance.

- Documentation is essential throughout the chain-of-custody process.

RECEIVED for review September 17, 1990. ACCEPTED revised manuscript August 17, 1991.

Practical Applications for Chain of Custody Within an Analytical Laboratory

Robert J. Pollock

Analytical Development Corporation, 4405 North Chestnut, Colorado Springs, CO 80907

The climate in the agrochemical industry has changed over the past two years from limited to total documentation of chain of custody. This change is the result of the Good Laboratory Practice Standards that were enacted in October 1989. Documentation of chain of custody is necessary to provide information concerning the handling of test substances, reference substances, control samples, and treated samples within the analytical laboratory. Chain of custody includes not only the receipt of a substance, but also from whom that substance was received and the condition of the substance upon receipt. Once a substance or sample is in the possession of the analytical laboratory, the storage conditions must be documented. Chain-of-custody documentation provides a "paper trail" that tracks the removal of these items from storage for any reason: weighing, mixing, spraying, sampling, processing, assay, or shipment. All documentation must identify the item, the time and location of the transaction, the person conducting the transaction,

2192-8/92/0095$06.00/0 © 1992 American Chemical Society

and why the transaction occurred. This chapter will provide procedures on how to document the chain of custody of substances and samples within an analytical laboratory from receipt through storage, analysis, quantitation, and reporting.

W ITHIN THE AGRICULTURAL PESTICIDE INDUSTRY, the requirement for documentation has undergone significant changes during the past few years. No longer is it enough simply to do the work. Test compounds cannot just be synthesized, mixed, and applied, and samples taken. The samples cannot just be shipped, received, stored, weighed, and assayed, and the results reported. Now each of these phases of a given study must be thoroughly documented as to what was done, how it was done, when it was done, where it was done, and who did it.

With the establishment of the reporting and record-keeping requirements in support of pesticide registration under the Federal Insecticide, Fungicide, and Rodenticide Act (FIFRA) and the Toxic Substances Control Act (TSCA) Good Laboratory Practice (GLP) Standards, documentation has become the key issue (*1, 2*). The establishment of a defensible "paper trail" is essential to the successful completion of every study. A study can be considered successfully completed when the Environmental Protection Agency (EPA) accepts the data and conclusions submitted in support of a pesticide registration. Who is responsible for compliance with GLPs? "Management"[1], study directors, and investigators are responsible. The quality assurance unit (QAU) must verify the organization's compliance with GLPs.

Management of samples is imperative from the field through the laboratory to the archives or destruction. Whether the organization is large or small, a formal approach to sample management is necessary. Forms designed by and for your personnel to meet your specific organizational requirements must be available. The forms that are developed must be consistently used. With complex programs, a computer-based data management system may be necessary to provide the test substance and sample tracking for chain-of-custody documentation.

In any given study, sampling is recognized as one of the most critical steps. All aspects of sampling (generation, collection, preparation, subsampling, analysis, and archiving or destruction) must be completed according to an approved protocol.

The practitioner of these studies must be able to document when the test substance, reference substance, and samples were generated or

---

[1]Management is defined here as the person or persons responsible for setting policy at a facility. For a description of the duties of testing facility management, see § 160.31 of the GLP Standards.

received. Documentation must not only cover when these items were received, but also from whom they were received, their condition when received, and the quantity received. Once these substances and samples are in the possession of the field or laboratory personnel, storage conditions and duration must be documented. Documentation is also necessary for the removal of these substances and samples from storage for whatever reason (weighing, mixing, application, sampling, shipping, processing, analysis, reporting, and disposal). How is the chain of custody of these substances and samples documented?

This chapter will provide a set of procedures to document the chronological history of substances and samples from the sponsor to the field and from the field to the analytical laboratory. The chronological history of substances and samples, commonly referred to as the *chain of custody*, provides integrity to sample collection, identification, shipment, analysis, and reporting.

SOPs have been written for every conceivable contingency, and the specific study protocol has been prepared and revised many times over. That very important study has been placed in the capable hands of the field personnel, and the analytical laboratory has been selected.

Analytical Development Corporation (ADC) is an independent laboratory specializing in chemical research and analysis. The following procedures are those that are used within our facility and are examples of how to provide sample, analytical reference standard, and test substance chain-of-custody documentation.

Preparing the Analytical Laboratory for Receipt of Field Samples

The analytical laboratory has been alerted via telephone or in writing that the study is under way. This notification may come from the study sponsor or from the field organization. The sponsor supplies the analytical laboratory with the analytical reference standard, technical data sheet, and Material Safety Data Sheet (MSDS). The technical data sheet should contain all the pertinent substance characterization information (purity, stability, method of analysis, and storage conditions). The analytical reference standard is logged into the master log for documentation of receipt.

Master Log. The master log (Figure 8.1) is set up to identify the material received, including protocols, analytical reference standards, and samples. Also identified are the compound or test substance and the quantity of each received. The location of storage for the material is specified with as much detail as possible in the space available.

ADC#	Material	Compound	Amount	Storage Location	SUBMITTED BY		DATE RECEIVED		Comments	Signature
					Company	Individual	Time	Date		

Figure 8.1. Master log.

An example of an entry for the storage location of an analytical reference standard would be "standard storage cabinet, refrigerator". The individual and organization submitting the material should be documented as part of the master log. The date and time of receipt must be specified as well as the condition of the material when received, such as whether it is frozen, thawed, or at ambient temperature. Each master log entry must be signed by the individual making the entry. Each entry must also be given a master log number. This number is a chronological transactional number containing the year and project number; an example is 90–001–1180 (the year, the log entry number, and the project number). The master log may be maintained by hand or by using a data base system on a computer. A master log number must be assigned to every incoming item (all written material, analytical reference standards, test substances, and samples).

Analytical Reference Standard Logs. The shipping document and the analytical reference standard are marked with the master log number, and the shipping document is placed in the main project file. The analytical reference standard is logged into the standard log (Figure 8.2) and placed into the proper storage as specified on the technical data sheet.

Compound: _____

Lot No.: _____

Percent Purity: _____

Master Log Number: _____

Storage Location; Freezer: _____ Refrigerator: _____ Ambient _____

Submitted By: _____

Amount Received: _____

Date: _____

Signature: _____

Disposition: _____

Figure 8.2. Analytical reference standard receipt log.

The standard log is simply an alphabetical listing of all analytical reference standards and test substances located at an organization. Each log entry contains the identification and lot number of the compound, percent purity, amount received, storage location, and date received. Each log entry is cross-referenced to the master log with the master log number and is signed by the individual making the entry.

Analytical reference standards and test substances may require storage under three separate types of conditions. Storage may be necessary under refrigeration, freezer, or ambient temperatures. The reference standard is logged into the appropriate standard storage log (Figure 8.3) at the time it is initially placed in storage. A separate log is maintained for each of the three different standard storage conditions. Each standard storage log contains the identification of each reference standard or test substance, the study project number, the date and time of entry or removal, and the signature of the individual performing the task.

When the reference standard is removed from storage for preparation, the key to the locked standard storage cabinet is obtained from the Materials Control Manager, and the cabinet is unlocked. The reference standard storage container is removed from its designated storage position and the analytical reference standard is removed. The storage container is returned to its designated position in the cabinet. The cabinet is locked, and the standard is logged out in the standard log. After weighing, the standard is returned to storage and logged back into the standard log. Every time the analytical reference standard or test substance is removed from storage the above procedure is repeated. The weighing is logged into the analytical balance log (Figure 8.4). This log cross-references the study project number, the project notebook and page, as well as the reference standard for identification of what is being done on the analytical balance. The balance log also contains information on the amount of substance weighed, the calibration weighings, and the date and signature of the analyst conducting the weighing. The analytical reference standard or test substance weight is also recorded in the study notebook for calculation purposes and is cross-referenced to the balance logbook. The analytical balance log is permanently associated with a particular balance by using the balance serial number on the front of the logbook along with the logbook number.

Shipping Field Samples to the Analytical Laboratory

The field personnel inform the analytical laboratory that samples have been shipped in dry ice via a nationally recognized carrier (examples are Federal Express via air and ACDS via land) with an estimated arrival date. Dry ice is not required when samples are shipped by freezer truck; how-

Project No.	Compound	Remove		Return		Initials
		Date	Time	Date	Time	

Figure 8.3. *Analytical reference standard storage log.*

Date	ADC Cross-Reference Book #	Page	#	Client	Compound	Standard Weight	Calibration Weights	Analyst	Verified

Figure 8.4. Analytical balance log.

ever, even in this case dry ice is recommended. When the sample shipment arrives, it will have an associated transportation or air bill of lading from the carrier. This bill of lading should be retained because it will be the documentation needed to verify who had custody of the samples, how long they had them, and where the samples were located when they were out of the field personnel's custody and prior to being in the custody of the laboratory. This form, along with the entry in the master log under comments, will provide information about sample transport and the condition of the samples upon receipt at the laboratory. The sample shipment should also contain a field sample transmittal (Figure 8.5) or tracking form, which was originated by the study sponsor or the field organization. This form should have a complete inventory of samples contained in the shipment; each sample is identified by the field sample number. The transmittal form should identify who shipped the samples, how they were shipped, and when they were shipped and should identify the study from which the samples were taken. The transmittal form should identify the organization the samples were shipped to, the air bill number, how many containers were in the shipment, and a location for the date they were received at the laboratory. The form should also have questions to be completed by the receiving laboratory, concerning the conditions of the samples upon arrival; examples are (1) frozen with dry ice, (2) frozen with no dry ice, (3) partly thawed, (4) thawed but cold, and (5) warm. The form must also contain a signature and date line for the individual shipping the samples from the field and the individual receiving the samples at the laboratory. The shipment is inventoried by the receiving technician, and the transmittal form is completed and returned to the sponsor. A copy of this form is kept in the master project file. The shipment is logged into the master log and is given a master log number, as described previously.

All documentation, methods, protocols, air bills, transmittal forms, MSDS forms, telephone messages, and any correspondence pertaining to the study are maintained in a project master file. When progress reports are prepared, copies are also placed in the master file along with the raw data.

The bill of lading and the sample transmittal form are stamped with the master log number for cross-referencing receipt of the samples. This same master log number is placed on each container of samples, thus cross-referencing the sample shipment and receipt. The samples are placed in the freezer, which is set at nominal −20 °C, and logged into the freezer logbook. The freezer log is set up with a project per page (Figure 8.6). Each page has space for identification of the sponsor, the master log number, sponsor sample number, reason for sample movement (such as for preparation, analysis, or shipment), time of entry (in or out), date, and initials of individual adding or removing the samples. The freezer is kept inside a locked building and access is limited to those individuals with

```
Shipped By:_____        Shipped To_____
                                              _____
Address: _____          _____
City/State/Zip:_____           _____
Courier: _____          Airbill:_____
Date Shipped: _____          Number of Boxes:_____
Study ID: _____          Date Received:_____
ADC Project #:_____
Logged in upon receipt? ____Yes ____No

Information to be filled out by the shipper[1,2,3]/recipient[4]: Nos. 1,2,3, & 4

Date [1]   Sample Identification [2]        Sampler [3]     Receiver [4]
Sampled    Number                           Initial         Initial
------------------------------------------------------------------------
```


```
Condition of samples upon arrival (please check one of the following):

(  ) Frozen with dry ice      (  ) Partly thawed
(  ) Frozen with no dry ice   (  ) Thawed but cold
(  ) Warm

Were any of the samples damaged upon arrival?  (  ) Yes   (  ) No
If yes, describe:


Please describe any other poblems with the shipment:


Signature of shipper: _____
Print Name: _____          Date: _____
Signature of recipient: _____
Print Name: _____          Date: _____

Please send this document back promptly to:  Analytical Development Corp.
                                              4405 N. Chestnut Street
                                              Suite D
                                              Colorado Springs, CO  80907
```

Figure 8.5. Field sample transmittal form.

ADC Project# _____   Sponsor _____

ADC Log No	Sample Numbers	Reason For Movement	Time	Date	Initials
			In		
			Out		
			In		
			Out		
			In		
			Out		
			In		
			Out		
			In		
			Out		
			In		
			Out		
			In		
			Out		

Figure 8.6. Freezer log.

study samples contained within. The freezer temperature is recorded by using a 7-day constant temperature recorder. These strip chart temperature recordings are maintained in the archives to support the storage conditions under which the samples were maintained during the period of time they are at the facility. When the samples are removed from the freezer for processing, they are signed out of the freezer log; all columns of the log should be completed. As the samples are processed, they are entered into a processing notebook that is specific to the project. The samples are entered into the processing notebook by using the sponsor sample number. The samples are then returned to the freezer to await analysis and are logged back into the freezer log.

Analyzing Field Samples

When the analyst retrieves the samples from the freezer for analysis, the samples are again logged out of the freezer in the sample log. The samples are then listed in the bound analytical notebook assigned to the project. Notebooks are assigned to specific studies rather than to an analyst. This procedure eliminates the possibility of a notebook containing more than one unrelated study for a particular sponsor. Also sponsor and study confidentiality is maintained by preventing an analyst from entering data from multiple sponsor studies in a single notebook. By assigning a notebook to a study, it is easier to track the study conduct. An alternative to bound notebooks is the use of worksheets. Each worksheet should contain the same information previously described for notebooks.

Generally, the field sample number or sponsor assigned sample number is too long to use as a working number within a laboratory setting. A simple procedure is to assign a unique laboratory number to each sample for assay. These numbers, the sponsor–field sample number and the laboratory sample number, are associated throughout the analysis and reporting by their presence in the analytical notebook. As samples are weighed for analysis, the weights are entered into the analytical notebook beside the sample numbers. The weighings are cross-referenced to the balance logbook by entering the balance logbook number and page in the project notebook. The individual balance logs are differentiated from each other by serial number of the particular balance to which they are assigned. This procedure is identical to that used for analytical balance weighings. The samples are assayed according to the validated methodology and prepared for quantitation. Each time a sample is transferred from one laboratory container to another its laboratory number is written on that container. This procedure ensures that each sample is tracked through the method of analysis.

Samples and analytical reference standards are then placed in automatic sampler vials and loaded onto the automatic sampler for injection. The laboratory sample number is written on the sample vial for identification within the injection sequence. The concentration of the analytical reference standards injected to generate the standard curve is written on each respective vial. The data acquisition system is prepared with a method for collection and data processing. An example system is the Nelson Analytical Series 3000 Multi-Instrument Data System using a Novell local network of IBM AT and XT computers to provide data acquisition. The acquired files are accessed by the analyst to print chromatograms, standard curves, and sample results.

To maintain sample tracking within the series of automatic injections, data acquisition, and reduction processes, the analyst assigns a file name to the sequence. Within this sequence is either the laboratory sample number and as much of the field sample number as the available space will allow or the standard concentration. These samples or standards are entered in the sequence that the analyst has loaded them into the automatic sampler. Included in the sequence is the sample weight, final volume, and any appropriate dilutions. The instrument parameters are also entered as part of the sequence. These parameters are obtained from the instrument log book.

The instrument log is maintained by the analyst using the instrument and contains information pertaining to each day's analyses (Figure 8.7, GC log; Figure 8.8, HPLC log). Each analytical system must have its own log. These logs reference the date of instrument use, the individual using the instrument, project number, column description, gases or mobile phase used, flows, detector type, temperatures, number of injections, and any information pertaining to maintenance. This maintenance is either routine or preventive.

The chromatograms obtained from the analytical analysis for each sample and each standard are transferred electronically from the instrument to the computer. As each injection is made, the electronic signal is transmitted through an analog-to-digital converter to the computer. The computer, using the Nelson software and the methods provided by the analyst, collects the chromatographic data. After the data are acquired, the system automatically assigns a cycle number to the injection. This cycle number is unique within the specified file, and a cross-reference is provided between the laboratory sample number and the chromatogram. Each chromatogram is automatically and serially numbered by the computerized data acquisition system with this cycle number. The analyst prepares the injection sequence for the computer according to the way the samples and standards are loaded into the automatic injector. As the samples and standards are injected into the instrument, they are automatically assigned a file number, which is an alphanumeric designation sequentially assigned by the data acquisition system. This file number cannot be

				Gases															
				Carrier						Detector									
				mL/min		Column				Make-up		H_2		Air		O_2			
								Head PSI											
Date	Init	Proj #	Column Description	Split Vent	Sept Purge	mL/min	Type	ISO	Prog	Type	Total Flow	mL/min	PSI	mL/min	PSI	mL/min	PSI	Detector Type	Flows Measured at what Oven Temp

Figure 8.7. Instrument log for a GC.

Purge System			Range/ Atn	Temperature °C					Run Time	Injection/Sample Information						Maintenance & Comments
Time On(min)	Time Off(min)	Signal		Detector	Inlet ISO	Prog	Oven ISO	Prog		Insert Type	Type	#	Vol	Auto	Solvent	Septum, Glasswool, Detector Clean., Etc.

Figure 8.7. Continued

Date												
Operator												
Project #												
Column Maker												
Description												
Packing												
Length/Diam.												
Serial #												
Temperature												
Precolumn												
Date												
Mobile Phase — Initial — A comp.												
Mobile Phase — Initial — B comp.												
Mobile Phase — Initial — C comp.												
Mobile Phase — Initial — hold												
Mobile Phase — Gradnt time												
Mobile Phase — Final — A comp.												
Mobile Phase — Final — B comp.												
Mobile Phase — Final — C comp.												
Mobile Phase — Final — hold												
return time												
equil time												
Flow												
Pressure												
Detector												
Serial #												
AUFS/Gain												
Range												
Wavelength												
Injector												
Serial #												
Inj. Vol/Loop												
# samples												
Type Samples												
Autoinjected												

Comments

Figure 8.8. Instrument log for an HPLC.

changed or repeated within the sequence. Each chromatogram is individually labeled with the correct sample number (laboratory number and an abbreviated sponsor number), project number, notebook and page reference for sample workup, date of injection, final volume, sample weight, and injection volume. The peaks of interest are identified, and the initials of the analyst are placed on each chromatogram. If the chromatogram is for a fortified sample, the fortification level is also identified. If the chromatogram is for a standard injection, the concentration of the solution is included. In addition, the instrument is identified along with the operating parameters, which are temperatures (injection, column, and detector), identification of the gases or mobile phase used along with the associated flows, the column identification, and detector information.

The chromatography is processed and the data is reported by using PR–86–5 (*3*) for the final report format. The final report and all the raw data are placed in the archives. After the report has been reviewed by the study director and the sponsor, the samples may be placed into archival storage for future use. If the samples are determined to be of no further use, they are disposed of in an appropriate manner. The sample archiving or disposal is documented by sample number, and the documentation is placed in the project file in the archives.

Role of the Quality Assurance Unit

Each of these various phases is monitored and documented by the QAU for accuracy and to ensure that the protocol that was so carefully prepared in the beginning of the study is being followed. If the protocol was not followed, the deviations must be documented. The implications of the variations must also be presented in the final report. The QAU reviews the final report for completeness and for accuracy of the data.

Conclusion

Many different forms have been presented in this chapter to provide documentation for analytical-reference-standard, test-substance, and sample chain of custody. The information contained in each form and the consistent use of the form are important; the actual format is unimportant. Each form should be designed by the individuals who are going to use it, with some general guidelines as to the information each form must contain. Each form must answer the basic questions: who, what, when, where, why, and how.

Providing complete chain-of-custody documentation is a formidable task. However, the establishment of chain-of-custody procedures that provide unbroken documentation of the location and use of the analytical reference standards, test substances, and samples will only add credibility to the data generated in any study. Chain-of-custody procedures provide the knowledge that an established system was in control throughout a study. The establishment of these procedures eliminates doubt as to the identity or character of the analytical reference standard, test substance, or samples.

Remember, *if you didn't write it down, you didn't do it!*

References

1. "Good Laboratory Practice Standards", *Code of Federal Regulations* Title 40, Pt. 160; *Federal Register* 54:34052 (August 17, 1989).

2. "Good Laboratory Practice Standards", *Code of Federal Regulations* Title 40, Pt. 792.

3. Akerman, J. W. "PR Notice 86-5, Notice to Producers, Formulators, Distributors, and Registrants", U.S. Environmental Protection Agency, Washington, DC, July 29, 1986.

RECEIVED for review September 17, 1990. ACCEPTED revised manuscript August 17, 1991.

Chapter
9

Test Substance and Analytical Reference Standard Characterization and Accountability

S. Rand Fuller and Willa Y. Garner[1]

Quality Associates Inc., Ellicott City, MD 21043

The identity, purity, composition, and other characteristics of test substances, control substances, and analytical reference standards used in GLP-driven studies must be determined before using them in those studies. Methods of synthesis of these substances must be documented by the sponsor or testing facility. The stability of test substances and analytical reference standards under storage conditions at the test site must be known. Use documentation of these materials is required. Compounds used in studies of greater than 4 weeks' duration must be archived. Reagents and stock solutions also must be handled in a defined manner. These and other topics will be examined for their impact on the scientific aspect of the conduct of field and laboratory studies.

A TEST SUBSTANCE IS DEFINED AS "a substance or mixture administered or added to a test system in a study," according to the Federal Insecticide, Fungicide, and Rodenticide Act (FIFRA) Good Laboratory Practice

[1]Corresponding author. Current address: Garndal Associates, 17485 Sierra Way, Monument, CO 80132

2192-8/92/0113$06.00/0 © 1992 American Chemical Society

(GLP) regulations (*1*). A control substance means any chemical substance or mixture that is administered to the test system in the course of a study for the purpose of establishing a basis for comparison with the test substance. A reference substance is any chemical, mixture, analytical standard, or material other than the test substance, feed, or water, administered to or used in analyzing the test system to establish a basis for comparison with the test substance. For the purposes of this chapter, reference substances refer to analytical reference standards. Except where specifically cited, the regulations are the same for all three types of substances.

The requirements, as set forth in the FIFRA GLP Standards, are designed to ensure the integrity of test substances and analytical reference standards from their shipment to and receipt at the testing facility through test use and disposal at the completion of the study. The test substance and analytical reference standards are critical components of any regulated study. To assess the validity of a study, specific parameters for each test substance and analytical reference standard must be documented. These parameters include the storage conditions under which these chemicals must be maintained throughout substance retention; chemical properties such as stability, purity, and strength; and the distribution and use of all these substances.

The study sponsor and the testing facility or test site have an obligation to ensure and maintain the integrity of the test substances and reference standards during shipment to as well as during storage and use at the test site. The following discussion outlines the requirements of the GLP Standards and offers suggestions for the handling and accountability of these substances.

Facilities for Handling Test Substances and Reference Standards

The facility requirements necessary for handling test substances and reference standards are listed in § 160.47 of the GLP Standards (*1*). Except for physical and chemical characterization studies as cited in § 160.135(b), this section requires separate areas, as necessary, to prevent contamination or mix-ups as follows:

- Separate areas must be available for the receipt, mixing, and storage of test substances and reference standards.

- Separate areas must be available for the storage of mixtures of test substances and analytical reference standards. Mixtures include stock solutions.

- These areas must be kept separate from the areas in which the test systems are maintained.

The areas must be adequate to preserve the identity, purity, strength, and stability of these substances and mixtures. The storage facilities should be designed so that access to the test substances and analytical reference standards is limited; that is, the storage areas should be secured with locks, and only authorized personnel should have access to these areas. Test substances and analytical reference standards must be kept separate from samples and stock solutions. Test materials labeled with radioactive isotopes should be maintained separately from those that are unlabeled.

Handling Test Substances and Analytical Reference Standards

The handling requirements for GLP test substances and analytical reference standards are stated in § 160.107. For all investigations that are conducted under the GLP Standards (most physical and chemical characterization studies are excluded), procedures must be established to ensure the following:

- There must be proper storage.

- The receipt and distribution of each batch must be documented. Such documentation must include the date and quantity of each batch distributed or returned.

- Distribution must be made in a manner designed to preclude the possibility of contamination, deterioration, or damage.

- Proper identification must be maintained throughout the distribution process.

The storage conditions should duplicate the conditions recommended by the supplier. Lighting, relative humidity, and temperature should be controlled and maintained within specified parameters for the duration of storage. Accurate records of the storage conditions must be maintained. If storage conditions such as temperature are manually recorded, the standard operating procedure (SOP) for this function should state the number of times per week these parameters are to be documented (i.e., on each working day or daily). The appropriate dispensing techniques employed in the distribution of test substances and analytical reference standards should be consistently followed for the duration of the study to prevent accidental contamination. The information on the container label should be verified each time a substance is used. Containers must be handled with care to prevent breakage and loss. Should the distribution of small amounts of these substances be required (i.e., the substances are removed directly from their original containers), use clean pipets or spatulas only. Substances that require refrigeration should be returned to such conditions as quickly as possible to prevent loss (i.e., volatilization) or degradation (i.e., thermal or photo). Frozen or refrigerated substances should be allowed to reach ambient environmental conditions before opening them for use. Placing the substances in a desiccator during this temperature equilibration period will minimize the effects of condensation.

The identity of these substances must be clearly maintained throughout the study. Use a consistent method for identification to prevent the misuse of a substance, the use of expired materials, or confusion over the use of multiple containers or lots of the same substance.

The receipt and distribution of the test substances and reference standards must be accurately accounted for prior to and during their use in a study. Figure 9.1 is an example of a "Chemical Receipt and Use Log" form. This form acts as a receipt log as well as a substance distribution log. These documents can be separate logs or combined as in the figure. This type of form is especially useful for a contract facility that must maintain client confidentiality. All label information should be recorded onto the log sheet from the container label and the shipping invoice, if applicable. Other pertinent information for the substance should also be recorded. The Material Safety Data Sheet (MSDS) should be maintained with the information for the test substance and analytical standard. The testing facility or test site should request a copy of the MSDS for each test substance and analytical reference standard it receives. The total amount of material received will be entered upon receipt, thus establishing the bulk inventory. Each distribution of the substance must be accurately recorded, and the amount withdrawn should be subtracted from the previous total amount. The reason for use should be recorded, as well as the identity of the person receiving the material. This practice is vital for an accurate record of substance usage. The disposition of the unused material or empty containers must also be documented.

Test, Control, and Reference Standard Characterization

The GLP Standards outline the requirements for proper characterization and retention of test substances and analytical reference standards in § 160.105. Most physical and chemical characterization studies are exempt from these requirements (with the exception of number 5), the contents of

PTRL RECEIPT RECORD AND INVENTORY LOG OF INCOMING TEST,
REFERENCE AND/OR CONTROL ARTICLES/SYSTEMS

Receipt Date: _____ PTRL Log Number: _____

Name on Container Label: _____ Formulation: _____

Active Ingredient: _____ Purity: _____

Batch/Lot No.: _____ Expiration/Re-analysis Date: _____

Other Label Info: _____

Shipper (Name, Address): _____

No. of Containers: _____ Gross Weight Received: _____

Received By: _____ Date: _____

Condition on Arrival: _____

Physical Description: _____/ Solid _____/ Liquid _____/ Solution _____/ Other

Storage Location: _____ Storage Temp: _____

Comments: _____

Shipper: _____Way Bill #: _____

INVENTORY LOG

Project #	Purpose	Gross Wt. ( )	Amount Dispensed ( )	Adjusted Gross Wt. Remaining ( )	Initials	Date	Balance I.D. No.

Disposed: _____ Returned To: _____ Date: _____

Initials: _____: Date: _____ Return Approved By: _____

DEV 4/90

Figure 9.1. Chemical receipt and use log. (Reproduced with permission from PTRL-East).

which are contained in the following list. The first four parameters in the following list should be determined and documented prior to the "experimental start date"[2].

1. Required general information on test substances and analytical reference standards:

    - The identity, strength, purity, composition, or other characteristics that define the test substances or analytical reference standards must be documented. Methods of synthesis, fabrication, or derivation must be documented by the sponsor or by the testing facility, and the location of these data must be specified in the study file. The source of these substances should be documented.
    - If a substance that was labeled with a radioactive isotope had an unlabeled substance added to it to reduce specific activity, the mixture must be considered as two test substances, and the required characterization information on both of the substances must be supplied to meet the GLP requirements.
    - Other information relevant to the study that must be documented includes the correspondence between the test substance and its commercial counterpart, if applicable.

2. The stability of each batch used must be determined prior to the start of the study or concomitantly with the study according to written SOPs that provide for periodic analysis of each batch. The solubility of the test substances or analytical reference standards must be determined by the sponsor or by the testing facility when relevant to the conduct of the study. Solubility is an important factor for such studies as hydrolysis, solution photolysis, and fish accumulation.

3. Each storage container for a test substance or analytical reference standard must be labeled by name, Chemical Abstracts Service (CAS) number or code number, batch number, expiration date (if any), and where appropriate, storage conditions that are necessary to maintain the substance's identity, strength, purity, and composition. Storage containers must be assigned for the duration of the study (until acceptance of the final report).

4. When the experimental portion of a study will last longer than four weeks, reserve samples from each batch or lot of test substance and analytical reference standard must be retained for as long as these substances afford evaluation (i.e., the ability to obtain reproducible results).

[2]The EPA defines "experimental start date" as the first date the test substance is applied to the test system.

5. The stability of the test substances and analytical reference standards under storage conditions at each test site must be known for all studies.

The identity of every substance must be confirmed prior to its distribution or use in a regulated study. This confirmation prevents the accidental misuse of a test substance or analytical reference standard. Purity or strength of these substances must be determined prior to the initiation of the study. These assays are usually conducted by the sponsor; however, the testing facility should verify the purity of all substances upon receipt. This is done to ensure the integrity of the test substances and analytical reference standards during transport from the sponsor or supplier to the test site. The purity of these substances, if stability was not determined concomitantly, should be reverified at the conclusion of the study to ensure the integrity of the substances for the duration of the study. Thin layer chromatography (TLC), gas chromatography (GC), high-performance liquid chromatography (HPLC), or infrared (IR) or ultraviolet (UV) spectral analyses may be employed by the testing facility to check substance purity. All methods used and the results must be documented in the study file. Each lot or batch used during the course of the study must be analyzed. Samples of test substances from field studies as well as those from laboratory studies and all analytical reference standards must be assayed for purity after their use in a study to ensure stability under storage conditions at the test site. Simple GC or HPLC analyses will be adequate for the re-assay of most organic substances. If the sponsor or supplier has generated stability data under GLP Standards and if the testing facility can show through documentation that the material was shipped and stored under the same conditions (i.e., same temperature and comparable duration), then requirement number five from the previous list will have been met (*2*).

Any impurities that occur in quantities greater than 5.0% in the test substances and analytical reference standards should be identified and quantified. For toxicological studies, impurities should be determined down to 0.1%, which is the same requirement for EPA product chemistry requirement guidelines, Subpart D. Test substances that are formulated, such as a 50% wettable powder, do not need impurity analyses of the inert ingredients. However, all components of the mixture must be documented. The methods of impurity identification should also be documented. Each lot or batch will be subject to impurity identification. The impact, if any, the impurities may have on the outcome of the study must be documented.

The testing facility must document all information on a test substance or analytical reference standard provided by the sponsor or manufacturer. Should any information be unavailable, the testing facility should employ appropriate techniques to determine the necessary characteristics or request that the sponsor obtain the information. The method of synthesis may be the manufacturing process employed by the sponsor or, in cases of fabrication or derivatization, may be the methods employed by the facility. If this information is not available in the testing facility's study file, the location of this information must be documented (e.g., "the methods of synthesis are the responsibility of the sponsor and are maintained in the sponsor's archive").

The stability of test substances and analytical reference standards is to be determined prior to the start date of the experiment. Stability can be determined concomitantly with the study. Additionally, the stability of these materials under the storage conditions at the test site must be known for all studies. These conditions must be documented, and all substances must be maintained under those conditions for the duration of the study. The test site is the location of the test system, which may be in the testing facility or may be one or more ancillary sites (as in a field study). The analytical methods used to determine or confirm the purity of these substances may be applicable to the stability determination; that is, purity confirmation conducted prior to study start and at study termination may be used to verify the stability of these substances. These methods must be documented.

Storage containers must be assigned to a particular test substance for the duration of the study. Containers are to be clearly and uniquely labeled. The information needed to identify the storage container properly must include name; Chemical Abstracts Service (CAS) number or code number; batch or lot number; expiration date, if any; and storage conditions where appropriate. Storage containers may be the original containers received from the sponsor or they may be the testing facility's container used to facilitate removal or use of the test substance (i.e., the shipping container may be too large for convenient dispensing of the test substance).

The EPA requires that these test substance containers be maintained by the facility until the acceptance of the final report, that is, through quality assurance review. Most laboratories or field sites maintain empty containers until the study director signs the final report. A proper identification and documentation trail should be established to prevent the accidental disposal of empty substance containers.

Maintaining empty containers can be very difficult for field studies where, frequently, hundreds of containers are involved. If the study director feels that maintaining these containers is a burden, an exception can be requested in writing from the EPA. For applicants granted an exception (3), the following information must be documented:

1. The following records must be maintained: (a) shipment information pertaining to each container leaving the storage site; (b) receipt records of the chemicals to be tested at each testing cite; (c) complete use logs of material taken from each container; and (d) a record of the final destination of each container, including the place and date of disposal or reclaiming.

2. An inventory of empty containers must be prepared before disposal, and this inventory must include information to identify these containers uniquely. Data on each container must be documented.

3. The identity of the location of facilities where test material is stored; where empty containers are stored prior to disposal; where records of shipment, use, and disposal of containers are maintained; and location where the test substance is used in studies (i.e., the testing facility) must be reported to the Office of Compliance Monitoring at the EPA within two weeks of receipt of notification of any pending inspection involving a study in which these containers were used.

4. A statement must be included with the statement of compliance or noncompliance required in § 160.12 describing that this exception to GLP Standards is in accordance with the conditions provided by the exception.

Should these conditions not be fully met, all of the provisions of the GLP Standards, including assignment of storage containers for the duration of the study, apply. Exceptions are granted on a case-by-case basis and must be requested each time a testing facility needs to dispose of empty test substance containers prior to the end of the study.

Experimental duration is the period from the start date of the experiment (first application) to the termination of the experiment (experiment termination date, the last date on which data are collected directly from the study, including analytical data). If the experiment duration is longer than four weeks, reserve samples from each batch or lot of test substance and analytical reference standard must be collected and retained as long as the quality of the material affords meaningful evaluation. Reserve samples should not be discarded until written sponsor approval to do so has been obtained by the testing facility. These samples must be archived in a manner to prevent deterioration and indexed to permit expedient retrieval by the responsible archivist.

Mixtures of Substances with Carriers

The GLP Standards that discuss the mixtures of substances with carriers are given in § 160.113 (*1*). Regulated studies, except most physical and chemical characterization studies, must adhere to the following requirements:

1. Determine the uniformity of the mixture, and determine, periodically, the concentration of the test substance or analytical reference standard in the mixture.

2. When relevant to the conduct of the study, the solubility of each substance in the mixture must be determined by the testing facility or the sponsor prior to the start date of the experiment.

3. The stability of the test substance or analytical reference standard in a mixture must be determined prior to the start date of the experiment, or concomitantly with the study, according to the facility's approved written SOPs, which should provide for periodic analysis of each batch.

Although not required by the GLP Standards, each batch of a mixture should be evaluated for homogeneity and concentration by appropriate analytical methods. All analytical methods must be documented. The testing facility has the primary responsibility for determining these parameters. The stability of the mixture should be determined under test site conditions.

Should any of the components of a mixture have an expiration date, that date must be clearly indicated on the container. Should more than one component have an expiration date, the earliest date must be documented on the storage container label. Expired mixtures must not be used.

If a vehicle is used in the preparation of the mixture to facilitate the mixing with a carrier, then data must be provided ensuring that the vehicle does not interfere with the integrity of the study. All data associated with the carrier should be documented.

Tank mixtures for field studies are considered a mixture of the test substance with a carrier, usually water. Analysis of tank-mix samples usually provides inconsistent data (i.e., similar results are not usually obtained). The agrochemical industry has requested that the EPA exclude the analysis of tank-mix samples from the GLP Standards. No response had been obtained from the EPA at the time this book went to press. Until the EPA renders a decision on tank-mix sample analyses, tank-mix samples should be taken and analyzed to meet the GLP requirements. The water used for tank mixtures and in all other analyses should be analyzed to ensure that there are no contaminants that would have an effect on the study results, especially if the quality of the water used is in question. A laboratory study conducted to show the stability and homogeneity of a mixture that was continuously agitated to simulate a tank mix under field conditions would meet this GLP requirement thus obviating the need for collecting tank-mix samples.

Physical and Chemical Characterization Studies for Test Substances and Analytical Reference Standards

Most studies that are conducted to determine the physical and chemical characteristics of a substance do not need to adhere to all of the GLP Standards. Those GLP Standards that do not apply are listed in § 160.135 (*1*). Several of these studies do fall under total GLP compliance, however. The characterization studies that must comply with all of the GLP Standards are

1. stability

2. solubility (aqueous and organic)

3. octanol–water partition coefficient

4. volatility (vapor pressure)

5. persistence, such as biodegradation, photodegradation, and chemical degradation studies

Items 2, 3, and 4 are used as benchmarks to predict the potential for leaching, bioaccumulation, and drift, respectively.

The following list contains studies that are not subject to full GLP compliance:

melting point	explodability
boiling point	viscosity
color and odor	miscibility
density	corrosion characteristics
dissociation constant	dielectric breakdown voltage
pH	flammability
physical state	oxidizing or reducing action

For these studies, separate areas are not required for handling, mixing, and storage of test substance and analytical reference standards; characterization is not required for test substances and analytical reference standards, except for stability under test site storage conditions; and characterization of mixtures of test substances and analytical reference standards with carriers is not required.

Reagents and Solutions

All solvents used for a regulated study, whether in a laboratory or in the field, must follow the GLP requirements as noted in § 160.83 (*1*). The following information must be determined for each reagent or solution and is to be clearly identified on the container label (Figure 9.2):

1. identity
2. titer or concentration
3. storage requirements
4. expiration date

Additional label information, such as preparation date or date received, date container was opened, and initials of the scientist, is helpful in tracking the usage of reagents and solutions (Figure 9.2). Reagents or solutions in wash bottles are also covered by the GLP Standards, and the wash bottles must have labels containing the information in the previous list.

Deteriorated or expired reagents and solutions cannot be used. Extremely stable materials may be reevaluated upon reaching a given expiration date to determine if the material is still usable. A new expiration date would then be assigned. An SOP should be written to define the expiration date of a reagent or solution if a date (month, day, and year) is

REAGENTS:

IDENTITY: _____
CONCENTRATION: _____
DATE OPENED: _____ INITIALS: ____
EXPIRATION DATE: _____
STORAGE CONDITIONS: RT/R/F/Other: _____

SOLUTIONS/MIXTURES:

IDENTITY: _____
CONCENTRATION: _____
PREP. DATE: _____ INITIALS: ____
EXPIRATION DATE: _____
STORAGE CONDITIONS: RT/R/F/Other: _____

Figure 9.2. Label formats for reagents and solutions.

not provided by the supplier. A reasonable expiration date should be assigned to each type of chemical (i.e., acids, bases, organic solvents, dry chemicals, or mixtures). For most reagents (except reagents with a manufacturer-specified expiration date, such as ether or pH buffer solutions), a color-code system may be used to identify expiration dates. In such a system, the last day of a particular year is the expiration date for all reagents. Colored stickers that correspond to a specific year (i.e., green, 1990; orange, 1991; and blue, 1992) are placed on each reagent bottle. On December 31, a technician could go through the facility and discard the appropriate chemicals (denoted by a certain color). When dates are written on bottles, outdated reagents may be overlooked during the discarding process and may be used, or worse, may show up during an EPA audit. Red stickers could be used to alert the technical to bottles with a more frequent expiration date (i.e., ether or buffer solution). This color-coding technique is suitable for outdated-reagent spot-checks only. All reagent and solution containers must have complete GLP labeling information (as previously listed) to be in compliance.

Conclusion

The FIFRA GLP Standards, 40 CFR Part 160, August 17, 1989, which became final on October 16, 1989, are intended to ensure the integrity of a regulated study. The GLP Standards provide specific information to standardize the handling, characterization, and use of test substances, control and analytical reference standards by the testing facility. In order to ensure the validity of a study from initiation to termination, the parameters defining the characteristics of test substances and analytical reference standards must be known. This information must be available prior to the first administration of the test substance or an analytical standard's use for identification or quantification. Chemical stability must be maintained throughout the duration of the study.

Proper identification, storage conditions, and distribution must be accurately recorded and maintained. Separate areas within the laboratory facility must be available for the receipt, storage, and mixing of the test substances and analytical reference standards. The responsibility of the study sponsor and the testing facility to properly maintain these materials is paramount to the success and acceptance of any regulated study.

References

1. "Good Laboratory Practice Standards Under the Federal Insecticide, Fungicide, and Rodenticide Act Final Rule", *Code of Federal Regulations* Title 40, Pt. 160; *Federal Register* 54:158 (August 17, 1989) pp 34067–34074.

2. "Interpretation of the Good Laboratory Practice (GLP) Regulations", GLP Regulations Advisory No. 23, U.S. Environmental Protection Agency, Washington, DC, November 7, 1990.

3. "Interpretation of the Good Laboratory Practice (GLP) Regulations", GLP Regulations Advisory Nos. 24, 25, and 26, U.S. Environmental Protection Agency, Washington, DC, March 22, 1990.

RECEIVED for review November 5, 1990. ACCEPTED revised manuscript August 17, 1991.

Chapter

10

Reporting Study Results

Judy H. Hochman[1,2] and Willa Y. Garner[1,2]

[1] Quality Associates Inc., Ellicott City, MD 21043

*The objective of the final report is to describe accurately the
materials and methods required by the protocol and documented in the raw data and to discuss the results and conclusions while meeting the reporting requirements of the
Good Laboratory Practice (GLP) Standards, the Data
Reporting Guidelines (DRGs), the addenda to the Pesticide
Assessment Guidelines, and the Pesticide Registration (PR)
Notices. These documents describe not only the topics that
need to be addressed but also how the entire report is to be
presented. The GLP Standards, being a generic document,
simply list the minimal requirements that ensure integrity
during study conduct. The DRGs provide guidance on what
scientific topics should be discussed, for example, soil characterization, analytical methods, and graphical presenta-
tions, as well as providing a format for presenting these
data. PR Notice 86–5 provides the format requirements for
individual studies, beginning with the title page, and for
entire submittal volumes.*

[2] Current address: Garndal Associates, 17485 Sierra Way, Monument, CO 80132

2192-8/92/0127$06.00/0 © 1992 American Chemical Society

Good Laboratory Practice Reporting Requirements

Good laboratory practice (GLP) reporting requirements are stated in §
160.185 of the GLP Standards (1). A final report must be prepared for
each study whether or not the study was completed. As listed in the GLP
Standards, final reports must include the following:

1. the name and address of the testing facility performing the study and
    the dates on which the study was initiated and completed, ter-
    minated, or discontinued

2. the names of the study director, other scientists or professionals, and
    all supervisory personnel involved in the study (Figure 10.1)

3. the location where all samples, raw data, and the final report are to
    be stored

4. objectives and procedures stated in the approved protocol, including
    any changes in the original protocol

5. the test substance and analytical reference standards identified by
    name, Chemical Abstracts Service number or code number, strength,
    purity, and composition, or other appropriate characteristics (Figure
    10.2)

6. stability and, when relevant to the conduct of the study, the solubil-
    ity of the test substance and analytical reference standards under the
    conditions of administration

7. a description of the test system used, and where applicable, the
    source of supply, the species, and the procedure used for identifica-
    tion

8. a description of the dosage, dosage regimen, route, and duration of
    administration

9. a description of the methods used

10. a description of the transformations, calculations, or operations per-
    formed on the data, a summary and analysis of the data, and a state-
    ment of the conclusions drawn from the analysis

11. statistical methods employed for analyzing the data

12. a description of all circumstances that may have affected the quality
    or integrity of the data

13. the signed and dated reports of each of the individual scientists or
    other professionals involved in the study, including each person who,

Study Personnel	Job Title	Signature	Initials
Jane Scientist	Study Director		
John Scientist	Sr. Research Scientist		
Rex Junior	Staff Scientist		
Kori Junior	Staff Scientist		
Fred Helper	Assistant Scientist		
Judy Sure	QA Specialist		
Frank King	QA Specialist		

Figure 10.1. Format for listing study personnel.

Chemical	Grade	Source	Date Received	Lab Number	Specific Activity or Percent Purity	Lot or Code	Expiration Date	Physical Appearance
Oxy	Analytical	X Chem. Co.	2/1/90	462	99.5	Y23	1/7/92	Beige solid
Mono	Analytical	Y Chem. Co.	7/6/91	973	97.8	90–4–L	9/1/92	Amber liquid

Figure 10.2. Suggested format for reporting information on test substances and analytical reference standards. (Reproduced with permission from EN-CAS Analytical Laboratories.)

at the request or direction of the testing facility or sponsor, conducted an analysis of samples or evaluation of data from the study after data generation was completed

14. the statement prepared and signed by the quality assurance unit as described in § 160.35(b)(7) (i.e., a statement showing the dates inspections were made and the dates they were reported to the study director and management) (Figure 10.3). Although not required,

Quality Assurance Final Report Statement

Chemical X: Magnitude of the Residues on Corn

This report has been reviewed by the Quality Assurance Unit in accordance with the EPA GLP Standards set forth in 40 CFR 160.35(b)(6) and (7).

The Quality Assurance Unit has conducted the following inspections during the conduct of this study and has submitted written reports of said inspections to the study director and management.

Date of Inspection	Type of Inspection	Date Reported
09/27/88	Protocol review	09/27/88
10/19/88	Test material preparation	10/19/88
10/19/88	Test material application	10/20/88
11/03/88	Sample collection	11/03/88
02/16/89	Sample collection	02/16/89
03/07–08/89	Data review	03/09/89
06/09/89	Data review	06/09/89
07/07/89	Sample analysis	07/07/89
08/15/89	Sample collection	08/17/89
08/15/89	Sample analysis	08/17/89
08/28/89	Data review	08/28/89
09/30/89	Final report review	10/01/89

Quality Assurance Officer   Date

Figure 10.3. Format for the quality assurance statement.

this statement should list not only the dates inspections were made but also the type of inspection that was performed on those dates

If a study is terminated or discontinued, a final report, although brief, must be prepared. The raw data generated, as well as the report, must be maintained in the archive for at least 2 years. The final report must be signed and dated by the study director, and a copy of this report and all amendments to it must be maintained by the sponsor and the testing facility. The study director's signature on the compliance statement will serve as the required signature.

Corrections or additions to a final report must be in the form of an amendment by the study director. The amendment must clearly identify the part of the final report that is being added to or corrected and the reasons for the addition or correction and must be signed and dated by the person responsible (*1*). Modification of the administrative pages of a final report to comply with the submission requirements of the EPA does not constitute a correction, addition, or amendment to a final report if this modification is executed before submitting the report to the EPA.

The EPA Office of Pesticide Programs (OPP) has verbally indicated that submission of amendment pages to the final report is unacceptable. All changes to the final report must result in a reissue of the entire report. The reissued report must have a new title page that states "Amended Final Report". The amended report must also contain a revised table of contents, a statement that describes the portions of the report that have been modified and the modifications and that documents reasons for these changes. The amended pages should be identified as "amended". The reissued report must be signed and dated by the study director and key personnel involved in the modification. A new quality assurance (QA) statement is also required because the amended report will have been audited by the QA unit (QAU). To address some of the generic requirements of the GLP Standards, a study identification page should be included in the report. The study identification page includes the study title; sponsor name and address; testing facility name and address; study director name; initiation, "experimental start date"[3], "experimental termination date"[4], and completion dates; the location of raw data, samples, protocol, and final report; and lists the telephone number where inquiries may be made.

---

[3]The EPA defines "experimental start date" as the first date the test substance is applied to the test system.
[4]The EPA defines "experimental termination date" as the last date on which data are collected directly from the study.

Pesticide Registration Notice 86–5

The purpose of this Pesticide Registration (PR) Notice is to provide a standard format for data submitted to the EPA. This notice applies to all data required for granting or maintaining pesticide registrations, experimental use permits, tolerances, and "related approvals under certain provisions of the Federal Insecticide, Fungicide, and Rodenticide Act (FIFRA) and FFDCA." The FFDCA is The Federal Food, Drug, and Cosmetic Act. Although the PR Notice contains requirements for organizing and formatting submittal packages, it does not address which study data are to be reported.

A submittal package consists of all studies submitted to the EPA at one time in support of a single regulatory action, along with a transmittal document and other related administrative material. Because this chapter focuses on reporting formats rather than submission requirements, the discussion will be centered primarily around the requirements for submission of individual studies.

Transmittal Document. The first item in a submittal package must be the transmittal document that identifies the submitters, the regulatory action that the package is intended to support, the transmittal date, and a list of all the individual studies included in the package.

Individual Studies. Each complete study must include all the pages in the following list, where applicable, in the order indicated:

- study title page
- statement of data confidentiality claims
- certification of good laboratory practices
- flagging statements (for reporting adverse results)
- body of the study
- study appendices (at submitter's option)
- cover sheet to "Confidential Attachment" (if confidential business information is claimed under FIFRA 10(d) (1) (A), (B), or (C))
- the "Supplemental Statement of Data Confidentiality Claims" (used only if confidentiality is claimed on a basis other than FIFRA 10(d) (1) (A), (B), or (C))

The Study Title Page. The title page (Figure 10.4) is required for each study and must contain the following information:

- *Study Title.* The title must clearly identify the substances tested, must correspond to the name of the data requirement as it appears in the testing guidelines, and should be as descriptive as possible. This title should be the same as that appearing on the protocol for study conduct.

- *Data Requirement.* This is the Pesticide Assessment Guideline (PAG) number of the specific requirement that the study is intended to satisfy.

- *Authors.* Cite only individuals with primary intellectual responsibility for the conduct of the study.

- *Study Date.* This is a single date for the study. Use only the date of the latest element collected in the study. This date has been interpreted to mean the date the final report is signed by the study director.

- *Performing Laboratory.* Include the name and address of each laboratory or field operation contributing to the reported study. If there are too many sites to list on the title page, a notation should be made to refer to the next page in the report where these laboratories are listed.

- *Laboratory Project Identification.* This is each performing laboratory's internal project number for the study.

- *Supplemental Submissions.* Include on the title page the preceding elements for the previously submitted study, along with the EPA Master Record Identifier number. Do not include supplements to more than one study under a single title page.

- *Facts of Publication.* If the study is a reprint of a published document, include on the title page the journal title, volume, issue, inclusive page numbers, publication date, and any other publication information.

- *Page Numbers.* Show the total number of pages in the complete study on the title page. The study may be paginated in one of two ways: (1) Give the total number of pages on each page (i.e., 1 of 75, 2 of 75, ... 75 of 75) or (2) use a company name or mark and a unique study number on each page of the study (e.g., Wereachemlab–3210–1 ... Wereachemlab–3210–75).

Statements of Data Confidentiality Claims. Each study must be accompanied by a "Statement of Data Confidentiality Claims". There are two alternative statements that apply to confidentiality based on FIFRA 10(d)(1)(A), (B), or (C) (*1*). One form asserts a claim of data confidentiality, and the other waives it. The "Statement of Data Confidentiality

STUDY TITLE
(Chemical name)—Magnitude of Residue on Corn

DATA REQUIREMENT
EPA Pesticide Assessment Guideline 171–4

AUTHOR
John C. Davis

STUDY COMPLETED ON:
January 5, 1990

PERFORMING LABORATORY
XYZ Contract Laboratory
940 Your Street
Anytown, USA 10001

LABORATORY PROJECT ID
XYZ–47–89

SPONSOR
Large Chemical Company
1234 South Street
Washington, CO 32148

Page 1 of ✕
or
Total Pages: 92

Figure 10.4. Sample title page.

Claims" must be signed and dated and must include the typed name and title of the official who signs it. See Figure 10.5 for an example of a No Confidentiality Claims statement.

Good Laboratory Practice Compliance Statement. The GLP compliance statement must be included in all studysubmissions. As specified in 40 *Code of Federal Regulations* (CFR) 160, one of the following statements must be used, and it must be signed by the applicant, the sponsor, and the study director:

- a statement that the study was conducted in compliance with 40 CFR 160

- a statement describing in detail all differences between the practices used in the study and those required by 40 CFR 160

- a statement that the person was not a sponsor of the study, did not conduct the study, and does not know whether the study was conducted in compliance with 40 CFR 160

According to an EPA Office of Compliance Monitoring Advisory, the EPA stated that more than one compliance statement could be submitted (i.e., one signed by the study director and one signed by the sponsor or applicant); however, all statements must be identical in content (2).

Under the GLP regulations, the compliance statement is a very significant document, and the wording therein should not be taken lightly. Any GLP deficiencies that cannot be addressed by a GLP mechanism must be listed in the compliance statement. Examples of deficiencies that cannot be addressed by a GLP mechanism (such as an amendment or a properly addressed protocol deviation) are water, soil, and feed analyses that were not conducted under the GLP Standards. In addition to the aforementioned analyses, mass spectroscopic analyses frequently are not conducted under the GLP Standards.

No claim of confidentiality is made for any information contained in this study on the basis of its falling within the scope of FIFRA § 10(d)(1)(A), (B), or (C).

Company:

Company Agent   Date

Figure 10.5. Format for "Statement of (No) Data Confidentiality Claims".

In addition to the requirements discussed in the preceding sections, the PR Notice lists physical format requirements for all elements in a submittal package. For example, frayed or torn pages and carbon copies cannot be included. All data must fit on an 8 1/2 ×11-inch page; oversized pages cannot be submitted (*3*).

EPA Pesticide Assessment Guidelines

The EPA PAGs describe both general and specific reporting and data evaluation requirements that apply to studies within each subdivision. Each study report submitted under a particular subdivision should satisfy the reporting requirements specified in that subdivision unless another document pertaining to the subdivision directs otherwise.

Each study report should include a comprehensive description of test methods and a discussion of the test results. It should also contain an abstract or summary of the study, an analysis of the data, sufficient data to allow for verification of calculated statistical values, and a statement of the conclusions drawn from the data analysis. The abstract or summary should be viewed as a "stand-alone" document and thus should contain sufficient detail to permit the reader to understand the objectives, study design, results, and conclusions.

The general information to be included in the report (as directed by the testing guidelines) must be in agreement with the GLP requirements. For example, study dates, name and address of the testing facility, test location, and names of principal investigators and senior scientific personnel are all required. Examples of more specific reporting requirements set forth by the testing guidelines are contained in the following list:

- Test equipment: The report should include a description of the test equipment used, including photographs or detailed descriptions of nonstandard equipment.

- Units of measurement: Reporting units should be in the metric system, but the English system may be used in addition. In no instance should the systems be mixed (e.g., kilograms/acre).

- Calculations and tabular or graphic information: Each report should contain the principal mathematical equations used in generating and analyzing data, as well as representative calculations using these equations. When rates of formation and decline of parent compounds or their degradation products are reported from any test, data should be expressed as amounts, concentrations, and corresponding percentages. Rate constants, when required, should be reported in conjunction with rate data. Tabulated data, as well as graphs for decline curves and soil sorption, should be submitted (*4*).

Data Reporting Guidelines

The Data Reporting Guidelines (DRGs) (*5*) are addenda to the EPA PAGs. These guidelines are study-specific and are designed to provide consistent reporting of data. The requirements of the DRGs are compatible with the PR Notice and GLP requirements; however, the DRGs often provide a great deal more detail than the testing guidelines as to the content of the report. For example, the "Material and Methods" section of the DRGs specifically spells out the amount of detail required to describe the test substance, the equipment used, and the test method. Additionally, the DRGs define how the report is to be formatted (i.e, abstract, introduction, materials and methods, results and discussion, conclusion, certification statement, tables and figures, references, and appendices, all inclusive). The certification statement should read "This report is a complete and unaltered copy of the report provided by the testing facility except for title page changes required by PR Notice 86–5." This statement must be signed by the sponsor.

The following table of contents is recommended as a sample to incorporate the formatting requirements of the documents previously described. The pagination is illustrative and may change if any of the items under the "Contents" heading spans more than one page.

Page	Contents
1	Study title
2	Statement of Data (No) Confidentiality Claims
3	GLP compliance statement
4	Certification by applicant that report is unaltered copy received from the testing facility
5	QAU final report statement
6	Scientific personnel signature
7	Table of contents, with tables and figures listed individually
8	Study identification
9	Abstract (with summary table if required by DRGs)
10	Body of report, followed by references and appendices (including protocol, amendments, and deviations)

Standard Evaluation Procedures

Standard Evaluation Procedures (SEPs) are guidance documents that are used to evaluate data submitted to the OPP and to provide interpretive policy guidelines when appropriate. The SEPs go hand in hand with the DRGs. The EPA scientific reviewer is asked to evaluate all the items that are to be reported according to the DRGs. Additional requirements may be raised by the SEPs; for example, in specific studies SEPs require that soils be sifted through a 2-mm sieve prior to use (*6*). The SEPs may be used to provide guidance for the performance of studies, but they do not supersede the requirements of the GLP regulations. Should there be a terminology difference between the GLP Standards and an SEP, the requirements of the GLP Standards take precedence (*7*).

Phase 3 Technical Guidance Document

The Phase 3 Technical Guidance Document is used for the accelerated reregistration of pesticides established by Section 4 of FIFRA as amended in 1988. Although the information required by the Phase 3 Technical Guidance Document is not used for report preparation, the "Acceptance Criteria" and "Guidance for Summarizing Studies" items should be addressed in the final report because these sections are a compilation of the most recent study requirements set forth by the EPA, and they encompass the testing guidelines and the SEPs (*8*).

Adherence to Guideline Documents

A study is the report of a single scientific investigation, including all supporting analyses required for logical completeness. A final report is a comprehensive document that must not only satisfy all of the GLP requirements but must also incorporate the requirements of the applicable EPA PAGs, PR Notices, SEPs, FIFRA reregistration Phase 3 Technical Guidance Document, and DRGs. These documents present either general report-formatting requirements, or study-specific requirements, or both, as set forth by the EPA. Because the final report presents a complete interpretation of the study results and is used in the support of pesticide registration, the aforementioned formats and the scientific requirements should be carefully followed.

References

1. "Good Laboratory Practice Standards Under the Federal Insecticide, Fungicide, and Rodenticide Act Final Rule", *Code of Federal Regulations* Title 40, Pt. 160; *Federal Register* 54:58 (August 17, 1989) pp 34067–34074.

2. "Interpretation of the Good Laboratory Practice (GLP) Regulations", GLP Regulations Advisory No. 9, U.S. Environmental Protection Agency, Washington, DC, April 10, 1990.

3. "PR Notice 86–5", U.S. Environmental Protection Agency, Washington, DC, 1986.

4. "Pesticide Assessment Guidelines, Subdivision N, Chemistry: Environmental Fate", U.S. Environmental Protection Agency, Washington, DC, 1982.

5. "Pesticide Assessment Guidelines, Subdivision N, Chemistry: Environmental Fate, Series 162–1, Aerobic Soil Metabolism Studies", Addendum 5 on Data Reporting, U.S. Environmental Protection Agency, Washington, DC, January 1988.

6. "Hazard Evaluation Division, Standard Evaluation Procedure, Aerobic Soil Metabolism Studies", Office of Pesticide Programs, U.S. Environmental Protection Agency, Washington, DC, June 1985.

7. "Interpretation of the Good Laboratory Practice (GLP) Regulations", GLP Regulation Advisory No. 18, U.S. Environmental Protection Agency, Washington, DC, October 30, 1990.

8. "FIFRA Accelerated Reregistration Phase 3 Technical Guidance", Office of Pesticides and Toxic Substances, U.S. Environmental Protection Agency, Washington, DC, December 24, 1989.

RECEIVED for review November 5, 1990. ACCEPTED revised manuscript August 17, 1991.

Quality Assurance Responsibilities

Chapter

11

The Quality Assurance Unit

The Master Schedule and the Archives

Patricia D. Royal

Environmental Sciences Division, Springborn Laboratories, Inc., 790
Main Street, Wareham, MA 02571

*The good laboratory practices (GLP) regulations define the
function of the quality assurance unit (QAU) in regulated
studies as that of ensuring managers that all aspects of the
facility, personnel, performance, record-keeping, and reporting are consistent and in compliance with the regulations.
The objective of the regulations is to ensure users of health
and safety information and of the standards of accuracy and
to ensure integrity of study conduct and reported results
according to specifications in the GLPs. This objective is
achieved through the development of quality assurance programs that systematically evaluate and monitor various
ongoing and completed studies, as well as the activities of
the facility and personnel. Evaluations of ongoing activities
are called* inspections, *and evaluations of completed studies
to ascertain accuracy of the final report are called* audits.
*Two additional requirements of the GLPs are the master
schedule and the archives. The master schedule identifies*

ongoing studies, test materials being used, and study schedules; and maintenance of the master schedule is described as part of the QAU function. Controversy has always existed over QAU involvement in maintaining the archives. The requirements and accepted practices relating to these activities are discussed in this chapter.

Quality assurance programs have a relatively short history of only 12 years, yet the growth in scope and impact of quality assurance (QA) programs over these years has been great. The first Good Laboratory Practice (GLP) Standards regulation was issued by the Food and Drug Administration (FDA) in 1978 (*1*). This GLP regulation mandated the independent QA function as we know it today. Although the initial description of the QA function has not changed since 1978, its impact and scope have changed. In 1983, the scope of the quality assurance unit (QAU) changed to include studies to be submitted to the Environmental Protection Agency (EPA) (*2, 3*). Under the Federal Insecticide, Fungicide, and Rodenticide Act (FIFRA), the scope of the QAU role was expanded to include toxicological, health-related studies, and the Toxic Substances Control Act (TSCA) included environmental studies in addition to traditional health-effects studies. This transition was easily accommodated; most laboratories conducting regulated mammalian toxicology studies under the EPA were already moving toward GLPs and were operating under quasi-GLP-compliance programs, and environmental studies mandated under TSCA programs were few. The World Health Organization then orchestrated the Organization for Economic Cooperative Development (OECD) GLP program. Since then, other nations, including Japan, England, and Italy, have also begun to implement GLP programs. More recently the EPA has amended its GLP program under both FIFRA and TSCA to include (with very few exceptions) all studies submitted to the EPA to support the registration of all chemicals and pesticides (*4*). Interestingly enough, even though the scope and application of QA programs have changed drastically, the verbiage described in the regulation, its purpose, and the responsibilities of QA programs have not changed. Through this process, a regulation initially developed for a contained laboratory environment was expanded to cover multisite operations of diverse disciplines in a very short time. The impact and responsibilities of QA programs in the agrochemical industry has truly expanded, and understanding those implications and responsibilities and meeting those challenges is the objective of this chapter and of the symposium upon which this book is based.

The purpose of an independent QA program, as it was written in 1978 and as it exists today, is to ensure data integrity and compliance with

GLPs. The definition of data integrity in 1978 assumed that a study was conducted at a facility and that all data generated from a study would be under the direct control of one person identified as the study director. The regulation also states that the QAU is responsible for conducting inspections and data audits to assure the "management"[1] of the facility of compliance with the regulation. This directive is the theoretical application of the QAU, and it has not changed. Recently there have been discussions about the definition of these terms and their impact on the conduct of a study and therefore compliance with the regulation. These debates will continue for some time; thus, the interpretation of compliance with the regulation will continue to evolve. In the past, if segments of a study were subcontracted to another facility, the lead QAU either worked with the subcontracted QAU to maintain a cohesive QA program and thus ensure the overall compliance of the study or implemented its own QA program to cover those activities conducted off site. Today, studies are routinely conducted at multisite locations, and the lead QAU must ensure compliance at several locations simultaneously. Because there are more off-site locations for which the lead QAU is responsible, in practical terms the responsibility for ensuring compliance for the study to management means different things to different people, and many groups depend on the QAU to ensure compliance of the study for different reasons and to keep them informed about different facets of study conduct. These groups are illustrated in Figure 11.1.

Figure 11.1. The responsibilities of the QAU extend to many different people.

---

[1]Management is defined here as the person or persons responsible for setting policy at a facility. For a description of the duties of testing facility management, see § 160.31 of the GLP Standards.

To fulfill this need, the QAU must orchestrate universal study mile-stones into time points that are then targeted for review. These mile-stones include signing the protocol and entering the study onto the master schedule, in-process activities, study termination, study completion, and archiving.

Inspections and Data Audits

The regulations state that the QAU must conduct inspections at a fre-quency that will ensure the integrity of the study. To meet this challenge, a strategic plan that systematically evaluates all parts of the regulation, the study, and the facility must be developed. The regulations state that each study must have data and report audits. Above and beyond this require-ment, the regulation is open-ended. The frequency of inspections is the decision of the QAU. This decision is a change from earlier versions of the GLP Standard in which inspections at 3-month intervals were required for long-term studies (2, 3). The regulation states that studies should be inspected often enough to ensure their integrity and compliance with the regulations.

Three general rules apply to the conduct of inspections and data audits:

- Data and report audits must be conducted on each study.

- Inspection schedules can be flexible and should be dependent on the duration and complexity of study.

- Targeted phases to be inspected can be selected by using one of the following methods.
  — all phases for each study
  — random selection (hit or miss)
  — phases selected off the master schedule
  — comprehensive series of rotating generic phases (frequency depen-dent on length or complexity of study)

A facility management decision must be made concerning the approach for scheduling inspections. Some laboratories attempt to inspect all phases of all studies. Although this approach may be comforting to laboratories in the early stages of GLP compliance, to inspect all phases of each study is burdensome, redundant, and unnecessary for laboratories with experienced GLP programs. Reports must be written for each inspection. If the same operation is reviewed too often, the process becomes dull, and important events may be overlooked. In a typical busy laboratory, the inspection of all phases of a study would be excessive and would not be a cost-effective use of personnel. Equally unacceptable would be to base selection of study phases for inspection on a hit-or-miss method with no forethought. This approach would not provide a comprehensive evaluation of overall compliance. The most efficient method for selecting phases for review is to use the master schedule if it is kept in enough detail so that phases can be realistically selected. This last approach uses the master schedule and the length and complexity of the study to target specific operations and to identify "generic" phases that all studies encompass. Thus, although the type of phase inspection may stay the same, the operations reviewed depend on the type and duration of the study. Suggested generic phases that should be targeted for inspection are given in the following list. Likewise, examples of how the generic phase can be tailored to specific study type are given in Table 11.I.

- *Facility*: location of test system and test substance, maintenance and calibration of equipment, location and availability of equipment, structure of the QAU and company management staff, master schedule, and archives

- *Pretest*: method validation, homogeneity, stability, pretest clinical chemistry (mammalian studies), health certification, diluter calibration (aquatic studies), and applicator calibration (field studies)

- *Initiation*: protocol, standard operating procedures, training, and first study event

- *Test substance*: receipt, distribution, accountability, dose preparation, dosing, and storage

- *Test system*: receipt of test system or culture records, water and feed analysis, randomization, and accountability

- *Ongoing test*: water quality and biological observations (fish), body weight, feed consumption (mammalian), growth, and length (plant)

- *Termination*: necropsy, harvesting test system, growth, mortality, transformation, and degradation

Scheduling and managing QA personnel to inspect and audit all required activities in a cost-effective manner is becoming more difficult as the number of locations and study types increases. Activities must be continuously reviewed to ensure compliance with the most recent interpretations of the regulations.

Master Schedule

The GLPs require that a copy of the master schedule be maintained by the QAU. Requirements for the master schedule as identified in §

Table 11.1. Examples of Generic Phases of Studies

Study Type	Initiation	Dose Preparation	Termination
Teratology	Mating	Gavage preparation	Necropsy
Plant growth	Seed planting	Blending with soil	Root and shoot weight and length
Field residue	Site selection Application of chemical	Tank mix	Residue analysis
Bioconcentration	Diluter calibration Randomization	^{14}C stock and specific activity	Tissue oxidation and identification

160.35(b)(1) are that the master schedule must be "indexed by test sub-stance and contain[ing] [the] test system, nature of study, date study was initiated, current status of each study, identification of the sponsor, and name of the study director."

The initial purpose of the master schedule was to assess work loads and to schedule equipment. However, in a laboratory the master schedule is rarely used for this purpose. The requirements for the master schedule as written have received a great deal of criticism. Federal agencies find the master schedule a convenient tool for scheduling future inspections, but the master schedule often changes for legitimate reasons, and thus the study status is usually very general. The study start date or the date of ini-tial application of the test material can change; likewise, the study director can change. Any change in the master schedule must leave an audit trail so that the change can be reconstructed and justified. The master schedule provides an immediate overview of ongoing activities, but many laboratories are reluctant to include nonrequired information about indivi-dual laboratory operations on the master schedule for fear of revealing too much information during review by either enforcement officers or clients. This point was reiterated recently by Goeke (5) in her chapter on the mas-ter schedule; she said that too much information in the master schedule is counterproductive and that often it is more sensible to integrate the mas-ter schedule with other types of scheduling documents. The initial pur-pose and intent of the EPA in requiring the master schedule was as a management tool to assess work loads and anticipate overcommitment on the part of available personnel and equipment. However, because of regu-latory restrictions, the master schedule is often maintained only to fulfill regulatory requirements. For a more detailed description of the master schedule, see reference 5.

Exceptions to this evaluation do exist. Computerized master schedules seem to be more adaptable and useful than handwritten master schedules. Computerized schedules can be sorted by different column headers, thus facilitating assessment of such entries as study type, study director work load, and needed test material. If included, nonrequired or sensitive information such as client name, identification of test substance, or budgeting can be purged from any printout. A computerized master schedule can be particularly useful for tracking field studies. Scheduling field programs is subject to change due to weather, and thus tracking these studies and scheduling inspections are difficult. The most efficient way to update the master schedule for field studies is in the field, where weather conditions and thus test initiation can be immediately assessed. Workers at each location could then be responsible for updating their portions of the master schedule on a modem that relays the information back to a mainframe computer where the overall master schedule is maintained. Likewise, in-house scheduling could be updated on the basis of pretest events entered into the master schedule by the task leaders responsible for those events. Computerized master schedules can facilitate scheduling QA inspections at distant locations as well as within one laboratory and can provide real-time assessment and time management of ongoing work.

The regulation states that the QAU must "maintain a copy of the master schedule" [§ 160.35(b)(1)]. Many individuals have interpreted this requirement to mean that the QAU must physically update all entries on the master schedule. To interpret this requirement, as written, to mean that the QAU and only the QAU may update the master schedule becomes a question of semantics when realizing that the GLPs also state that the QAU must "maintain a copy of all protocols" [§ 160.35(b)(2)]. The EPA never said that the QAU must physically write the protocol.

Another set of requirements for the master schedule needs clarification. These requirements are for entering and removing studies on the initiation date and test completion date. These dates are important milestones that must appear on the master schedule. By definition, *test initiation date* is the date the study director signs the protocol and *test completion date* is the date that the study director signs the final report. The preamble of the GLP Standards states that a study must be entered onto the master schedule no later than at test initiation, and that all data must be archived at test completion. At that time, the study may be removed from the master schedule. As some people have suggested, these requirements do not mean that studies may not be entered and removed prior to and after these dates, rather that studies must be included on the master schedule for the duration of the study and that these dates must be entered onto the master schedule. When considering the originally intended function of the master schedule, clearly, it is advantageous from a management point of view to enter a study as soon as is reasonable to facilitate managers' assessment of work loads.

Archives

At test completion, all raw data, correspondence, specimens, an aliquot of the pure test material necessary for the reconstruction of the study, the final report, and the protocol must be archived. Currently the EPA interprets test completion as the date the final report is signed and issued by the study director. The location of archived data must be included in the final report.

Archiving requirements specified in the GLP Standards include facility requirements, security (including logs and indices), longevity, and personnel. Most of the archive requirements are identified in § 160.190 of the FIFRA GLP Standards, but a careful review of the entire GLP Standards and the preamble is necessary to comprehend fully archiving requirements.

Archive Facility. Requirements specify that a secure location that provides limited access to archived material must be set aside for archiving purposes. Occasionally, separate facilities may be necessary for paper and wet specimens, depending on study type and optimal storage conditions for these data. The archives should have adequate fire protection and controlled temperature and humidity to optimize storage. However, extraordinary storage procedures, such as −20°C freezers, are not necessary. Sections 160.190 (b) and 160.195 (c) of the GLP Standards specify the following requirements:

> Conditions of storage shall minimize deterioration of the documents or specimens...for the time of their retention....

> Wet specimens; samples of test, control, or reference substances; and specially prepared material, which are relatively fragile and differ markedly in stability and quality during storage, shall be retained only as long as the quality of the preparation affords evaluation. Specimens obtained from mutagenicity tests; specimens of soil, water, and plants; and wet specimens of blood, urine, feces, and biological fluids do not need to be retained after quality assurance verification.

The archives should be of sufficient size and location to retrieve data conveniently when needed. Although often customary, this requirement does not mandate that an entire room be designated for archive use only. The requirement is that a separate, secure area be assigned as the archives. Thus, a locked freezer or locked safe with adequate fire protection or designated office space with locked file cabinets and fire protection could fulfill the facility requirements as long as all other requirements were also fulfilled. Archive requirements include limited access and that an archivist be designated and in control. All data entered into the archives should be indexed, and the material and the date archived should be indicated. A logbook should be established that documents any material leaving the archives by type of material, individual removing the material, and dates removed and returned. All entries into the index should be signed and dated.

Data Retention. Data-retention requirements are also defined in § 160.195. This section should be carefully reviewed because specifications differ according to type and use of data. Data should be retained for whichever of the following periods is longest:

- For a study used to support a marketing permit approved by the EPA, data must be retained for the period during which the sponsor holds the marketing permit to which the study pertains.

- Full-fledged study data must be retained for at least 5 years following the date on which the results were submitted to EPA in support of an application for a research or marketing permit.

- For studies for which the results do not lead to the submission of the study in support of an application for a research or marketing permit, data must be retained for a period of at least 2 years following the date on which the study is completed, terminated, or discontinued.

- Facility data, such as maintenance and calibration records, training records, culture logs, and facility environmental conditions, should be retained indefinitely.

In addition, all records are subject to the following requirements:

- Records may be retained either as original records or as true copies such as photocopies, microfilm, microfiche, or other accurate reproductions of original records.

- All copies of raw data substituting the original record must be identified as verified copies and must be signed and dated at the time of copying.

Interestingly, 40 *Code of Federal Regulations* (CFR) Part 169, entitled "Books and Records" also specifies data-retention requirements for research data. These specifications are broader than those given in the GLP Standards. 40 CFR Part 169 specifies that "data must be retained for as long as the registration is valid and the producer is in business". The difference here is that liability for data integrity does not end with product ownership. Presumably, both 40 CFR Part 160 (which was amended in 1989) and 40 CFR Part 169 are regulations written to enforce FIFRA. FIFRA does not specify data-retention requirements; it specifies only that the administrator may write specifications for the retention of records. This discrepancy has led to confusion over data-retention requirements. Often contract laboratories are not notified when a sponsor submits specific studies or applications for registration to the EPA for approval, a practice that compounds the problem. Contracted studies can be completed years prior to the submission of the application. Thus, retention requirements need careful review. Generally, data (with the exception of certain specimens) supporting a registered product should be archived indefinitely until additional clarification is obtained from the EPA.

Designated Archivist. The GLP Standards specify that the archivist be a designated individual. Most laboratories have interpreted this to mean that the archivist must be identified by name. The naming of a designated individual is the only requirement of this type. A long-standing controversy has existed as to whether or not a member of the QAU can also function as the archivist. In a small laboratory, designating a member of the QAU as archivist would facilitate data security, personnel requirements, and efficiency. However, occasionally using a member of the QAU as the archivist has been interpreted by the EPA and the FDA as a potential conflict of interest. Although this practice is not specifically disallowed, most laboratories prefer to keep these functions separate and thus do designate an individual as an archivist. Many archivists have a library sciences degree and a general background in science. Managing an archive well requires special organizational skills that should not be overlooked.

Because of space limitations and regulatory repercussions, many laboratories have resorted to using commercial archiving facilities, or in the case of contract laboratories, charging sponsors an archiving fee. In the past, many contract laboratories provided archiving services to their clients free of charge, but this service is quickly disappearing. Likewise, because of liability, many sponsors are requesting that all raw data be archived at the sponsor's facility at the termination of any study.

During the FIFRA 88 reregistration process, many manufacturers discovered that the EPA may require that raw data be available indefinitely. As part of the FIFRA 88 reregistration package, a signed statement confirming the availability of raw data must be included. Additional specifications for data retention and archiving are given in FIFRA Parts 169 and 169.2 "Books and Records of Pesticide Production and Distribution and Maintenance of Records" (*6*).

Conclusion

QA personnel, study directors, and corporate managers are well aware of the current changes in the GLP Standards and their impact. These changes result in increases—increases in the types of regulated studies, in the personnel and training required to conduct these studies, and in budgets to run the studies. Likewise, the recent change in the position of the EPA from one of simply expecting conformance with the regulations to one of active enforcement of the regulations, with potential penalties for noncompliance has raised many questions and uncertainties. In many laboratories, the master schedule has changed from a rudimentary hand-recorded document to a sophisticated computerized system. Archiving requirements have also changed, along with liability for maintaining records for long periods of time. These changes mean new challenges for

QA programs that are only just beginning to be realized. The functions of the QAU have remained the same since the original GLPs were promulgated, but the scope and responsibilities of the QAU have expanded. This expansion is most readily apparent in study types and logistics under the recently amended FIFRA GLP Standards. To fulfill the mandate given to the QAU by the EPA, QA managers must develop a strategic plan to assess many unrelated study types of different disciplines in different locations and in an atmosphere of evolving interpretations of the regulation.

The challenges of developing new compliance programs by QA managers in the future must be coupled with an understanding of the diverse scientific disciplines and with a knowledge of the original intent of the regulation. Keeping those issues in mind (diverse disciplines and the original intent of the regulation), will lead both the EPA and the regulated community on the same path—a path where the integrity and quality of a study can be judged.

References

1. "Good Laboratory Practices", *Code of Federal Regulations* Title 21, Pt. 58; Food and Drug Administration, Washington, DC, December 22, 1978.

2. "Good Laboratory Practices", *Code of Federal Regulations* Title 40, Pt. 160; Environmental Protection Agency, Washington, DC, November 29, 1983.

3. "Good Laboratory Practices", *Code of Federal Regulations* Title 40, Pt. 792; Environmental Protection Agency, Washington, DC, November 29, 1983.

4. "Good Laboratory Practices", *Code of Federal Regulations* Title 40, Pt. 160; Environmental Protection Agency, Washington, DC, August 17, 1989.

5. Goeke, J. "Good Laboratory and Clinical Practices; Techniques for the QA Professional". In *The Master Schedule*; Carson, P.; Dent, N., Eds.; Heinemann Newnes: London, 1990.

6. "Books and Records of Pesticide Production and Distribution", *Code of Federal Regulations* Title 40, Pt. 169; Environmental Protection Agency, Washington, DC, 1985.

RECEIVED for review September 17, 1990. ACCEPTED revised manuscript August 17, 1991.

Facility Inspection Conduct

Quality Assurance Perspective for Field Studies

David Johnson and Jesse Burton

Stewart Agricultural Research Services, Inc., P.O. Box 509,
Macon, MO 63552

*To ensure that the testing facilities are adequate for con-
ducting field studies according to good laboratory practices
regulations, five basic areas need to be inspected: The testing facility must have a sufficient number of substantiated,
qualified personnel for performing studies, and these personnel must use proper sanitation and health practices; facilities must be of sufficient size and suitability for the proper
conduct of studies, with functions and activities to preserve
the identity, integrity, and control of the samples; equipment
must be appropriate, properly maintained, and calibrated
with documentation on its use; current and historical
records on standard operating procedures must be written,
available, and followed by the personnel; and test substances
must be properly labeled and maintained with documentation on their receipt, storage, and use.*

Toxicological testing in a laboratory setting was the first
research area to come under Good Laboratory Practice (GLP) Standards
(*1*). The regulated tests were performed at one site by a single group of
scientists. The scope of the regulations has now expanded to include field
studies.

Adapting the regulations to field studies has caused considerable confusion and is more difficult than adapting the regulations to other laboratory studies. Consequently the interpretation of GLP regulations is quite

2192-8/92/0155$06.00/0 © 1992 American Chemical Society

varied among quality assurance units (QAU) throughout the agricultural chemical industry.

The purpose of the GLP regulations is "to ensure the quality and integrity of data submitted" (§ 160.1(a) of the GLP Standards). The regulations outline procedures for ensuring data quality and integrity. Quality assurance units are "responsible for monitoring each study to assure management that the facilities, equipment, personnel, methods, practices, records, and controls are in conformance with the regulations" (§ 160.35). The regulations specify that the QAU will

- conduct inspections

- maintain appropriate records

- maintain a copy of the master schedule

- maintain a copy of study protocols

- inspect each study at adequate intervals

- maintain written records of each inspection

- submit reports to the manager and the study director

- determine that deviations and amendments are properly authorized and documented

- review the final study report to ensure accuracy

- sign a statement in the final report stating the dates of inspection and reporting to the manager and the study director

These regulations apply to both field and laboratory portions of studies. The specific procedures used to ensure the quality and integrity of data may differ between the laboratory and the field. In the field portions, the primary concerns are minimizing the chances of contamination, ensuring that samples are of the highest possible quality, ensuring an orderly arrangement of activities, and ensuring complete documentation of all activities.

In some cases, such as prestudy testing facility evaluations, field inspections conducted when no study was in progress or inspections conducted during other studies may provide necessary assurances. This evaluation method is acceptable as long as the facilities and personnel are representative of those to be used in the study. However, each study must still be inspected at least once during execution.

During an inspection, the QAU must physically observe the facilities to ensure adequate separation of incompatible activities and suitability for the study; review documentation to determine adequacy; review the methods, practices, and records that were used and in place when the study was being executed to ensure compliance with protocols and standard operating procedures (SOPs); and judge the adequacy of the facilities as determined by the standards adopted by the manager.

According to the regulations, "the facilities shall be of sufficient size and construction to facilitate the proper conduct of studies" (§ 160.41). The term sufficient must be defined. Facility "management"[1] is responsible for determining what is sufficient within the confines of the regulations. The QAU is then responsible for observing and evaluating the facilities to ensure that they conform to the regulations and management directives. The results of the QAU inspection are reported to the managers and to the study director for possible action.

During a facility inspection, sites that are obviously sufficient are relatively easy to identify. Facilities that are sufficient have adequate equipment and space, qualified personnel, good SOPs, and orderly and complete documentation of all activities.

Facilities that are obviously insufficient are also relatively easy to identify. These facilities will commonly lack sufficient qualified personnel for performing the duties, data will not have been recorded currently, the field investigator may not have control over the test site or sample storage facilities, or adequate facilities may not be available for processing samples or handling excess spray and rinsing solutions. These are the sites that need to be identified quickly because they could jeopardize the quality and integrity of the studies.

Some requirements are spelled out very clearly in the regulations and must be followed. Other items are not specifically spelled out in the regulations but are practices generally accepted by the agrochemical industry, and they should be followed. Examples of both types of requirements are presented throughout this chapter.

Requirements that are specific to individual sponsors must also be considered. A sponsor's practices may not be accepted across the industry, and although this usually is not a problem within a sponsor's own facilities, it can result in confusion and frustration for the investigators at contract facilities because of disagreements and inconsistencies in interpreting the GLP Standards among sponsors.

Field facilities can be divided into five physical areas: office, equipment, storage facilities, operation areas, and field sites.

---

[1]Management is defined here as the person or persons responsible for setting policy at a facility. For a description of the duties of testing facility management, see § 160.31 of the GLP Standards.

Office

Office space should provide room for personnel plus space for archiving and storing raw data and supporting documentation prior to transfer to the sponsor.

The regulations require that all raw data be archived [§ 160.190(b)] but do not specify the location of the archive. In the agrochemical industry, raw data associated with each test are usually transferred to the sponsor for archiving. The items to be archived at the field facility might include items that are specific to the test site as opposed to a particular study. These items include weather data, maintenance logs, calibration logs, chemical receipt and shipment logs, personnel records (curricula vitae, training records and job descriptions, and organizational charts), and SOPs.

The regulations do not specify the type of storage facilities to use, although some type of fire-proof, lockable container (such as a fire-proof filing cabinet) is commonly used. This is not the only way of keeping records safe. The important point is that storage should be adequate for safe retention of the data so it can be retrieved in future years if necessary.

The archive should also be organized "for orderly storage and expedient retrieval of all raw data" [§ 160.190(b)]. This need for an organized archive requires a system for entering data into and retrieving data from the archives as explained in an SOP. An index to the archives is also required.

No regulations are specified for the handling and short-term storage of raw data prior to transfer to the sponsor for archiving. However, a system should be in place for this purpose. Raw data should be well organized and secure during the time they are in the possession of the field investigator. This security can be achieved by storing the data in a safe place or sending verified copies of the raw data to the study director or project monitor as the data are recorded.

The facilities inspection should include an examination of the records to ensure that personnel records are complete and up-to-date. The organizational structure should allow for proper separation of the QAU from those who execute the field trials. Maintenance, calibration, and shipping logs should be located where they are used, and it should be evident that they are being used. SOPs should be located where they are easily accessible to personnel. Other items that need to be checked include Material Safety Data Sheets and proper clothing and first aid facilities to ensure proper sanitation and health for the personnel (§ 160.29). SOPs for the activities that occur at a particular site should be immediately available at that site. Any QAU that is functioning at a field site should have a copy of the master schedule, copies of all relevant SOPs and protocols, and documentation on inspection and audit dates and reports to managers and study directors (§ 160.35).

Equipment

The equipment used for residue studies varies in both type and size. It may range from large equipment such as tractors to small sampling equipment such as hand clippers.

Each field facility must be organized so that equipment can be easily obtained by the personnel. The facilities should provide for the operation, cleaning, maintenance, and storage of equipment in a manner that will ensure proper operation of the equipment and freedom from contamination.

Using equipment that is maintained in good working order is important to ensure that the tests will be executed properly with minimal potential for problems with applications of chemicals or sample collection.

Freedom from contamination requires having facilities and procedures for cleaning equipment after use. The dirt and dust adhering to equipment can contain contaminants. In some cases cleaning can be done at a special facility, and in other cases it may be accomplished in the field. Section 160.43 requires "provisions for the collection and disposal of contaminated water, soil, or other spent material." The cleaning facilities should conform to applicable laws and regulations dealing with waste disposal.

Another important aspect of reducing contamination is the physical separation of equipment and supplies from chemicals or other sources of contamination such as dirt. Even large equipment used solely for land preparation should be kept clean and separated from equipment and supplies used for applying test substances and collecting samples.

The design and capacity of the equipment should be adequate for the normal functioning of the studies. The equipment should either be standard commercial equipment or equipment that accurately simulates standard operations and provides quality samples (such as backpack sprayers or research-sized threshers).

"Equipment shall be adequately inspected, cleaned, and maintained" (§ 160.63). Some type of documentation is required to verify these actions. Usually, some type of maintenance log is used, although other techniques are acceptable as long as the documentation verifies that these operations were performed. Number-generating equipment (e.g., thermographs, weighing scales, and pH meters) and standards used in calibration (e.g., scale masses and thermometers) should have documentation verifying their accuracy.

Records on the cleaning and maintenance of equipment used for applying chemicals and collecting samples are especially important because these procedures directly affect study results. The need for records on support equipment such as tractors and tillage implements is less obvious.

Storage Facilities

Storage facilities for field sites should be adequate to maintain equipment, supplies, samples, and chemicals without danger of cross contamination.

Supplies for a field test include the containers used for collecting and storing the samples. These include bags, bottles, plastic tubes and caps (for soil samples), pads, and shipping boxes. It is wise to check with the analytical lab to ensure that the intended containers will not cause a problem with sample analysis. Supplies should be stored to minimize the chances of contamination from other stored chemicals and chemicals that could be present on dust particles.

The regulations do not specifically address sample storage, although samples are usually stored in some type of freezer. These are usually chest-type home freezers or walk-in freezers. Either is acceptable as long as it is free from potential contamination from chemicals and dirt and will maintain the samples in a frozen state while in storage.

Some sponsors require that check and treated samples be stored in separate freezers and packed in separate boxes for shipment. Most sponsors allow storage and shipment of check and treated samples together as long as the check samples are double bagged or check and treated samples are physically separated in the freezer or box. The exact requirements are dictated by individual sponsors rather than by the regulations. As verification of the storage conditions, a system is required to ensure that the freezers operated properly and that the samples were not subject to thawing. Maximum–minimum thermometer readings will accomplish this verification. The regulations do not specify the recording times; the best system is one with continuous temperature monitoring because the exact time and temperature conditions will be known if and when a freezer shuts down.

To prevent contamination or mixing, separate areas are needed for storing test, control, and reference chemicals. The storage area must be separate from areas used for the storage of equipment, supplies, and samples.

The best areas for storing the different chemicals are ones with limited access and a system for logging the receipt, use, and shipment of chemicals. Maintaining a complete paper trail on the chemical from time of receipt to its final disposal or shipment is important, particularly for nonregistered compounds.

The storage area does not require complete temperature control, although it is preferred to have some means of ensuring that temperatures remain within the range required for maintaining chemical stability. The temperature in the storage area must be monitored to ensure that the ideal temperature range is not exceeded. In this case, the use of maximum—minimum thermometers is generally considered to be sufficient.

Each chemical must be maintained in the container in which it was received and must be labeled with the test and chemical identification, concentration, lot and batch numbers, and expiration date. If containers are maintained at the field site until the studies are completed, then facilities must be available for storing these containers.

Operation Areas

Adequate space must be available to carry out the activities associated with the execution of field studies. This includes areas for the orderly receipt and shipment of chemicals and supplies, the preparation of chemicals for application, calibration of equipment, and performance of special procedures.

In many cases the preparation of chemicals for application and calibration of application equipment are performed in the field. It is important that these processes be performed in a manner and at a site that will not interfere with the test area.

Special procedures may include the handling or processing of samples after they are removed from the field (e.g., cutting soil cores into incremental lengths and threshing grain). A contamination-free area should be available whenever the special procedure is a part of the facility's SOPs or is required by specific protocols.

Field Sites

The regulations do not specifically address the field sites. For most field residue studies, the test system includes land either for direct use or for growing the test crop. The land area must be sufficient for containing the plots and buffer areas as required in the protocols. The test area must be separated from potential problems such as neighboring fields where there is no control over what pesticides are used or where traffic from livestock, people, railroads, and unpaved roads could create contamination from dust and spraying.

The test area should be typical of land used for crop production and free of steep slopes and variability in soil types. The soil should be free from physical and chemical problems that could be adverse to crop production. The soil characteristics and past crop and pesticide history must conform to the requirements specified in the protocols.

The field investigator must have complete control over the field site for the duration of the study. If the site is subject to the activities of farmers or growers, their activities may not be compatible with the objectives of the study.

Reference

1. "Federal Insecticide, Fungicide, and Rodenticide Act (FIFRA); Good Laboratory Practice Standards Final Rule", *Code of Federal Regulations* Title 40, Pt. 160; *Federal Register 54*:158 (August 17, 1989).

RECEIVED for review September 17, 1990. ACCEPTED revised manuscript September 13, 1991.

Chapter
13

Analytical Phase Inspection of Residue Chemistry and Environmental Studies

Frances A. Dillon[1] and Jeffery L. Harris[2]

[1]Stewart Pesticide Registration Associates, Inc., 2001 Jefferson Davis Highway, Suite 603, Arlington, VA 22202
[2]Spectralytix, Inc., 200 Girard Street, Suite 204, Gaithersburg, MD 20877

With the promulgation of Good Laboratory Practice Standards in October 1989, residue chemistry and environmental fate studies became subject to facility inspections. Inspection programs may need to be modified or developed to accommodate this type of laboratory work. This chapter provides specific information and guidance for the establishment of an inspection program. To have an effective inspection program, comprehensive standard operating procedures (SOPs) are required. The SOPs define frequency of inspection, phases to be inspected, procedures for conducting the inspection, and documentation and reporting of the inspection and findings. The quality assurance auditor performs the inspection by comparing the work in progress against the study protocol, analytical method, and applicable SOP documents. Professionalism and objectivity are essential to the success of the inspection process. Findings made by the quality assurance unit during the inspection are documented in reports to managers. Managers must then ensure that corrective actions for any findings are taken and appropriately documented.

INSPECTION OF STUDIES IS A FUNCTION assigned to the quality assurance unit (QAU) by the Good Laboratory Practice (GLP) Standards. Section

2192-8/92/0163$06.00/0 © 1992 American Chemical Society

160.35 of the Federal Insecticide, Fungicide, and Rodenticide Act
(FIFRA) GLP Standards (1) and Section 792.35 of the Toxic Substances
Control Act (TSCA) GLP Standards (2) define the responsibilities of the
QAU for study inspection. The purpose of the study inspection is to
assure "management"[3] that the study is being performed in compliance
with GLPs. Adherence to GLPs, study protocol, and internal standard
operating procedures (SOPs) is monitored when actual facilities, equipment, personnel, methods, practices, records, and controls are checked.
    The success of the laboratory inspection program is contingent on
managerial support of the program. The inspection and corresponding
quality assurance (QA) reports provide management with current, nonbiased information on study status. In effect, the program becomes an
information-gathering management tool. Management support for the
inspection program starts with (1) establishing a separate QAU that is
independent of study managers and department heads, (2) careful selection and training of QAU auditors, (3) involvement in defining the inspection program SOPs, and (4) routine follow-up with study directors regarding inspection findings.
    This chapter will provide guidance on how to set up and implement
an analytical-phase GLP inspection program for environmental fate and
chemistry studies. The following topics will be addressed: the QAU auditor, GLP requirements for inspection programs, the inspection program
SOP, phases to be inspected and frequency of inspection, procedures for
conducting the inspection, topics to include in an inspection, documentation of the inspection, and records available to the Environmental Protection Agency (EPA) or the Food and Drug Administration (FDA)
representatives.

The QAU Auditor

To perform an objective evaluation, the QAU auditor must be independent from personnel engaged in the study direction or conduct. Therefore, the QAU auditor may not participate in any study-related activities
other than QA functions. The QAU auditor designated to inspect
environmental-fate and chemistry studies must understand the GLP Standards, chemistry, and quantitative analytical techniques. The auditor must
be trained to perform a GLP inspection and must be familiar with the
laboratory SOPs and the study protocol. Underqualified staff members
should not be placed in the position of QA auditor. Underqualified QA
auditors may overlook serious problems, and operations may be needlessly
disrupted by unnecessary questions.

---

[3]Management is defined here as the person or persons responsible for setting policy at a facility. For a description of the duties of testing facility management, see
§ 160.31 of the GLP Standards.

GLP Requirements for Inspection Programs

Before defining or modifying the laboratory GLP inspection program, the minimum requirements outlined in the GLP Standards should be considered. These requirements provide the basis for building an inspection program that must be incorporated by all laboratories.

- The GLP Standards state that the QAU must inspect each study at intervals adequate to ensure the integrity of the study and must maintain written and properly signed records of each periodic inspection. The records must show the date of the inspection, the identity of the study, the phase or segment of the study inspected, the person performing the inspection, any findings or problems, the action recommended and taken to resolve existing problems, and any scheduled date for reinspection.

- Any problems that are likely to affect the study integrity during the course of an inspection must be brought to the attention of the study director and management immediately.

- The QA auditor must periodically submit to management and the study director written status reports on each study, noting any problems and the corrective actions taken.

- The responsibilities and procedures applicable to the QAU, the records maintained by the QAU, and the method of indexing such records must be maintained in writing. These items, including inspection dates, the identity of the study inspected, the phase or segment of the study inspected, and the name of the individual performing the inspection, must be made available to authorized employees or representatives of the EPA or the FDA.

- Authorized employees or representatives of the EPA or the FDA must have access to the written procedures established for the inspection and may request the testing facility manager to certify that inspections are being implemented, performed, documented, and followed in accordance with 40 *Code of Federal Regulations* (CFR) parts 160.35 or 792.35.

The Inspection Program SOP

An SOP should be established to define the inspection program. This SOP is typically prepared by the QAU with input from managers, and it should include the following information:

- scope of the inspection program

- a statement that the QAU is responsible for the performance of the inspection and maintenance of required records

- phases of each type of study to be inspected

- frequency of inspection

- discussion of the actual procedures for the inspection (checklists may be included if the laboratory uses them)

- responsibility of the QAU to notify management immediately if there are problems that affect study integrity

- records to be retained by the QAU and the methods used for indexing these records

- content of inspection reports to be issued by the QAU

- instructions on distribution of inspection reports to the project supervisor, management, and the study director

The remainder of this chapter provides information that may assist in the development of an inspection SOP.

Phases To Be Inspected and Frequency of Inspection

As discussed previously, GLP Standards require inspection of each study at intervals adequate to ensure the integrity of the study. The interpretation of "adequate intervals" can vary widely. In practice, the decisions about what to inspect and how often inspections are performed are influenced by the type of study, the length of the study, the frequency with which that type of study is performed in the laboratory, and staff turnover.

The first step is to review the protocol and to identify the critical phases of the study. Examples are provided in the following list.

- Hydrolysis

  — buffer preparation
  — sterilization of glassware and solutions
  — test-substance-solution preparation and addition to buffer
  — test solution reaction mixture incubation and sampling
  — extraction and analysis

- Photolysis (aqueous)

  — buffer preparation
  — sterilization of glassware and solutions

- test-substance-solution preparation and addition to buffer
- light exposure and sampling
- extraction and analysis

• Soil leaching (^{14}C-labeled test substance)

- soil column preparation
- test-substance preparation and application
- leaching and leachate collection
- soil core collection
- extraction, radiochemical analysis, and soil combustion

• Soil metabolism (^{14}C-labeled test substance)

- test-substance preparation and application to soil
- aerobic incubation and sampling
- anaerobic incubation and sampling
- microbiology and moisture monitoring
- radiocarbon assay
- metabolite characterization, extraction, and thin-layer chromatography

• Residue analysis of crops, soils, or water

- sample receipt, chain of custody, and storage
- spiking standard solution and analytical standard solution preparation and storage
- sample preparation (subsampling, chopping, or grinding)
- extraction or digestion
- extract cleanup
- analysis

Inspecting every phase of every study performed in the laboratory is neither desirable nor necessary. The process would be economically disastrous and could potentially interfere with laboratory operations. However, some studies may require inspection of all critical phases. An example is a study that requires the use of a methodology performed only once or twice a year by a particular laboratory. In such a case the QA time required may be greater than that needed for more routine studies.

Studies that are routinely performed in a laboratory with low to moderate staff turnover provide an opportunity for a more cost-effective inspection schedule. If a laboratory concurrently conducts multiple tests using the same generic protocols, the laboratory may choose to inspect one phase of each study as long as all phases of the protocol are inspected during a 3- or 6-month period. For example, a laboratory conducting several soil metabolism studies could inspect the test compound preparation and application to soil of one study, the aerobic incubation conditions and sampling of a second study, and different phases of the remaining studies. In this example, the inspected phase is rotated so that all phases are monitored. This approach requires a sufficient volume of similar studies performed within a suitable time frame.

Frequency of inspection is not mandated in the current GLP regulations. Laboratories should consider periodic monitoring of long-term studies. Additional inspections are warranted whenever there are changes in laboratory personnel, SOPs, or protocol.

Procedures for Conducting the Inspection

The GLP inspection is a compliance inspection rather than a technical review (the technical review is the responsibility of the supervisors and the study director). The QA auditor is responsible for evaluating compliance of laboratory operations with the GLP Standards, the study protocol, and the internal SOPs.

Prior to visiting the laboratory, the QA auditor should take appropriate steps in preparation for the audit. The auditor should review the protocol, methods, and SOPs. All necessary QA paperwork such as checklists and reporting forms should be in order. This preparation will enable the auditor to perform the inspection in an efficient and thorough manner and will minimize employee distraction from the technical work being monitored.

Inspections may be either announced, with notice to staff, or unannounced. The laboratory policy regarding this matter should be spelled out in the inspection SOP. Unannounced inspections are effective if the laboratory schedule is well defined with few disruptions. The auditor can observe all activities in the laboratory at that time to obtain a true indication of laboratory operations. However, the unannounced audit is not practical when schedule changes are frequent or a study phase is performed infrequently. The study director, laboratory supervisors, and QAU personnel need to work cooperatively. The QAU certainly should not dictate to the laboratory when selected project phases need to be performed but, at the same time, study directors need to keep the QAU informed of any last-minute schedule changes that might have an impact on a planned inspection.

The QA auditor is an observer when visiting the laboratory for an inspection. The auditor should not dictate or influence events and should maintain a professional, objective demeanor. The auditor may watch, listen, open notebooks or bench drawers, and ask questions as necessary to perform the evaluation. Technicians being watched frequently feel uncomfortable in this situation. A cordial attitude makes the procedure more enjoyable for all parties. If the technicians performing work under inspection are new, the auditor should explain what will be checked and what will be done with any findings. Employees being audited should respond openly to questions from the auditor. The project supervisor should be available in case it is necessary to discuss questions or problems encountered.

If a noncompliance issue or other problem is found, the auditor should take the following steps. First, the suspected error should be rechecked. Questions should be asked to clarify what has happened. The auditor should talk with the study personnel about the problem. This communication keeps study personnel informed and also may provide the auditor with additional information or explanations. Second, the matter should be brought immediately to the attention of the project supervisor. The situation may be resolved at this point. If it is determined that there is no problem, the matter should be dropped. If the problem is confirmed, management and the study director should be informed promptly, whether or not corrective action is taken. The report from the QAU should indicate if the project supervisor has taken or proposed corrective action.

Topics To Include in an Inspection

The QAU is charged with monitoring facilities, equipment, personnel, methods, practices, records, and controls for conformance with the GLP Standards. When done properly, this task is not easy. Thoroughness is an essential trait in the performance of a QA auditor. Many auditors perform an inspection by using a checklist. The advantages of checklists are (1) the list provides a reminder to the auditor of topics to be included, (2) there is better consistency among QA auditors within the same QAU, and (3) the checklist is a good training tool for new auditors. However, auditors must realize that the checklist is simply an inspection tool and not a substitute for careful investigation or attentiveness.

Having a separate study-specific checklist for each study performed in the laboratory is not always practical. If this is the case, the checklist should be written in a generic fashion. Suggestions for topics to be reviewed during inspections of environmental fate and chemistry studies are provided in the following list.

1. Qualifications of study personnel

   • Are the training files for the study director and laboratory personnel complete?

2. Protocol

- Does the protocol comply with GLP requirements? Does the copy of the protocol in the QA file agree with the protocol used in the laboratory?

- Do the study participants have a copy of or access to the protocol?

- Have protocol amendments been authorized in writing by the study director and sponsor?

- Are protocol amendments complete, including study title, test substance, effective date, amended protocol section, justification for the change, and dated sponsor and study director signatures?

- Have unplanned protocol deviations been documented in the data by the study director and reported to the sponsor? (For most laboratory environmental fate studies, temperature fluctuation outside the range specified by the protocol is an example of an unplanned protocol deviation that could influence the acceptability of the study.)

3. Test substances and analytical standards (reference substances)

- Are there separate secure areas for receipt and storage of test substances?

- Are the receiving records complete? Was the shipping paperwork (air bill or bill of lading) retained?

- Does the test substance label information, including the test substance identification, lot or batch number, and expiration date, match that given in the protocol and transmitting paperwork? Are the storage conditions proper?

- Are distribution and usage logs maintained and complete?

- Have retention samples been taken?

- Are the preparations of stock solutions and dilutions from stock solutions completely documented? Are the solution-concentration calculations correct? The documentation system must be trackable, and therefore, a number code for each solution is desirable. Check that the record shows the name and lot number of the source material, quantity used, final diluted volume, solvent, date of preparation, signature of preparer, expiration date, and storage location. Prepared solutions should be labeled with solution number, compound identification, concentration, solvent used, storage requirements, date of preparation, and expiration date.

- Are proper techniques used in the preparation of solutions and in the spiking of samples? Is the glassware used for measuring and storage of the appropriate type? Are Eppendorf-type pipets restricted to aqueous solutions only?

4. Sample receipt and storage

This information is pertinent to residue samples and any samples that are stored prior to analysis.

- Was the condition of the samples acceptable (frozen, undamaged) upon receipt? Are sample receipt records complete? Are samples stored as specified in the documentation?

- Have required temperature controls been maintained in the laboratory?

- Are samples secured when not in use? Are transfers documented completely? Are sample storage areas sufficiently well organized to avoid mixups?

5. Laboratory operations

- Do actual operations agree with procedures specified in the protocol and internal SOPs?

- Are the SOPs adequate to ensure the quality and integrity of data generated in the study? List in the inspection report any SOPs that need to be written or revised.

- Are data recorded promptly as the work is performed? Do the data pages have project identification? Are entries signed and dated?

- Is there sufficient laboratory space for the operation under inspection? Is the area kept clean?

- Are there sufficient laboratory supplies available? Are supplies stored and maintained to avoid cross contamination?

- Are all solutions and reagents labeled with identity, concentration, storage requirements, and expiration date? Are solutions stored appropriately? Are any expired materials in the laboratory?

- Are samples, subsamples, and extracts adequately labeled with test system, study identification, and nature and date of sample collection? Are samples properly stored?

- For samples being analyzed, does the sample identification listed in the data agree with the labels on the actual samples?

- Are calibration and maintenance records for all equipment used current and complete? Are all logbook entries signed and dated? Does the logbook identify the equipment by model and serial number?

6. Facility controls

- Have exhaust hood efficiencies in the chemistry laboratory been checked? Is there an SOP for laboratory hood checks and certification?

- Are operations that should be performed in a hood actually performed in a hood?

- Is the laboratory water source free of potential study contaminants? Water quality should be monitored routinely according to an SOP, and records must be maintained.

Documentation of the Inspection

All inspections should be documented in a report to the study director and management. The report should be signed and dated by the QA auditor and should be identified by the project number, sponsor name or code, type of study, test compound, phase of the study inspected, and the inspection date. Findings and actions recommended should be reported in an objective, complete manner. The discussion should be concise but also understandable to persons not present during the inspection. The report should also include a date for reinspection or follow-up, if required.

After the inspection report is written, a copy should be retained in QAU files. The original report should be sent to the project supervisor. The project supervisor is responsible for recording action taken to address any problems noted in the report. The report should have space available for recording the project supervisor's comments. Commitments for corrective actions should include proposed completion dates. The project supervisor should also be objective and professional when recording comments on an inspection report. Anger and sarcasm do not belong in the document. Insufficient responses will simply require additional follow-up time and increase costs for the study. The project supervisor is allowed to provide clarifications and explanations.

After the project supervisor has completed the response, the report is forwarded to management and the study director for review. Management should follow up with the project supervisor to ensure that corrective actions are taken. Management can show a commitment to a high-quality laboratory operation by reviewing the inspection reports carefully; addressing problems with staff, equipment, or other resource availability; and enforcing GLP requirements. Management's support is critical to the success of the laboratory inspection program.

Records Available to the EPA or the FDA Representatives

The inspection findings and corrective actions taken do not need to be shown to the EPA or the FDA investigators unless requested by subpoena. The GLP Standards require that the QAU make the following laboratory inspection reports available to the EPA or the FDA investigators:

- a record of the inspection dates, the study inspected, the phase or segment of the study inspected, and the name of the individual performing the inspection

- the SOPs describing the responsibilities and procedures of the QAU (including those for inspection) and the records maintained by the QAU and the method of indexing these records

- a signed statement (to be included in the final study report) that specifies the dates inspections were made and the findings that were reported to the manager and the study director

References

1. "Good Laboratory Practice Standards under the Federal Insecticide, Fungicide, and Rodenticide Act (FIFRA); Good Laboratory Practice Standards Final Rule", *Code of Federal Regulations* Title 40, Pt. 160; *Federal Register* 54:158 (August 17, 1989) pp 34051–34074.

2. "Good Laboratory Practice Standards Under the Toxic Substances Control Act (TSCA) Final Rule", *Code of Federal Regulations* Title 40, Pt. 792; *Federal Register* 54 (August 17, 1989) pp 34033–34050.

RECEIVED for review September 17, 1990. ACCEPTED revised manuscript August 17, 1991.

Chapter
14

Data Audits for Field Studies

Del W. Huntsinger

Consumer Products and Life Sciences Division, BASF
Corporation, Agricultural Chemicals, 26 Davis Drive,
Research Triangle Park, NC 27709–3528

*Field phases that are carried out as part of studies con-
ducted under the Federal Insecticide, Fungicide, and Rodenticide Act Good Laboratory Practice (GLP) Standards, as
with all other study phases, must be audited. The intent of
these audits is to assure management of the integrity of the
data that are generated. In order to fulfill the GLP requirements of the audit, the auditor must be independent of the
study activities and must be knowledgeable of the GLP and
individual study requirements. Cognizance of the requirements and a systematic approach to auditing will result in
an effective audit and in a study that conforms to the regulations.*

To GAIN AN UNDERSTANDING OF THE GOALS of field data audits, the
need for and requirements of these audits must first be understood.

    Under the Federal Insecticide, Fungicide, and Rodenticide Act
(FIFRA) Good Laboratory Practice (GLP) Standards (*1*), Code of Federal
Regulations (CFR) Title 40 § 160.35(b)(3), the quality assurance unit
(QAU) is responsible for monitoring each study to ensure that, among
other things, the methods, records, data, and procedures are in confor-

2192-8/92/0175$06.00/0 © 1992 American Chemical Society

mance with the regulations contained in this part of the CFR. One way to provide these assurances is to perform audits of the study during its in-life phase (the time during which the study is conducted) and of the corresponding raw data that have been generated.

Under 40 CFR § 160.35(b)(6) of the GLP Standards, the QAU must review the final report of the study to ensure that the report accurately describes the methods and standard operating procedures (SOPs) used during the study and that the reported results accurately reflect the raw data of the study.

In addition, because the Environmental Protection Agency (EPA) may eventually contact a company and request to audit a study report that has been submitted to support a registration requirement (or some other data requirement), both "management"[1] and the study director will benefit from having had someone who was not actually involved in conducting the study to audit the work that was performed. The focus in this chapter is on auditing raw data for the field component of a study.

Preparing for Data Audits

What are raw data with regard to studies performed in compliance with the FIFRA GLP Standards? The EPA, under 40 CFR § 160.3, "Definitions", describes raw data as

> any laboratory work-sheets, records, memoranda, notes, or exact copies thereof, that are the result of original observations and activities of a study and are necessary for the reconstruction and evaluation of the report of that study. In the event that exact transcripts of raw data have been prepared (e.g., tapes that have been transcribed verbatim, dated, and verified by signature), the exact copy or exact transcript may be substituted for the original source as raw data. "Raw Data" may include photographs, microfilm, or microfiche copies, computer printouts, magnetic media, including dictated observations, and recorded data from automated instruments.

During the audit of the field data, whether during an active study or at the completion of the study, four basic areas of concern should make up the nucleus of the QAU investigation:

---

[1]Management is defined here as the person or persons responsible for setting policy at a facility. For a description of the duties of testing facility management, see § 160.31 of the GLP Standards.

1. adherence to the approved study protocol
2. adherence to company SOPs
3. compliance with internal company policies
4. compliance with the FIFRA GLPs

The goal of audits is to ensure that the data that have been generated support the aforementioned categories, that the data are accurately recorded, and that at the end of the study the data are accurately reflected in the final report.

To perform an effective data audit, the QAU must adequately prepare for the task at hand. A clear idea of the scope of the audit and the requirements of the study being audited needs to be defined. As a starting point, the protocol and any amendments that exist should be reviewed. In order to perform an effective audit, preparing a list of the data required by the protocol may be helpful. The pertinent SOPs for the study need to be reviewed to obtain an understanding of how the data are to be recorded and the nature of the data that will be audited (i.e., pre-printed forms, lists, or data logs). These protocols, amendments, and SOPs must also be readily available during the study for reference by the study personnel. Training records and curricula vitae (CV) for each person involved in the technical conduct of the study should be up to date. Many people consider training records and CVs part of the raw data package for the study, especially if the work is being performed by a field cooperator.

The types of field data that are audited are generally determined by the appropriate EPA Office of Pesticide Programs (OPP) guideline that has been developed for a study or study type. For example, a terrestrial field dissipation study being run to fulfill the EPA OPP Pesticide Assessment Guideline, Subdivision N, Chemistry: Environmental Fate § 164–1, "Field Dissipation Studies for Terrestrial Uses" (2) should have a protocol written to ensure that the information specifically required by this guideline will be collected. Residue studies for raw agricultural commodities would be designed to contain at minimum the requirements set forth in the EPA Pesticide Assessment Guideline, Subdivision O, Residue Chemistry § 171–4(h) (3).

Data from the field phase of a study that are typically audited include

- test substance chain-of-custody documentation, identification (name and Chemical Abstracts Service or code number), purity, stability, synthesis or manufacturing documentation, storage conditions, and solubility data. This information must be available to the testing facility prior to the use of the test substance in the study.

- tank mixing (test solution preparation), calibration of the application equipment, and application documentation for the test substance, including the application rate. The application pressure, the amount applied, and the exact area to which the test substance mixture is applied can also be important pieces of information depending on the specific nature of the study.

- field plot location (e.g., maps, diagrams, and photographs), history, preparation, maintenance, and description of soil type.

- environmental records such as temperature (air and ground), relative humidity, cloud cover, wind direction and speed, slope of plot, depth to water table, time from last rainfall prior to and after application, and supplemental irrigation. The EPA in some instances requires that pan-evaporation data be collected during the field phase of the study.

- soil characterization. Whether this has been done can be critical for such studies as field-dissipation or prospective-groundwater studies. Recent interpretations by the Policy Branch of the Office of Compliance Monitoring indicate that if the soil characterization is part of the study then it must comply with the GLP Standard. More laboratories specializing in soil-characterization analysis are now operating under GLPs.

- sampling methods. Any sampling must be performed in accordance with the protocol and SOPs. Samples must be properly identified, and records of such information as number, weight, crop part, location in plot, and treatment-to-harvest interval must be kept. The time between sampling and storing, especially if the sample requires freezing, must be documented. If the sample is stored, the conditions of storage must be documented. The temperature records for the time of storage should be available. If the samples have been shipped, documentation as to the method and conditions of shipment must also be recorded and available. Samples that have been shipped should have an appropriate chain of custody from their origin (usually the field site) to their final destination.

- activity logs. An activity log should be kept on the events that have taken place at the site. Unforeseen occurrences that might have an effect on the study are critical to the study's raw data package.

The actual review or inspection of the raw data involves not only checking for its mere presence but also other factors. These other factors will be discussed in the following section.

Performing the Field Data Audit

When you have gathered the data in front of you, whether during the study or in preparation for the final report audit, start with the basics. A good way to start is by going through the data and checking for improper entries. That is, verify that the data are recorded in ink (for manual entries), that they are legible, and that any data that has been modified has been done so properly (i.e., the changed or corrected data entries have a single line drawn through them, the reason for the change is stated, and the changes are signed and dated), and that there are no missing entries on the forms being used.

The next step is to assemble the data to match the sequence of events as they are outlined in the protocol. Check the data to ensure that the protocol was followed and to check whether any deviations from the protocol have occurred; if deviations exist, verify that they have been appropriately documented and approved by the study director. You should be able to trace any data point that has been generated back to the piece of equipment used to generate this point, even to the use or maintenance log for that particular piece of equipment. When calculations are made (for example calibration of spray equipment), the equation and the base point used should be available. At least 20% of the data that have been reported in the final report, depending on your internal policy, should be verified by the QAU by following these data points back to their origin. If there are significant errors or missing data points, the report should go back to the study director for a thorough review and corrections.

Benefits of Data Audits

Data audits in the field during the in-life stage of a study are an extremely useful tool. If performed properly, these data audits can serve not only to satisfy agency requirements but also to help the study personnel to avoid problems by discovering potential errors that, if left undetected until study completion, might lead to an unacceptable study. If detected during the study, deviations and errors might possibly be rectified so that the validity of the study will not be jeopardized.

These in-life field audits are useful when performed in conjunction with a contract facility inspection to help in the overall assessment of a facility. Performing both inspections at once enables an auditor to see firsthand how a facility records the vital study data, handles changes to the data, and complies with the study protocol, SOPs, and GLP regulations.

At the end of a study, these same issues apply with regard to the field data collected. The field data must be compared to what is contained in the final report. The report must be an accurate reflection of the raw data package. Any inconsistencies or incorrect findings must be reported to the study director. When the final report has been signed by the study director and the complete package is ready to be archived, the QAU in many cases will have one last look at the package to see if it is complete.

With the advent of the FIFRA 1988 reregistration standards, the QAU auditor's skills are being used to audit studies (including field phases of studies) that may be several years old. Because the audit goal is basically to ascertain whether these studies meet today's quality standards, the same basic rules are followed. The EPA, in its Phase 3 Technical Guidance publication (4), has furnished (in Appendix H of the guideline) what it believes are the minimum raw data requirements for these older studies. Appendix H can make a good checklist when reviewing a study or placing a study into the archive as to what, at minimum, needs to be included as part of the archive data package.

Data audits are an integral part of the QAU and the overall GLP compliance program for any organization. Effective QAU audits are an important management resource; the audits result in high-quality reports that the study sponsor and the EPA can rely on to make their assessments of the products that these studies support.

References

1. "Federal Insecticide, Fungicide, and Rodenticide ACT (FIFRA): Good Laboratory Practice Standards Final Rule", *Code of Federal Regulations* Title 40, Pt. 160; *Federal Register* 54:158 (August 17, 1989) pp 34051–34074.

2. "Pesticide Assessment Guidelines, Subdivision N. Chemistry: Environmental Fate", U.S. Environmental Protection Agency, Office of Pesticide Programs, Hazard Evaluation Division, Washington, DC, October 18, 1982.

3. Schmitt, R. D. et al. "Pesticide Assessment Guidelines, Subdivision O. Residue Chemistry", U.S. Environmental Protection Agency, Office of Pesticide Programs, Hazard Evaluation Division, Washington, DC, October 1982.

4. "FIFRA Accelerated Reregistration Phase 3 Technical Guidance", U.S. Environmental Protection Agency, Pesticides and Toxic Substances (H-750C), Washington, DC, December 24, 1989.

RECEIVED for review September 17, 1990. ACCEPTED revised manuscript August 17, 1991.

Laboratory Data Auditing

Harry L. Hyndman

Monsanto Agricultural Company, 700 Chesterfield Village Parkway, St. Louis, MO 63198

Good laboratory practice (GLP) regulations require the quality assurance unit to be responsible for assuring management (management is defined here as the person or persons responsible for setting policy at a facility. For a description of the duties of testing facility management, see § 160.31 of the GLP Standards) that study records are in conformance with the GLP regulations. Aspects that are audited include adherence to the protocol and standard operating procedures (SOPs), accountability of all documentation, completeness of the raw data, and validation of data points. The test, control, and reference substances and samples must be identified and characterized, and their chains of custody must be verified. Analytical methods must be available and the data must be unambiguously labeled. Record-keeping procedures for ensuring the integrity of the data, the correction of errors, and the archiving of the data must be verified. Final study reports must contain specified items and must accurately describe the methods and SOPs used, and the reported results must accurately reflect the raw data.

THE PROCEDURES TYPICALLY USED BY THE QUALITY ASSURANCE UNIT (QAU) in conducting an audit of the analytical laboratory data associated with a field residue or environmental fate study will be given in this chapter. The following discussion and the checklists in the appendices to this chapter are intended to provide assistance to investigators, authors, study directors, and others who are preparing and organizing the study

2192-8/92/0181$06.00/0 © 1992 American Chemical Society

data package and writing or reviewing a report that will be submitted to the QAU. Understanding these procedures will help the investigator to better prepare the data package and report for auditing, thereby making the process of auditing faster and easier for all concerned.

Procedures for similar audits of the field data associated with these types of studies are described in Chapter 16 and, although the field data would normally be reviewed in the course of an audit, these data will not considered further here. Likewise, other types of required data accompanying a laboratory study, such as personnel records, standard operating procedures (SOPs), facility records, and equipment maintenance and calibration logs are generally reviewed in the course of facility inspections and are not included in this discussion.

Purpose

The QAU is charged by the Good Laboratory Practice (GLP) Standards [40 *Code of Federal Regulations* (CFR) 160.35(b)] with (1) determining that no deviations from approved study protocols or SOPs were made during the study without proper authorization and documentation and (2) reviewing the final study report to ensure that the report accurately describes the methods and SOPs and that the reported results accurately reflect the raw data of the study. These responsibilities are fulfilled by conducting data audits of studies in progress and of final study reports. Other items relative to report quality are reviewed in addition to those items required for ensuring the quality and integrity of the study as required by the GLP Standards. Such items include general readability, clarity, and completeness with respect to the appropriate "Addenda for Data Reporting" document and "Standard Evaluation Procedure".

Procedure

In the following sections, the typical steps that are taken when conducting a data audit will be discussed. Although these steps are presented in the more or less logical and sequential order of a review of a final study report, these review steps may be conducted either simultaneously or at various other times in the course of a study.

Data Transfer. The first step in the review of a final study report is to receive notification from the report author or study director that a final draft of the report is ready for review. This final draft should be typed, proofread, and peer reviewed; it should be ready, to all intents and purposes, for the study director's signature. Study personnel should be responsible initially for verifying data transcription and accuracy of the results and for proofreading the report.

When the QAU is ready to audit the report, it will call for the draft report and data package to be submitted. Up to this point the data package should remain in the custody of study personnel. When the QAU requests the final draft report, the data package is transferred to the custody of the QAU and is placed in a "working archive". The report is no longer available to the study personnel, except through established procedures for checking out materials from the archive. The data package now includes all of the data, including field and processing data, and documentation.

To accomplish this transfer from study personnel to QAU more effectively, a checklist (Appendix 15.I) is used by the report author to ensure that all appropriate data and documentation have been included in the data package. Upon receipt of the final draft report, the QAU reviews the checklist form for completeness and for the signature of the study director. The checklist is further verified for accuracy during the course of the audit. The presence of the required quality assurance (QA) statements with reports and data from other parts of the testing facility should be confirmed.

Protocol Review. At the start of a data audit, the reviewer rereads the protocol, along with any amendments that may have been issued, to see what was intended to be accomplished by the study, what specific data and records were to be maintained, and how the study director had stipulated that the study was to be performed. Many of the items called for in the protocol are specifically required by government regulations or by company policy. Such items are therefore a part of every study. Sometimes, however, the study director will also specify in the protocol additional requirements not normally done in a typical study of that particular type. When reading the report, the reviewer should pay particular attention that these requirements are in the report or in the data package. The protocol should be accompanied by all of the amendments that may have been issued.

Data Review. The next general step, the review of the actual data, is the most critical and time-consuming step of the review. Several aspects of this step are discussed in the following sections. The record-keeping practices are generally reviewed at the same time as the various other pieces of the data package. Current data are also reviewed during the course of in-progress data audits.

Record-Keeping Requirements. The review of record keeping occurs throughout the audit and verifies that all internal or company- and GLP-specified requirements for error correction, document signing, and adequate document labeling have been met. The record-keeping review includes verification of the accuracy and adequacy of the following items:

- the initialing, dating, and annotating of the correction of errors in the data package

- the identification of each piece of documentation with the study number

- the identification of each sample on a chromatogram by sample number or description and of the peaks of interest within the chromatogram (If identified by hand, doing so once on the chromatogram of an external standard is usually adequate.)

- the identification of the automated analytical report and documentation of the chromatographic conditions

- cross-referencing and the dated signatures of the people recording the data in laboratory notebooks and worksheets

- the identification of instrumentation used to collect the data

Traceability Audit. The most important part of a report is the results that have been obtained from the raw data. There must be a clear and complete trail from any reported value back to the original raw data. This requirement means that sample identification must be clear and consistent throughout the report and the data package. If samples are identified by a code, that code must be explained. The reviewer should be able to select any value from a table or figure in the report and then trace that selected value back to the original raw data with a minimum of difficulty. Establishing traceability will be much easier if the author has arranged the data package in the same order in which the results are presented in the report. Worksheets and computer reports included in the raw data package should contain appropriate cross-references to notebook pages for the analytical method, standards used, and any problems that may have been encountered. The author can assist in the review process by annotating references to the source of the data on all of the report tables and figures of the draft copy of the report.

The records of the source, type, treatment, and collection date of the field samples are reviewed for consistency with the other documentation. The sample preparation records are verified for dates and procedures for the grinding, homogenization, and storage of the analytical sample. The transcription of the analytical data and results from the analytical reports to the report table is carefully checked.

To ensure that the conclusions of the report are justified, the raw data are also tracked forward into the corresponding summary tables. The summarized data are also checked in the "Discussion and Results" and "Conclusion" sections of the report itself. Finally, summary data and individual data reported in the report abstract are verified against the summary and raw data tables.

Depending on the number of data records in the raw data and summary data tables, a complete audit or an audit of a randomly selected subset of the data may be conducted. For the random selection, a computerized version of the MILSTD-105D (*1, 2, 3*) procedure is used.

This procedure provides, in addition to the randomly selected subset, the maximum allowable number of errors to estimate, with 95% confidence, that the data table will have fewer than 5% errors. If the number of errors found is less than the maximum allowable number, those errors are corrected and the table is approved. If the number of errors found is greater than the maximum allowable number, the table is returned to the author for a complete and thorough re-examination and correction of all errors. Upon return of the corrected table to the QAU, a modified version of the program is used for the re-examination. This modified version allows even fewer errors for acceptability. Copies of the BASIC programs for subsample selection are included in Appendix 15.II.

Calculations Audit. In the course of the aforementioned traceability audit, the calculations performed with the raw data are also verified. The following calculations are included:

- the analytical result from the detector response and the calibration factors

- the various correction factors of averages and standard deviations or other statistical analyses

This review requires documentation of the calculation equation itself. Typically, an example calculation is provided to demonstrate how the equation was applied to the data. The calculations themselves are spot-checked to ensure that the algebra and arithmetic are correct. This checking will be done more extensively for hand calculations, where the error rate is likely to be higher, and less extensively when a computer program was used. The statistical analyses used are reviewed for conformance to the protocol. In cases in which a computer program was used, the documentation of the computerized system, software name and version, and input files are verified.

Analytical Method. The analytical procedure used for analysis of the samples must be documented. This documentation may be done by reference to a previously published method or to a research notebook and should be done for each sample set. The research notebook or worksheets should contain enough information to verify how the steps in the method, such as sample preparation and weighing, extraction, cleanup, and dilution, were carried out. The documentation of the analytical method should generally include the following:

- a complete copy of the analytical methods, including identification of the sources of those methods

- a discussion of any problems encountered with the procedures and a description of any deviations in reagents, procedures, instrumentation, or operating parameters from those specified in the analytical methods

- a description of the instrumentation used (e.g., type, make, and model of the instruments; type and specificity of detectors; column size and packing materials; carrier gases; and operating parameters such as flow rates, temperatures, and voltage)

- storage conditions to be used for samples and standards throughout the analysis (e.g., location, temperature, light or dark, and containers)

- calibration procedures for the instrumentation used, including procedures for preparing the calibration, fortification standard stock, and working solutions

- adequately labeled, representative instrumental output (e.g., chromatograms for each matrix analyzed, including solvent blank, lowest standard, check sample, treated sample, and fortified check sample at the lower limit of method validation)

- control measures or precautions taken to ensure the fidelity of the analytical results (e.g., blank, control, and fortified samples; fortification levels; and sample and standard injection sequences), including any method validation work performed prior to the analysis of any study samples

- any other additional information that the analyst considers appropriate and relevant to provide a complete and thorough description of the analytical procedures followed

Analytical Reference Standards. The analytical reference (calibration) standards used are a crucial part of the documentation of the analytical data. The identity, source, purity, and characterization data for these standards must be documented for each analytical sample set. All such documentation should include a code number that uniquely identifies that particular batch of reference standard.

The characterization of the standard includes both qualitative analysis of the chemical structure (to confirm that the container is accurately labeled) and quantitative analysis of purity.

Field Data. The field notebooks and data sheets for the field data are also examined for completeness and accuracy. When a contractor QAU has been involved in monitoring the field portion of the study, the presence of its QA statement is confirmed, and the content is examined. Because this topic is discussed in Chapter 16, the procedures and techniques used for auditing the field data will not be elaborated on any further here.

Processing Data. When field samples are processed into food or feed commodities, this process must also be documented in the study data package. Specific documentation should include process description and flow diagrams and sufficient notation to verify how these specific samples were produced. If a contract facility is used for this audit, the presence of the QA statement for that part of the testing facility is verified.

Test Substance Identification and Characterization. The test substance identification and characterization data are crucial to the study. These data are typically reviewed together with the field data. The identity, source, purity, and characterization data for the test substances must be documented. All such documentation should include a code number that uniquely identifies that particular batch of test substance.

Characterization of the test substance includes both qualitative analysis of the chemical structure of the active ingredient to confirm that the container is accurately labeled and quantitative assay of the active ingredient so that the application rate can be verified. Records of the formulation procedure should be included or cited to establish the identity of the test substance.

Chain of Custody. No mention has yet been made of the documentation of chains of custody. The chain-of-custody documents are crucial pieces of data and are verified throughout the course of the audit. This documentation includes the signature of the custodian, dates of custody, storage location, and storage conditions for each sample. Materials for which the chain of custody must be documented include the test substance and reference standards, the field samples, the processing samples, and the analytical samples. The record of custody should cover the life of the sample from its origination to its use in the study, return to storage, and ultimately to its final disposition.

Final Report. After carefully reading the report, the reviewer re-examines the protocol to determine if there were any deviations from the protocol instructions. If deviations from the protocol occurred during the study, documentation of those deviations must be included in the data package and noted in the report itself. The report is reviewed to ensure that all methods and SOPs are adequately and accurately described. The content of the final study report is also reviewed to ensure that all topics required by the GLP Standard (40 CFR 160.185) and others indicated by the appropriate "Addendum on Data Reporting to the Pesticide Assessment Guidelines" and by the appropriate Environmental Protection Agency standard evaluation procedure are adequately discussed. Compliance with other routine corporate requirements for report formats is verified for all testing facility reports included in the final study report. A checklist to assist the QAU in this evaluation is included in Appendix 15.III.

Quality Assurance Report. Upon completion of the review, the auditor documents the findings of the data audit. These findings, with a cover memo, are reported to the study director and the study manager for any necessary corrective action. The findings are also discussed with the report author. The report author will typically initiate corrections, which are then reviewed and approved by the study director. When the necessary corrections have been made, the report is resubmitted to the QAU for a follow-up review against the previous review findings. Verification of the corrections is made on the QA report by the reviewer. The report is then ready for signature.

Data Package Archiving. The study data package must be archived at the conclusion of the study, which has been interpreted as the signing of the final study report by the study director. The study data package was transferred to the working archive of the QAU upon submittal for review and was considered archived at that point. The QAU organizes the data package for intermediate storage when the archive copy of the final study report is received. The entire data package may then be microfilmed for backup security.

For both the intermediate and long-term storage archives, the exact storage location of the data package is recorded in a computerized data base for search and retrieval purposes. In the data base, studies are indexed by report number, report date, title, authors, analysts, study director, protocol numbers, storage location, and microfilm reference.

The data package and report generally remain in this intermediate storage for 2–3 years, depending on the volume of records to be stored. When space in intermediate storage has been exhausted and interest in the data package for a particular study has subsided, the data package is transferred in sealed boxes, marked "restricted access", to a corporate archive facility for long-term storage.

Conclusion

Three to five man-days of effort are required, on the average, for the audit of a field residue or environmental fate final study report that requires minimal correction and rework. The benefits to the laboratory that result from this overall auditing process include more uniform and systematic procedures, better quality data, better understanding of results, smoother and more reliable data retrieval, fewer incidences of lost data, improved documentation and record keeping for verification of results, and decreased risk of rejection of studies due to misleading or misrepresented results.

References

1. "American National Standard Sampling Procedures and Tables for Inspection by Attributes", ANSI Z1.4–1971, American National Standards Institute: New York, 1971.

2. Allen, G. O.; Hirsch, A. F.; Leidy, H. *Drug Info. J.*, April/June, 1980, pp 65–68.

3. "Military Standard Sampling Procedures and Tables for Inspection by Attributes", MILSTD–105D, April 29, 1963, Superintendent of Documents, U.S. Government Printing Office: Washington, DC 20402.

RECEIVED for review September 17, 1990. ACCEPTED revised manuscript August 17, 1991.

Appendix 15.I: Raw Data Transfer Statement and Checklist

Raw Data Transfer Statement

I hereby certify that all raw data and associated documentation (see attached checklist) for the report listed below has been transferred to the archives.

Report Number: _____

Study Director—   Signed: _____

                    Title: _____

                    Date: _____

Quality Assurance Officer— Signed: _____

                            Title: _____

                            Date: _____

Raw Data Transfer Checklist

Report Number: _____

Yes N/A No (with comments)

___ ___ _____   1) Original protocol, amendments, and deviations.

___ ___ _____   2) Laboratory notebooks.

___ ___ _____   3) A sample encoding/encryption key if samples were encoded or
                               encrypted during the study.

___ ___ _____   4) Affixed together for each analysis set, all chromatograms,

    worksheets, injection sequence information,
calibration, and analytical result reports.

__ __ _____ 5) All raw and supplemental data generated by outside laboratories including worksheets, chromatograms, final report, QA statement reference standard logs, and any magnetic storage media used to transfer raw data.

__ __ _____ 6) Reports or documentation covering the field plot locations. This would include soil characterization data, weather data, field notebooks, field final reports, and field data sheets.

__ __ _____ 7) Sample preparation or processing records including process flow diagrams, mass balance records, worksheets, QA statement, and final report.

__ __ _____ 8) Sample and reference standard shipping and receipt records.

__ __ _____ 9) Reference to the appropriate documentation for all reference standards used.

__ __ _____ 10) Test substance shipping and receipt records or reference thereto. Records should include identity, quantity, and lot or batch number.

__ __ _____ 11) Reference to test substance characterization documentation. These records should include the lot or batch number, synthesis procedure, and the characterization and assay data.

__ __ _____ 12) Intermediate data tables that support tables in the final report.

__ __ _____ 13) Names of computer systems and programs used to generate raw data, tables and figures that are found in the final report or that support results and conclusions presented in the final report. Reference computer programs to the specific tables and figures.

__ __ _____ 14) Memoranda, letters, and other communications pertaining to the study.

__ __ _____ 15) Signature sheets.

__ __ _____ 16) Photographs or videotapes taken at the field plots, test sites, contract analytical laboratories or other support facilities.

__ __ _____ 17) Any other raw or supplemental data that supports or documents the results and conclusions presented in the final report.

N/A means not applicable

Appendix 15.II: BASIC Programs for Audit Subsample Selection

Program for Initial Subsample

```
00010 PRINT "ENTER FILENAME FOR OUTPUT:"
00020 INPUT B$
00030 OPEN B$ AS #5
00050 PRINT #5, " RANDOM 1 - USED FOR FIRST AUDITS OF DATA."
00060 PRINT #5
00070 PRINT #5, " GENERAL INSPECTION LEVEL II."
00080 PRINT #5, " NORMAL INSPECTION."
00090 PRINT #5, " AQL = 1.0."
00100 PRINT #5
00110 PRINT #5, " DATA FILENAME = ";B$;"."
00120 PRINT #5
00130 PRINT "HOW MANY NUMBERS ARE IN THE TABLE?"
00140 INPUT B
00150 REM DETERMINE THE SAMPLE SIZE AND ACCEPTANCE NUMBER.
00160 IF B>15 THEN 220
00170 PRINT #5
00180 PRINT #5
00190 PRINT #5, " POPULATION SIZE = ";B;""
00200 PRINT #5, " DO A 100% AUDIT--REJECT THE TABLE IF ONE ERROR IS FOUND"
00210 GO TO 830
00220 IF B>150 THEN 260
00230 LET N=13
00240 LET A1=0
00250 GO TO 430
00260 IF B>500 THEN 300
00270 LET N=50
00280 LET A1=1
00290 GO TO 430
00300 IF B>1200 THEN 340
00310  LET N=80
00320 LET A1=2
00330 GO TO 430
00340 IF B>3200 THEN 372
00350 LET N=125
00360 LET A1=3
00370 GO TO 430
00372 IF B>10000 THEN 380
00374 LET N=200
00376 LET A1=5
00378 GO TO 430
00380 PRINT #5
00390 PRINT #5
00400 PRINT #5, " POPULATION SIZE = ";B;""
00410 PRINT #5, " YOU'VE GOT YOUR WORK CUT OUT FOR YOU."
00420 GO TO 830
00430 REM GENERATE THE RANDOM NUMBERS.
00440 DIM M1(200), M2(200)
```

```
00450 FOR A=1 TO N
00460 RANDOMIZE
00470 D1=INT(B*RND)
00480 REM CHECK TO SEE IF THE RANDOM NUMBER JUST GENERATED IS A NEW
ONE.
00490 FOR C=1 TO N
00500 IF D1=M1(C) THEN 460
00510 NEXT C
00520 M1(A)=D1
00530 NEXT A
00540 REM SORT THE RANDOM NUMBERS.
00550 FOR D=1 TO (N-1)
00560 C1=0
00570 FOR E=1 TO (N-D)
00580 IF M1(E)<=M1(E+1) THEN 620
00590 N1=M1(E)
00600 M1(E)=M1(E+1)
00610 M1(E+1)=N1
00620 C1=1
00630 NEXT E
00640 IF C1=0 THEN 780
00650 NEXT D
00660 REM CALCULATE INTERVAL BETWEEN RANDOM NUMBERS
00670 FOR K=1 TO N-1
00680 M2(K)=M1(K+1)-M1(K)
00690 NEXT K
00700 REM PRINT RANDOM NUMBERS AND INTERVALS
00710 PRINT #5
00720 PRINT #5
00730 PRINT #5
00740 PRINT #5, " POPULATION SIZE = ";B;""
00750 PRINT #5, " SAMPLE SIZE = ";N;""
00760 PRINT #5
00770 PRINT #5, " RANDOM NO.", "INTERVAL"
00780 FOR J=1 TO N
00790 PRINT #5, M1(J), M2(J)
00800 NEXT J
00810 PRINT #5
00820 PRINT #5, " REJECT THE TABLE IF MORE THAN ";A1;" ERRORS ARE FOUND"
00830 PRINT " DO YOU WISH TO ENTER ANOTHER NUMBER?"
00840 INPUT A$
00850 IF A$="Y" THEN 130
00860 END
```

Program for Repeat Subsample

```
00010 PRINT "ENTER FILENAME FOR OUTPUT:"
00020 INPUT B$
00030 OPEN B$ AS #5
00050 PRINT #5, "RANDOM 2 - USED FOR RE-AUDITS OF DATA."
00060 PRINT #5
00070 PRINT #5, "GENERAL INSPECTION LEVEL II."
```

```
00080 PRINT #5, "TIGHTENED INSPECTION."
00090 PRINT #5, "AQL = 1.0."
00100 PRINT #5
00110 PRINT #5, "DATA FILENAME = ";B$;"."
00120 PRINT #5
00130 PRINT "HOW MANY NUMBERS ARE IN THE TABLE?"
00140 INPUT B
00150 REM DETERMINE THE SAMPLE SIZE AND ACCEPTANCE NUMBER.
00160 IF B>25 THEN 220
00170 PRINT #5
00180 PRINT #5
00190 PRINT #5, "POPULATION SIZE = ";B;""
00200 PRINT #5, "DO A 100% AUDIT--REJECT THE TABLE IF ONE ERROR IS FOUND"
00210 GO TO 790
00220 IF B>150 THEN 260
00230 LET N=20
00240 LET A1=0
00250 GO TO 390
00260 IF B>1200 THEN 300
00270  LET N=80
00280 LET A1=1
00290 GO TO 390
00300 IF B>3200 THEN 340
00310 LET N=125
00320 LET A1=2
00330 GO TO 390
00340 PRINT #5
00350 PRINT #5
00360 PRINT #5, "POPULATION SIZE = ";B;""
00370 PRINT #5, "YOU'VE GOT YOUR WORK CUT OUT FOR YOU."
00380 GO TO 790
00390 REM GENERATE THE RANDOM NUMBERS.
00400 DIM M1(200), M2(200)
00410 FOR A=1 TO N
00420 RANDOMIZE
00430 D1=INT(B*RND)
00440 REM CHECK TO SEE IF THE RANDOM NUMBER JUST GENERATED IS A NEW ONE.
00450 FOR C=1 TO N
00460 IF D1=M1(C) THEN 420
00470 NEXT C
00480 M1(A)=D1
00490 NEXT A
00500 REM SORT THE RANDOM NUMBERS.
00510 FOR D=1 TO (N-1)
00520 C1=0
00530 FOR E=1 TO (N-D)
00540 IF M1(E)<=M1(E+1) THEN 580
00550 N1=M1(E)
00560 M1(E)=M1(E+1)
00570 M1(E+1)=N1
00580 C1=1
00590 NEXT E
00600 IF C1=0 THEN 740
00610 NEXT D
```

```
00620 REM CALCULATE INTERVAL BETWEEN RANDOM NUMBERS
00630 FOR K=1 TO N-1
00640 M2(K)=M1(K+1)-M1(K)
00650 NEXT K
00660 REM PRINT RANDOM NUMBERS AND INTERVALS
00670 PRINT #5
00680 PRINT #5
00690 PRINT #5
00700 PRINT #5, "POPULATION SIZE = ";B;""
00710 PRINT #5, "SAMPLE SIZE = ";N;""
00720 PRINT #5
00730 PRINT #5, "RANDOM NO.", "INTERVAL"
00740 FOR J=1 TO N
00750 PRINT #5, M1(J), M2(J)
00760 NEXT J
00770 PRINT #5
00780 PRINT #5, "REJECT THE TABLE IF MORE THAN ";A1;" ERRORS ARE FOUND"
00790 PRINT "DO YOU WISH TO ENTER ANOTHER NUMBER?"
00800 INPUT A$
00810 IF A$="Y" THEN 130
00820 END
```

Appendix 15.III: Final Study Report Format and GLP Requirements Checklist

Report Number: _____   Date: _____

       GLP

<u>Yes No N/A Req'd</u>

        I. <u>TITLE PAGE (Cover page)</u>

___ ___ ___   1.   Company/Division/Department/Location

___ ___ ___   2.   Type of Report

___ ___ ___   3.   Report Number

___ ___ ___   4.   Job/Project Numbers

___ ___ ___   5.   Date

___ ___ ___   6.   Descriptive Title

___ ___ ___   7.   Author(s)

___ ___ ___   8.   Study Director (Group Leader)

___ ___ ___   9.   Abstract

___ ___ ___   10.   Access Restrictions

II. DISTRIBUTION PAGE

___ ___ ___   1. List of Full Reports

___ ___ ___   2. List of Abstracts

___ ___ ___   3. Abstract (If continued from Title page)

III. PROJECT COST ESTIMATES

___ ___ ___   1. Period Covered

___ ___ ___   2. Man-months

___ ___ ___   3. Contract Facility Costs

IV. EPA PR 86-5 FORMATTED TITLE PAGE (Page 1)

___ ___ ___   *   1. Descriptive Title

___ ___ ___   2. Data Requirement

___ ___ ___   3. Author(s)

___ ___ ___   *   4. Study Completion Date

___ ___ ___   *   5. Performing Laboratory

___ ___ ___   6. Laboratory Project ID

___ ___ ___   7. Page 1 of nn

V. BLANK PAGE FOR NON-CONFIDENTIALITY STATEMENT (Page 2)

___ ___ ___   1. This page left blank for administrative purposes

VI. GLP COMPLIANCE STATEMENT (Page 3)

___ ___ ___   1. Dated Signature of Submitter

___ ___ ___   2. Dated Signature of Sponsor

___ ___ ___   *   3. Dated Signature of Study Director

Report Number: _____   Date: _____

GLP

<u>Yes No N/A Req'd</u>

VII. QAU STATEMENT (Page 4)

___ ___ ___   *   1. Dates and Types of Inspections

___ ___ ___   *   2. Dates Inspections Reported to Management

___ ___ ___   *   3. QAU Signature

___ ___ ___   *   4. Date of QAU Approval

VIII. BLANK PAGES FOR NOTES TO REVIEWERS (Pages 5 & 6)

___ ___ ___   1. These pages left blank for administrative purposes

IX. TABLE OF CONTENTS (Page 7)

___ ___ ___   1. Listing of Essential Elements with Page Numbers

X. BODY OF THE REPORT

___ ___ ___   *   1. Summary/Introduction

___ ___ ___   *   2. Materials

___ ___ ___   *   3. Methods

___ ___ ___   *   4. Results and Discussion

___ ___ ___   *   5. Conclusions

___ ___ ___   6. Certification Statement

___ ___ ___   7. Tables and Figures

___ ___ ___   8. References

XI. APPENDICES

___ ___ ___   *   1. Protocol

                    2. Raw Data

___ ___ ___   *   i. Explanation Of Tables

___ ___ ___   ii. Master Summary Table(s)

___ ___ ___   iii. Raw Data Table(s)

___ ___ ___   iv. Sample Field History

___ ___ ___   v. Recovery Table(s)

___ ___ ___ 3. Significant Dates

___ ___ ___ 4. Field Trial History and Crop History

     5. Test, Control, and Reference Substance Characterization

___ ___ ___ * i. Test Material Assay

___ ___ ___ * ii. Reference Standard Characterization

___ ___ ___ 6. Soil Characteristics

___ ___ ___ 7. Climatological Data

___ ___ ___ 8. Guidelines for Field Residue Plots

___ ___ ___ 9. Guidelines for Sample Preparation

___ ___ ___ * 10. ASOP(s)

Report Number: _____ Date: _____

    GLP

Yes No N/A Req'd

   XI. APPENDICES (cont'd)

    11. Example Chromatograms (each matrix)

___ ___ ___   i. Standard

___ ___ ___   ii. Check

___ ___ ___   iii. Fortification (at LLMV)

___ ___ ___   iv. Treated Sample

___ ___ ___ * 12. Supporting Reports

___ ___ ___ * 13. Contract Facility Reports

    14. Project Personnel

___ ___ ___ * i. Study Director

___ ___ ___ * ii. Investigators

___ ___ ___ * iii. Contributors

___ ___ ___ * iv. Contract Facilities

    15. Storage Location

___ ___ ___   i. Specimen, Raw Data, and Final Report Storage

      Location Statement

___ ___ ___   ii. Sample Analysis Notebook Pages

___ ___ ___   iii. Test, Control, and/or Reference Substance

                              Characterization Notebook Pages

___ ___ ___   iv. Reference Standards Preparation Notebook Pages

___ ___ ___   v. Sample Preparation Notebook Pages

XII. ADDITIONAL GLP REQUIREMENTS

___ ___ ___ *   1. Statistical Methods Employed for Data Analysis

                        2. Test, Control, and Reference Substances Identified By:

___ ___ ___ *   i. Name

___ ___ ___ *   ii. CAS or Code Number

___ ___ ___ *   iii. Strength

___ ___ ___ *   iv. Purity

___ ___ ___ *   v. Composition

___ ___ ___ *   vi. Other Appropriate Characteristics

                        3. Test, Control, and Reference Substances

___ ___ ___ *   i. Stability Under the Conditions of Administration

___ ___ ___ *   ii. Solubility Under the Conditions of Administration

___ ___ ___ *   4. Description of the Dosage, Dosage Regimen, Route of

                           Administration, and Duration

___ ___ ___ *   5. Description of all circumstances that may have affected

                         the quality or integrity of the data

Computer Validation

Working Models of Computer System Validation and Verification

Phillip M. Buckler

ABC Laboratories, Inc., 7200 East ABC Lane, P.O. Box 1097,
Columbia, MO 65205

*Over the past several years, ABC Laboratories has
developed many computerized systems to aid its scientists in
their work. From the beginning, validation of computer-
generated data has been addressed. As an example of what
one laboratory is doing, ABC's computerization system will
be discussed to show where it started, where it is today, and
where it may be in the future. The following topics will be
discussed: (1) systems used in the past and how validation
of systems evolved, (2) two systems used at present (one system was purchased and one was developed in-house) and
how validation differs for the two systems, and (3) what systems may be used in the future, including speculation about
necessary validation for future systems. The role of the
quality assurance unit will be included in all discussions.*

THE COMPUTER HAS BECOME A MANDATORY TOOL in the modern business world. Larger, faster systems are constantly being developed to make

2192-8/92/0203$06.00/0 © 1992 American Chemical Society

our lives and jobs easier. In the contract laboratory business, computer systems have allowed scientists to generate data calculations more efficiently and accurately and in a more presentable form than the manual methods used in the past.

Two of the primary concerns when generating data by computer are whether the system will perform the required function and whether the results will be accurate. The first concern can be answered by validation of the system software and hardware. The second may be addressed by verifying that the hard-copy print is a reliable representation of the information originally sent to the computer. The combination of validation and verification ensures the credibility of computer-generated results.

Since ABC Laboratories began using computers in 1979, computer system validation and verification have been addressed. From the early microcomputers that took several minutes to calculate results to current personal computers (PCs) that take only seconds for the same calculations, system validation and data verification have been and will continue to be necessities.

There are many techniques in use today to validate computer systems. We have developed methods that suit our laboratory's particular applications. Although the procedures described in this chapter have worked very well, they should be used only as a guide. Each business must determine its own needs and develop a system that applies specifically to the procedures being performed. At ABC, many individuals, including scientists, quality assurance personnel, data processing personnel, and computer vendors, have assisted in developing the procedures used today. The knowledge available from these participants allowed for the development of a solid validation system that substantiates reported results.

The development of the validation programs has been an interesting one. The first computer used consisted of a microprocessor, a disk drive, a keyboard, a printer, and a black-and-white television for a monitor. This system was used to calculate LC_{50} (the concentration of test material at which 50% of the test organisms die) results obtained from static acute tests in aquatic toxicology. Even at that time, a validation system was used to test the purchased software. Each time the program was used, a set of test data was run to verify that the software was calculating the results properly. The successful validation was then documented in a computer logbook.

Although the preceding example is simple in theory, it indicates the importance of confirming the computer system as well as the results. The degree of validation or verification depends on the sophistication of the software and hardware. The aforementioned system requires minimal documentation, but a large data acquisition mainframe may require a complex program.

The following sections describe two computer systems and their validation and verification procedures. The first is a system that was developed internally using purchased software language. The second is a software–hardware system that was purchased intact.

Standard Procedures

One of the primary items that must be addressed with any validation program is consistency. The procedures used must be performed the same way each time regardless of the person executing the process. Therefore, a standard operating procedure (SOP) must be written along with the development of the validation and verification program.

The SOP should be a concise document that outlines all procedures in detail. The following items should be included in an SOP: the frequency that validation procedures are to be performed, the training of personnel to perform the validation procedures, the specific parameters to be evaluated, the specific data to be generated, the parts of the system to be evaluated, and the specific steps that must be accomplished for each segment.

The list of items to be included in an SOP is long. However, the important thing to remember is that the SOP must be a comprehensive document that is detailed enough to allow a third-party individual to follow the procedures involved. If this is accomplished, the result will be a well-defined plan and a reliable validation system.

Validation of an In-House Data-Base System

Many laboratory studies are conducted with radiolabeled compounds. The calculation of results can become quite difficult for a study consisting of even 50 samples. Calibration curves and samples were formerly calculated by using a calculator. The development of a scintillation program on a PC cut the calculation time significantly. The components of this system include a scintillation counter and a data collector, which is usually a PC or data buffer (Figure 16.1). The collected data is subsequently dumped to a disk on a PC.

The development of this system involved choosing the correct software language, then writing, testing, and implementing the program. Deciding which software to use was probably the most difficult task. A

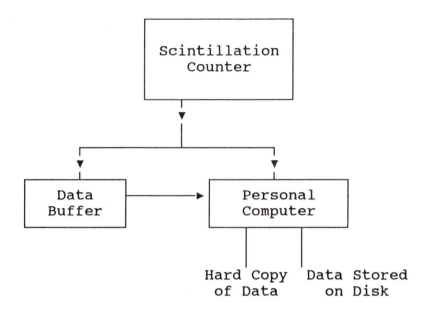

Figure 16.1. Data collection system for a scintillation counter.

committee consisting of laboratory, data processing, management, and quality assurance personnel was formed to discuss the various available software options. Two options were considered: spreadsheets and data base. Spreadsheets tend to be more user-friendly than data-base systems, but a data-base system is easier to validate, and the data are easier to verify at the end of an analysis set. A data-base system can be set up so that users cannot access the program, eliminating the possibility of accidental changes. A spreadsheet system does not have this option because program cells can be accessed by the user. On the basis of the aforementioned characteristics, the data-base system was chosen.

Once the type of software was defined, the committee discussed what parameters would be necessary for the program (i.e., parts per million, parts per billion, counts per minute, or disintegrations per minute). All of these units are used at various points in the calculation of scintillation results. The proper commands were written into the program to perform the appropriate arithmetic functions to the raw counts from the scintillation counter. One person was basically responsible for writing the program so that there were no coordination problems like those associated with larger systems. Once the programs were written, test data were used to ensure that the equations would provide the required results.

Figure 16.2 shows a sample of the printout obtained when the program is run. The counts-per-minute (CPM) number is recorded in a file and is accessed during the calculation process. The computer takes the CPM value, subtracts the designated background value, and multiplies it by a correction factor. The correction factor consists of the specific activity of the radioisotope, sample volume, scintillation counter efficiency, and dilution factor. This calculation gives an end result of parts per million, parts per billion, or as in the example, micrograms.

Verification of the data is relatively easy. Because a data-base system is used, the program and equations cannot be changed by the user. Therefore, if one line of the data is checked by hand and verified, then all lines will be correct. Laboratory personnel are required to check one line per set of data generated and to indicate which line was checked and who checked it. The correction-factor values mentioned previously are entered manually by the operator. Therefore, these entries are verified 100% by laboratory personnel.

The only portion of the package that must be validated is the link between the scintillation counter and the data-collection device. Validation is accomplished by comparing the CPM obtained from a printer (connected to the scintillation counter) to the CPM obtained from the collection device (Figure 16.3). If the numbers are the same, the collection system is verified as working. This exercise is conducted once a quarter during counting of the standard quench curve.

This fairly simple method of validation is adequate for the scintillation counter system because the computer is actually being used as a calculator. The inaccessibility of the data-base system has aided in the validation process, and accurate generation of data has been ensured. All methods described in the preceding sections, as well as operating instructions, are written in an SOP. The SOP ensures that all groups are performing procedures consistently.

Validation of a Computer Acquisition System

The data acquisition system is used to acquire data from gas chromatographs or high-performance liquid chromatographs. This software system was purchased after all development and initial testing had been performed, therefore the laboratory was not involved in validation during program development. However, procedures used in validating the programs were discussed with the vendor prior to purchase to ensure that a credible validation system was used during software development. The vendor also provided the source codes for the software. The source codes could be used at the laboratory to perform further validation or to assist

Solution Radioanalysis
Sample Set:
Analyst:
Study Sponsor:
Data Entered By:
Data Modified By:
User #:
ID:

Study Number:
Date of Analysis:
Test Material:
Date Entered:
Date Modified:
Collected:

Backgrounds:

Pos. #	Sample Code	Vol (ml)	Dilution Factor	CPM	H Factor	Efficiency
14-7	B Control Water	1.0000	1.0000	25.60	77.30	0.90812
14-8	B Control Water	1.0000	1.0000	27.00	76.70	0.90837
14-9	B Control Water	1.0000	1.0000	27.60	77.30	0.90812
Average:		1.0000	1.0000	26.733		0.90820

Samples:

Pos. #	Sample Code	Vol (ml)	Dilution Factor	CPM	NET CPM	H Factor	Efficiency	DPM	Sp. Act. (dpm/µg)	UG	Rep. Mean
14-10	B Sample Day 0-1	1.0000	%1000	112530	112503	142.00	0.88373	127305	152000	8.38D+02	
14-11	B Sample Day 0-2	1.0000	%1000	113647	113620	142.30	0.88360	128588	152000	8.46D+02	
14-12	B Sample Day 0-3	1.0000	%1000	113213	113186	143.00	0.88328	128143	152000	8.43D+02	8.42D+02

Reviewed By: _____   Date: _____

Figure 16.2. Sample data from a scintillation data-base system.

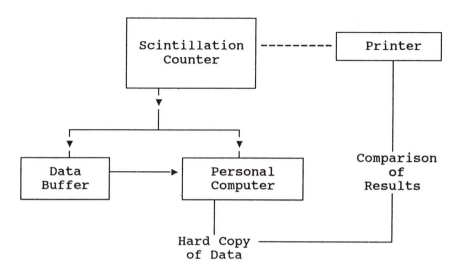

Figure 16.3. Data collection system for a scintillation counter with a verification printer.

in solving problems that occur during program execution. Usually, the source code is not used. A more direct approach to system validation and verification has been developed. The validation of this system is more complicated than the scintillation program because the data are manipulated more extensively. A brief explanation of the system components and the manner of obtaining data will help in describing the validation procedures.

Figure 16.4 shows the computer acquisition system (CAS). The signal is sent from the analytical instrument to an analog-to-digital converter. The converter translates the analog signal voltage into a digital format that is readable by the computer and sends it to the central processing unit (CPU) via a buffer multiplexer for storage on disk until needed for calculations. The chemist designates various parameters that are compound and study dependent. These parameters determine how base lines and other chromatographic controls are to be handled. After the parameters are keyed in, the data are manipulated, and the final results are calculated. Results are calculated by using a standard curve either in the linear or quadratic mode. A signal is sent to the printer from the CPU, again via the buffer multiplexer. This printout is considered the raw data. After the data have been printed, the file stored on the disk can be archived. Data accumulated on the disk are transferred to tape once a week and stored in a secure area.

The data processing staff and laboratory personnel operate as a team in validating the CAS. Maintenance of this system is an ongoing endeavor

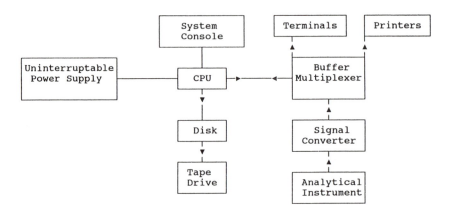

Figure 16.4. The computer acquisition system.

to ensure the soundness of the generated data. Each step of the computer process, from the collection of the signal produced by the analytical instrument to the backup and archival procedure, is monitored.

During the planning process for validation, many questions were considered: What parts of the system should be validated and what methods should be used to verify the system? The decision was made that a standardized set of test data would be developed to show that the calculations were being performed properly. The test data would also be stored on tape for assessment of the archival and retrieval system.

The next step was to determine how the test data would be generated. There were several options: Use genuine data obtained from a study, derive a set of data manually, or use a peak generator to create a set of data. The first two options leave gaps in the validation. The genuine data could be used to show consistency from one day to the next but would not prove that the system was operating properly at the beginning. The manually derived data would evaluate neither the signal from the instrument nor the area-calculation function of the system. The peak generator method was chosen because of its overall usefulness in the validation process. It addresses the above items in a format that is concise and easy to follow. The total system can be validated with one set of generated data instead of several. A device that generates a triangular peak of known area is used to simulate the signal from the analytical instrument to the computer. The device currently in use is capable of generating peaks at 25, 50, 75, and 100% of full scale and at a level of 0.001–10 V.

Several instruments of this type are on the market. The important thing to consider is that a peak should be generated within the range of the assay system.

To ensure that the peak generator was precise, a set of five replicates was generated at one voltage setting. The areas of these five peaks were calculated manually and the values of the manually calculated areas were compared to the values obtained by the computer. All five values were within one or two percent of the total average. Figure 16.5 shows an example of the chromatogram obtained from the calibrator. The y axis is in millivolts and the x axis is in minutes. The units of the x axis indicate retention time only. The area calculated by the computer is in millivolt-seconds (mV-s) because of the parameters entered for calculation of area. The theoretical area can therefore be calculated as 1/2 base width in seconds multiplied by the height of the peak in millivolts. In the example in Figure 16.5, the area can be calculated as follows: ramp rate = 1 s per side of peak, base = 2 s, 1/2 base = 1 s, height = 1 mV, and thus the theoretical area is 1 mV-s. The computer-calculated area was 1.003 mV-s (Figure 16.5). The percent difference between the two values is −0.3, a result that shows good agreement. This example shows that the signal sent from the instrument was received and quantitated correctly by the computer. The area reading is stored in the CPU in a nonaccessible file.

The total system is routinely checked on a monthly basis. Two levels of less than or equal to 1 V full scale are used, for example, 0.001 V and 1.0 V. If larger peaks are obtained in real data situations, higher levels are verified again as needed. Three replicates of each deflection amount (25, 50, 75, and 100%) are generated at two ramp rates to simulate packed- and capillary-column assays. Two replicate sets are used as standards, and the third set is used as the unknown. The results are calculated by using both linear and quadratic regressions. These data sets are generated during a peak usage time to reflect real use. This test helps ensure that the system is handling data consistently during both light and intense applications.

The area counts can be calculated further to determine parts per million by using a predetermined dilution factor. This step in the sequence will verify the accessibility of the raw data file and the subsequent transfer of the data into the file that calculates results. This is an important step because it proves that the values can be manipulated by the system successfully. The printed results are placed in a bound logbook to document the monthly validation. The results are cross-checked with hand-calculated results to show that the computer has performed the calculations correctly. The data are stored in the computer for one month, then retrieved to verify that the system has retained the data in the same form in which it was generated. After this verification, the data are transferred to tape and archived. At the end of another month, the data

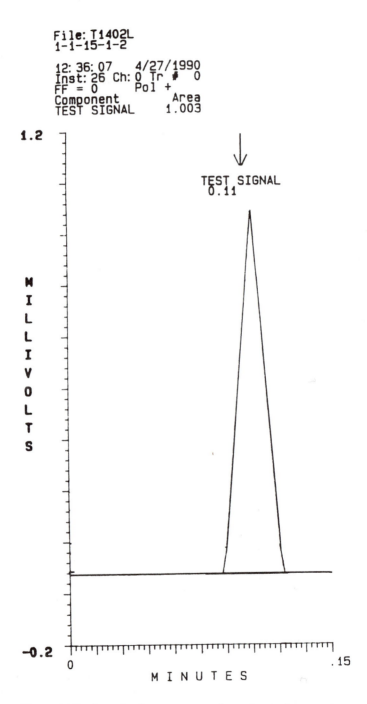

Figure 16.5. Sample chromatogram from the peak generator.

set is retrieved and verified again. This exercise proves that the system's storage capabilities are sound.

The one-month time frame is adequate to ensure that the system is working properly because many parameters are checked on a daily basis. Each day the previous day's events are printed. Items such as conditions controlling the total system, heading errors on data forms, and items performed during the past 24 hours (i.e., purging files) are reviewed by data processing personnel. This printout is stored in the archives for future reference.

Each set of data that is manipulated with this computer system contains quality control samples that indicate whether the analytical method is working properly. Routine acceptable recoveries also indicate that the system is acquiring and calculating data correctly. A set of standards prepared at a minimum of three levels is injected with each set of samples. Reviews of the standard curve results (y intercept, slope, and correlation coefficient) indicate whether the instrument run was acquired successfully. These subtle indicators give the chemist a guide that can be used to perform a miniature validation on each set of data analyzed.

Some individuals may be concerned that the computer will give erroneous results because of an electrical spike or other such uncontrollable phenomena without indicating that errors are present. Errors such as these would probably not go unnoticed. Errors would probably appear as garbage that does not fit the rest of the data set. We have never encountered such a problem with our system. However, if such a problem were to occur, the entire set of data would be reanalyzed. The entire computer system would probably not be revalidated unless the problem appeared widespread. However, the cause of the problem would be investigated.

Quality Assurance

As mentioned earlier, the quality assurance unit (QAU) assisted in the validation planning process. The QAU also has the responsibility of ensuring that the validation procedures have been properly executed. Each system has an SOP to outline what validation is required and how the validation will be accomplished. The QAU will check the system logbooks to verify that the appropriate procedures outlined in the respective SOP are being followed. This check is accomplished through periodic laboratory inspections of the procedures performed by the data processing and laboratory personnel.

Results of the computer runs from each study are reviewed by the QAU to check for any unusual occurrences that would indicate a potential problem. All generated data are evaluated by laboratory personnel and are reviewed in detail by the study director before being sent to the QAU. The quality assurance reviewer who checks the raw data will check random sections of data by hand to verify that the system was operating properly on that day. The QAU's job is made easier by the fact that a concise SOP has been developed for this system. The QAU will also check employee training records to ensure that personnel are properly trained to use the systems.

Security

Security plays an important role in the validation and verification of computer data. Data must remain in the form in which it was originally generated. Therefore, security involves not only the archival procedures but also how the data are treated before being archived. Several types of security apply to both the scintillation system and the data acquisition system.

Control of the data with the scintillation program consists of placing data from each study on a separate diskette. Because this system is PC driven, individuals can take their data to a secure area such as a locked file or cabinet. Unauthorized people cannot gain access to data easily. When a study is complete and the report has been sent to the QAU, the data diskettes are archived with the study report. A backup is stored in a separate secure location. The data are further protected by the data-base system, which does not allow changes unless they are made by the data processing unit.

Security of the data acquisition system is more involved than that of the scintillation system. One of the main reasons this is true is because the data acquisition system is a network, accessed by many individuals in the company. Security codes are used to protect specific user files and raw data files. User codes should not be readily recognizable to another person. For example, personal initials are not good user codes because they are easy for someone to ascertain. Individuals should select a code that is unique and should keep it confidential. This confidentiality will prevent accidental entries into files.

The raw data file is inaccessible to everyone—even the data processing personnel. Once the particular instrument run is made, only parameter-oriented changes that affect the way the results are calculated can be made. Examples of parameter-oriented changes are changing the way baselines are drawn, the threshold setting, or the sampling rate during acquisition.

Security of the data acquisition system hardware is of significant concern. The CPU and tape drive are in a limited-access room. A member of the data processing group is stationed in this room not only to maintain the system but also to monitor use of the terminals in this area. The computer room is locked after hours. The computer system is equipped with a security call-back modem that can be accessed by data processing personnel to check the system after hours. A password is required to access the modem system. When the caller reaches the system, the password is entered, the caller hangs up, and the modem calls a number that corresponds to the code that was entered. This prevents random access to the modem system by unauthorized personnel.

All computer data, both hard copy and tape, are placed in an archive area for long-term storage. The archive area is an important part of the security system. The archive room consists of two-hour-rated fire walls and smoke and heat detectors. The fire-alarm system is connected to a phone-calling system that will notify key employees in an emergency situation. A security guard is also on duty from 6 p.m. to 6 a.m. to monitor all areas of the facility.

The archive area is locked at all times, even during working hours. Access to this area is limited to archives personnel only; all other employees must be escorted by archives personnel.

All areas that house computers are temperature controlled. Heating and air conditioning units maintain an acceptable humidity range for the computer.

The PC-driven systems use a battery backup system to guarantee that data transferred from the scintillation counter to the PC are not lost if the electrical current is interrupted. The data acquisition system is connected to an uninterruptible power source that keeps the CPU operational if the electricity goes off.

The Future

This is a general overview of how one laboratory is addressing the subject of computer validation and verification. The current systems provide the satisfaction that the data being generated are accurate. However, methods are continually being evaluated for new ways to improve these systems.

The concept of validating data stored on diskette after archiving must be considered. This applies to data generated on a PC such as the scintillation system described previously. Data that have been stored on magnetic media for more than 5 years should be reviewed to verify that the information is still intact.

Where does the future lead? Many systems are being used and developed that are much more involved than the systems discussed in this chapter. These sophisticated systems create many challenges in the validation process. However, the overall procedures are still the same. Programs must be written to accomplish each step of the validation process. Deciding what components of the system should be validated will be an important consideration. How the validation of these specific components will be accomplished must also be considered. These are difficult questions, but they must be answered if we are to continue to strive for the ultimate in data generation: to improve how we conduct studies and to provide more accurate data.

Acknowledgments

This chapter was written in conjunction with the operations conducted at ABC Laboratories, Inc. The validation of software and hardware systems at ABC Laboratories is an ongoing process. The procedures described for the data-base and acquisition systems were developed at ABC Laboratories. The following persons assisted with the development of this document: L. Franklin, T. Spurgeon, E. Harper, K. Ediger, and B. Bowman.

RECEIVED for review November 9, 1990. ACCEPTED revised manuscript August 16, 1991.

Validation of Laboratory Computer Software

J. Drew Watson

Strategic Information Systems, EPL Bio-Analytical Services, Inc.,
P.O. Box 1708, Decatur, IL 62525

Accuracy in acquiring, processing, and reporting laboratory data is of paramount importance, especially when computers are used to perform these functions. This chapter discusses how one analytical laboratory is approaching the validation of programs used by the facility to minimize human- or machine-induced errors. Packaged software and user-written software alike are discussed as well as problems encountered during the evolution of computer-related standard operating procedures.

COMPUTER SOFTWARE IS THE BASIC TOOL that makes computers useful. As with all tools, the end users assume that the tools are properly crafted and designed to perform the required function. This chapter defines the processes necessary to establish the validity of that assumption.

2192-8/92/0217$06.00/0 © 1992 American Chemical Society

Validation Process

Validation is the process of evaluating computer software at the end of the software development process (Figure 17.1) to ensure compliance with user and system requirements. The user's requirements are functional, and the system's requirements are both regulatory and practical.

In our laboratory, the functional requirements are defined by the scientific staff in accordance with standard operating procedures (SOPs). An SOP is a document that details the acceptable method for accomplishing a given function. The SOP must contain adequate detail to assure facility managers that a particular function is being performed in a manner

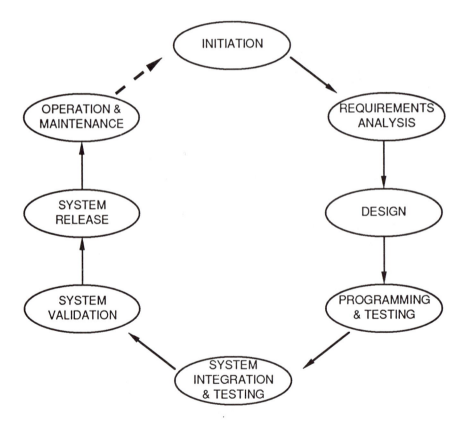

Figure 17.1. Software development cycle. (Reproduced with permission from Computerized Data Systems for Nonclinical Safety Assessment, *Drug Information Association, 1988.)*

that ensures the quality and integrity of the study. Inclusion of the scientific staff in developing SOPs ensures that a complete and accurate description of the needs and expectations of the end users is presented for analysis. This requirement means that SOPs stipulating the format for requesting development or purchase of computer software must be written and implemented before program development.

Many system requirements are based on SOPs that were written to ensure compliance with regulatory standards. These standards are described best in "Good Automated Laboratory Practices", a document now available in draft form from the Environmental Protection Agency. Based on these standards, we have prepared nine mandatory SOPs:

- application software system development

- application software documentation

- application software revisions

- programming standards

- system software installation and maintenance

- computer equipment maintenance

- computer backups

- computer system and data-base security

- application software validation and verification

In addition to the regulatory requirements, practical concerns must be considered. Primary among these are the use of hardware resources and data security. Analytical data by its sheer volume can rapidly overwhelm available storage capacity and challenge the number-crunching ability of even the largest machines. Programs must be designed within hardware limits.

Security poses special problems because of the diversity of available hardware. Many commercial hardware systems have built-in security designed to meet the stiff requirements of the Department of Defense standards. Other hardware systems have security attributes that meet less stringent guidelines. Among the latter category are personal computers; that is, those computers using the disk operating system.

The problem of security is compounded by the fact that much of today's commercially available software was written to run on the least secure computer systems. Addressing this contradiction in laboratory needs awaits regulatory guidance.

The validation process is also based on an SOP: "Application Software Validation and Verification". This SOP provides four tools to validate software. As shown in Figure 17.2, validation begins with a performance audit we call the "black box". The program is challenged with acceptable and unacceptable input, and the output is evaluated. The program must reject invalid input and properly handle valid input, as shown in Figure 17.3. This process is preceded by verification of each program

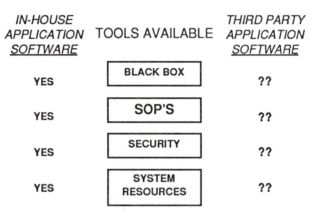

Figure 17.2. Validation of software.

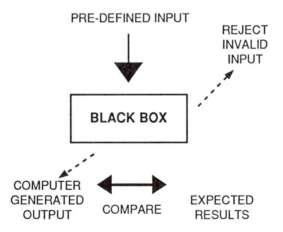

Figure 17.3. During the performance audit step of the validation process, the "black box" must check program outputs for given inputs and must reject invalid input.

segment or module if the program was written in-house. Rarely will third-party vendors make their program listings available for such an examination.

Successful completion of the performance audit is followed by an SOP review, documentation of security, and analyses of resources. The SOP review is a product of both the quality assurance unit and the "management"[1]. The review encompasses the technical requirements of the SOPs, including documentation of the program and training needs. Security and resource analyses are the responsibility of the computer support staff.

Problems with the Validation Process

As illustrated in Figure 17.2, a serious problem in the validation process is the lack of cooperation by third-party vendors. State-of-the-art computer software, especially in the area of chromatography, is available only from third parties. Generally, these vendors are unwilling to share the software source code or to comply with the regulatory requirements stipulated by the Good Laboratory Practice (GLP) Standards. Although the desire to protect such valuable property is understandable, some accommodation must be made so that laboratories using purchased software can ensure that the data generated or processed by that software is acceptable to regulatory agencies. Attaining this goal of GLP-acceptable software will require the cooperation of the regulatory agencies, the laboratories, and the vendors.

Conclusion

The validation process begins with well thought out, properly implemented SOPs. Four validation tools that can be used after the verification process are a performance audit (the "black box"), SOP review, security check, and resource utilization analysis. Software that was developed in-house can be fully verified and validated. Third-party application software can rarely be verified and can be validated only with the cooperation of the vendor.

---

[1]Management is defined here as the person or persons responsible for setting policy at a facility. For a description of the duties of testing facility management, see § 160.31 of the GLP Standards.

Bibliography

1. "Automated Laboratory Standards: A Guide to EPA Requirements for Automated Laboratories", U.S. Environmental Protection Agency, Office of Information Resources Management, Research Triangle Park, NC, 1990.

2. "Automated Laboratory Standards: Evaluation of the Use of Financial Systems Procedures", U.S. Environmental Protection Agency, Office of Information Resources Management, Research Triangle Park, NC, 1989.

3. "Automated Laboratory Standards: Survey of Current Automated Technology", U.S. Environmental Protection Agency, Office of Information Resources Management, Research Triangle Park, NC, 1989.

4. "Survey of Laboratory Automated Data Management Practices", U.S. Environmental Protection Agency, Office of Information Resources Management, Research Triangle Park, NC, 1989.

5. "Automated Laboratory Standards: Current Automated Laboratory Data Management Practices", U.S. Environmental Protection Agency, Office of Information Resources Management, Research Triangle Park, NC, 1989.

6. "Automated Laboratory Standards: Good Laboratory Practices for EPA Programs", U.S. Environmental Protection Agency, Office of Information Resources Management, Research Triangle Park, NC, 1990.

7. "Computerized Data Systems for Nonclinical Safety Assessment", Drug Information Association, Maple Glen, PA, September 1988.

8. "Planning for Software Validation, Verification, and Testing", U.S. Department of Commerce, National Institute of Standards and Technology, Gaithersburg, MD, November 1982.

9. "Validation, Verification, and Testing of Computer Software", U.S. Department of Commerce, National Institute of Standards and Technology, Gaithersburg, MD, February 1981.

10. *Draft Standard for Software Verification and Validation Plans*, IEEE Computer Society Press: Los Alamitos, CA, 1985.

11. *Draft Standard for Software Reviews and Audits (P1028)*, IEEE Computer Society Press: Los Alamitos, CA, 1985.

RECEIVED for review November 5, 1990. ACCEPTED revised manuscript October 28, 1991.

Specialty Studies

Product Chemistry
from the Formulation Perspective

John F. Wright

Pest Control Specialties, FMC Corporation, 2000 Market Street,
Philadelphia, PA 19103

Product chemistry is a necessary component of a pesticide registration application defined by the Federal Insecticide, Fungicide, and Rodenticide Act in 40 Code of Federal Regulations *158, Subdivision D. Each formulation research and development department deals with these regulations for end-use products. Based on the latest Good Laboratory Practices (GLP) Standards, the product-chemistry package for end-use products must be generated under GLPs. This chapter describes the product-chemistry requirements and gives a positive approach to integrating GLPs into laboratory activities with minimum disruption. Key recommendations are to provide a good audit trial to validate the data with minimum paperwork and to improve the work flow with proper planning. GLP is a state of mind that begins as a dynamic, grass-roots movement embraced by management to improve productivity without hindering progress (management is defined here as the person or persons responsible for setting policy at a facility. For a description of the duties of testing facility management, see § 160.31 of the GLP Standards).*

THE PHILOSOPHY OF DEVELOPING A QUALITY ASSURANCE PROGRAM
under the Good Laboratory Practices (GLP) Standards was the theme of
the first symposium on the topic of GLPs sponsored by the Division of
Agrochemicals of the American Chemical Society (*1*). The second symposium dealt with the practical applications of GLPs for laboratory and field
studies, the subject of this book. Industry was first introduced to GLPs by
the Food and Drug Administration (FDA) in 1978 (*2*) to ensure the quality and integrity of all data from nonclinical trials submitted to governmental agencies. The Environmental Protection Agency (EPA) introduced GLPs for health-effects trials of agricultural products in 1983 (*3*).
During the 1980s, the GLP concept evolved from a specific focus on trials
to a generic philosophy. The EPA has since instituted generic GLPs for
all data submitted under the Federal Insecticide, Fungicide, and Rodenticide Act (FIFRA) to support experimental-use permits and registration
applications (*4*).

Data Requirements for Product-Chemistry Studies

FIFRA's Section 3(c)2(A) (*5*) states that "the administrator shall publish
guidelines on information required to support registration of a pesticide."
This requirement manifests itself in 40 *Code of Federal Regulations* (CFR)
Part 158 as the Data Requirements for Registrations. One key element in
the data requirement focuses on the product or formulation being
registered. Historically, the formulation product-chemistry package was a
brief, single-page description of the pesticide product, generally the end-
use product (EP) or formulation being registered by the EPA. In many
instances, the technical material did not require a specific registration,
especially if it was not isolated in terms of the manufacturing scenario of
the EP. By the start of the 1980s, the EPA began to formalize the
product-chemistry package to standardize the data into several areas.
These general categories were

- product identity (Guideline Series 61)

- analysis and certification of product ingredients (Guideline Series 62)

- physical and chemical characteristics (Guideline Series 63)

- other product-chemistry requirements (Guideline Series 64)

The third item was generally the content of the product-chemistry
package, in which the physical and chemical properties of the product are
highlighted, and the other requirements dealt only with sample submissions to the EPA archives.

Registrants paid little attention to the product-chemistry area because the EPA was more concerned with the quality and integrity of the health-effects data and later with the environmental fate, residue, and metabolism data. In the mid-1980s, the EPA's concern was directed toward a toxicology data base on the inert compounds or "excipients" used in the formulations. The data in the product-chemistry package, especially in the Confidential Statement of Formula with certified limits for each component, became the mainstay of information. This evolution of product chemistry to its current state is described in Subdivision D of FIFRA, "Product Chemistry". The latest revisions to Parts 158.150–158.190 were published May 4, 1988, and generally are referred to as the "maxi regs" (*6*).

The purpose of product chemistry is to identify the composition of the product and the physical and chemical properties necessary for the EPA reviewer to determine or assess potential hazards. In addition, the generic GLPs were superimposed on product-chemistry studies (*4*). In the 1990s, product-chemistry acronyms will become commonplace as happened with the other technical groups addressing registration submissions. The acronyms PAGs, DRGs, SEPs, GLPs, SOPs, and QAU will be facts of life. For product-chemistry studies, as with other data requirements, the appropriate documents required to complete the technical task must be identified. The minimum requirements are contained in the following list.

- Pesticide Assessment Guidelines (PAGs) (*7*), Product Chemistry

- Data Reporting Guidelines (DRGs), Addendum 1 (*8*), Product Chemistry

- Standard Evaluation Procedures (SEPs) (*9*), Product Chemistry

- Pesticide Registration (PR) Notice 86–5 (*10*)

- Data Reporting Summaries (*11*), National Agricultural Chemicals Association (NACA) Standard Format (*12*)

- Internal GLP protocol for initiating product chemistry

The PAGs supply the details of the specific data requirements needed for the product-chemistry package. General or specific test procedures and the data required for a technical or end-use formulation are described in the PAGs. The DRGs augment the details of the initial PAG, especially to clarify technical issues. The SEPs explain what information will be reviewed by the EPA. For product-chemistry studies, the key items are the test compound and its manufacturing process. The test compound must be identified as a technical-grade active ingredient (TGAI), a manufacturing-use product (MP), or an EP. Test compound purity or the concentration of active ingredients must be clearly disclosed. If the EP is a single active ingredient or a blend of active ingredients, this information must be identified. If excipients are used, they must be clearly identified, and their exemption from the requirements of tolerance when used as excipients in formulations under 40 CFR 180.1001 must be stated. If the active ingredient source is registered with the EPA, its registration number must be identified. In addition, the sources of the active ingredients and their potential production locations must be highlighted. The proper test chemical must be employed for each requirement of the product-chemistry guideline.

For the manufacturing process of an EP, the reviewer must be supplied with enough mechanics to follow the synthesis or compounding of the product. Of concern are the impurities in the starting materials and the compounds likely to contribute to the formation of toxic ingredients. If multiple processing schemes are employed, all of the impurity profiles must be identified. All confidential business information should be placed in a confidential appendix. The decision on proprietary information is governed by company policy. Guidelines for reporting proprietary information are in Section 10 of FIFRA (5). If you wish to expand beyond these guidelines, a different format must be used as identified in PR Notice 86–5.

PR Notice 86–5, which is a press release issued by the EPA with specific instructions on the physical formatting of final reports (including location of a certification of GLP and other items), is critical to the product-chemistry package. A specific paragraph indicates the number of reports into which product-chemistry information should be earmarked. This number is a very important distinction.

- All product-chemistry data in support of EPs from a registered MP should be bound in a single study under a single title page.

- Product-chemistry data on a TGAI, an MP, an experimental-use product, or an EP produced from an unregistered source of active ingredient should be bound as a single study for each guideline series.

Therefore, PR Notice 86–5 specifies one report for certain uses and up to four reports for other uses. The data reporting summaries (11) are guidelines required by the EPA for the release of critical information allowing interested parties to assess potential hazards of a product without disclosing confidential business information. As part of the product-chemistry package, the formats suggested by NACA may be used to comply with this requirement.

Using GLPs for Product-Chemistry Studies—An Example

The GLPs must be superimposed on these data requirements. Working with GLPs is really a state of mind accompanied by a need to validate the integrity of the data being generated. Other groups have struggled with the concept of GLPs over the past 10 years. Today's scientist may learn from the growing pains of other disciplines trying to implement the GLP concept. After detailed internal discussion with the research and development department and consultation with the formulation chemist, FMC developed the strategy described in the following sections to fulfill the product-chemistry data requirements. As with any interpretation or implementation of regulations, the process is not static, but dynamic. It may evolve to the situation in which the analytical chemist is the study director and the formulation chemist is equally effective as a contributing or principal scientist.

In most organizations, the key elements being sold are the EPs or formulations. The formulation chemist plays a critical role in the commercial development of products being registered with the EPA. There are approximately 3000 active ingredients that are used in 44,000 formulations.

The formulation chemist is a good candidate for study director because he or she is well versed in all phases of the product. The sponsor is the manager of formulations. The sponsor generally receives a request from a development manager or registration specialist and assigns the formulation chemist as the study director to prepare the product-chemistry data package. The sponsor ensures that the study director has the technical support and facilities to accomplish the requested objective. As study director, the formulation chemist selects the test substance, initiates the protocol, assigns work activities, and writes the final report (including the data from the analytical chemist). The test substance is the formulation. The formulation chemist (study director) initiates a protocol to start the GLP project. Samples representative of the formulation to be registered, which should be available from commercial production or semi works (pilot plant), are selected. In the design efforts to bring out new products, the formulation chemist categorizes the technical (active ingredient) and understands its strengths and weaknesses especially with regard to its inherent chemical stability and what excipients are functionally and chemically compatible with it.

The formulation chemist is also aware of the final physical form of the product and how it will be stored (i.e., packaged to the consumer). From the laboratory to the pilot plant to the commercial product, the formulation chemist is involved. The product-chemistry package has many sections in which the formulation chemist will contribute this understanding of the entire process involved in preparing and packaging the product.

The test system may be defined as the test substance for product chemistry (e.g., the chemical analysis, the storage stability test, and the physical properties to be determined). The formulation chemist develops a series of standard operating procedures (SOPs) to convey the physical properties. The analytical chemist provides SOPs for chemical properties. The analytical work is generally not conducted by a formulation chemist but rather is requested internally or subcontracted out.

In most companies, a quality assurance unit (QAU) has been established. The generic concept of GLPs for a product-chemistry study is a good fit in the product-chemistry series.

GLPs can be used to the benefit of the researcher to control the destiny of research projects rather than having the projects control the scientist. The best approach to understanding and using the GLPs is a grass roots one. Researchers should read the regulations and build activities around them. By identifying needs in the PAGS and SEPs, SOPs can be generated to fit the tasks, including actual test procedures, but also formatting of reports, paper trail, and work flow. Researchers can develop a protocol request system that will be adaptable for each objective. The GLP concept is not a static issue but a dynamic one. Researchers should start by preplanning and scheduling, especially with the QAU. Use GLPs as a management tool.

The rate-controlling factor for a product-chemistry package is the storage stability data guideline (63–17) (6). The formulated product must be analyzed for its active ingredient at time zero and during 1 year of storage. Storage shall be at ambient warehouse conditions of temperature and humidity and in the product's commercial package or, if necessary, in a smaller size package of similar construction. The EPA will, however, accept accelerated-storage data for a product-chemistry package on an interim basis for a registration. The accelerated-storage data may be used to obtain a conditional registration from the EPA. This registration will be upgraded to an unconditional full registration pending the EPA's review of the 1-year data report. If the appropriate analytical methods are available, the interim product-chemistry package with 12 weeks of accelerated-storage stability data can be generated within approximately 24–28 weeks. This schedule allows approximately 6–8 weeks to put in place the paper audit trail, draft protocol, and final protocol and to establish analytical coordination. The set-up period is followed by 12 weeks of analytical data generation. The remaining 6–8 weeks are used to prepare the final report and to have it audited and ready to issue. Priority projects may require time frames to be compressed, and shortening the schedule to about 18 weeks is possible if it is justifiable. The interim package should be followed by the final report 36–48 weeks later, with the completion of the room-temperature data going out for 52 weeks of actual in-life time. The interim report is based on using the accelerated data in lieu of the 1-year data at room temperature, and the final report, which is released after 60–76 weeks, contains the 1-year room-temperature data.

Conclusion

The formulation chemist designs and develops the EPs sold to the consumer. The formulation chemist must know how the product will perform, the product's strengths and weaknesses, any unique hazards that are associated with the formulation, and how the product will be produced and packaged. The product-chemistry requirements for the EPA registration must be understood. Therefore, the formulation chemist is the ideal candidate to function as the study director. The study director accepts the conditions of the GLP Standards to ensure the quality and integrity of the data being generated and integrates the GLPs as a management tool. By understanding the philosophy and mechanics of the GLP Standards, the study director develops a good audit trail with minimum paperwork to minimize the length of time necessary to provide a high-quality product-chemistry package that conforms to FIFRA guidelines for registration purposes.

References

1. *Good Laboratory Practices: An Agrochemical Perspective*, Garner, W. Y.; Barge, M. S. Eds.; ACS Symposium Series 369; American Chemical Society: Washington, DC, 1988.

2. "Final Rule for Good Laboratory Practice Regulation Under the Federal Food, Drug, and Cosmetic Act"; *Code of Federal Regulations* Title 20, Pt. 58; *Federal Register* 43 (December 22, 1978) pp 59986–60025.

3. "Final Rule for Good Laboratory Practice Standards Under the Federal Insecticide, Fungicide, and Rodenticide Act", *Code of Federal Regulations* Title 40, Pt. 160; *Federal Register* 48: pp 53948–53969.

4. "Final Rule: Good Laboratory Practice Standards: Federal Insecticide, Fungicide, and Rodenticide", *Code of Federal Regulations* Title 40, Pt. 160; *Federal Register* 54:158 (August 17, 1989) pp. 34051–34074.

5. "The Federal Insecticide, Fungicide, and Rodenticide Act as Amended", U.S. Environmental Protection Agency, Office of Pesticide Programs, U.S. Government Printing Office: Washington, DC, 1989, 617-003/84346, p 16.

6. "Final Rule: Pesticide Registration Procedures; Pesticide Data Requirements", *Code of Federal Regulations* Title 40, Pts. 153, 156, 158, 162, and 163; *Federal Register* 53:86 (May 4, 1988) pp. 15952–15999.

7. "Pesticide Assessment Guidelines", Subdivision D, Product Chemistry, U.S. Environmental Protection Agency, October 1, 1982, National Technical Information Service, Springfield, VA, PB83–153890.

8. Perfetti R.B. "Pesticide Assessment Guidelines", Subdivision D, Product Chemistry, Series 61, 62, 63, and 64, Addendum 1 on Data Reporting, U.S. Environmental Protection Agency, National Technical Information Service, Springfield, VA, PB88–191705.

9. Perfetti R.B. "Hazard Evaluation Division, Standard Evaluation Procedure", Product Chemistry, U.S. Environmental Protection Agency, National Technical Information Service, Springfield, VA, PB88–243191.

10. "PR Notice 86-5, Notice to Producers, Formulators, Distributors, and Registrants", U.S. Environmental Protection Agency, U.S. Government Printing Office: Washington, DC, July 29, 1986.

11. *Code of Federal Regulations* Title 40, Pt. 152.50(4)C., pg. 13, revised as of July 1, 1990, U.S. Environmental Protection Agency, U.S. Government Printing Office: Washington, DC, 1990.

12. McCarthy J.F., to Registration Committee, Maxi-Regs Summaries, National Agricultural Chemicals Association, Washington, DC, February 16, 1989.

RECEIVED for review September 17, 1990. ACCEPTED revised manuscript September 3, 1991.

Compliance for Routine and Nonroutine Field Studies

James P. Ussary[1]

Agricultural Products Group, ICI Americas, Inc.,
Goldsboro, NC 27533–0208

*Good Laboratory Practice (GLP) Standards compliance
requirements for routine and nonroutine agricultural chemical products field studies are identical; however, the procedures used to obtain and monitor compliance for each
type of study may be quite different. Routine studies are
often done at several sites by only one or two scientists at
each site, and nonroutine studies may be done at single sites
with several people doing the work. This chapter describes
some of the GLP compliance methods used for each type of
study.*

REGISTRATION OF AGRICULTURAL CHEMICAL PRODUCTS requires a wide
variety of field studies that must be documented according to the Federal

---

[1]Current address: Ussary Scientific Associates, 1614 Boyette Drive,
Goldsboro, NC 27534

Insecticide, Fungicide, and Rodenticide Act Good Laboratory Practice (GLP) Standards. Some of these studies, such as magnitude of residues, crop rotation, and soil dissipation studies, are required for almost all new products, and these studies have become quite routine for field scientists as well as for the quality assurance unit (QAU). Thus, GLP compliance procedures for these studies are often somewhat automatic and predictable. Other studies, such as mesocosm and groundwater contamination studies and studies of worker exposure during product application, are required only under special conditions. Because these studies are not routine, each study may require planning and monitoring to ensure compliance with the GLPs. This chapter describes some of the GLP compliance problems that may occur in these two types of studies and how the QAU may need to be involved differently in each even though the GLP regulatory requirements are identical.

Protocols

The protocol, the backbone of any study, must include the entire study plan from initiation of scientific work to the location where data and the final report will be archived. The protocol for a complex nonroutine study may need to contain more details for scientific activities than a protocol for a routine study. For example, collection of sediment samples from a specific pond in an aquatic study would need to be described in more detail than would collection of tomatoes from a magnitude-of-residues study. Sediment samples may need to be collected from specific sections of the pond with a different number of cores from each section. The protocol for collecting tomatoes in the magnitude-of-residues study may specify only sample size and the requirement that the sample be collected from randomly selected sites throughout the test plot. The protocol must give enough instruction, either directly or by referring to standard operating procedures (SOPs), that the method of conduct of the study is quite evident. Furthermore, for a large nonroutine study, a summary of critical dates or study event sequences may be helpful to study personnel as well as to the QAU.

The QAU should be alert for protocol amendments, especially in large nonroutine studies. These studies are usually complex and are important to and visible within the organization. The study director will likely feel pressure to conduct the study in such a manner that there will be no doubt that the data will meet the study objectives and that the conclusions in the final report are well supported; therefore, the study director may feel compelled to make protocol changes during the study. Amendments are often made either early or late in a study. Early in a study, the protocol may need to be changed to solve some previously unforeseen technical problem, or a new study objective may be added. At the planned end of the study, the study director may wish to continue collecting samples beyond the time specified in the protocol, to make another test chemical application and continue the study, or to make some other major change. The QAU should ensure that the protocol has been properly amended and that the amended protocol is at the study site. The QAU inspection schedule will usually need to be changed when a protocol is amended.

The QAU should also be alert for protocol deviations. In field studies, a substantial number of deviations can occur if the protocol has not been written to allow for weather unpredictability and seasonal variations. Deviations must be thoroughly documented and approved by the study director. If the study director is not at the study site at the time of the deviation, the approval can be obtained with a telephone call from the field scientist to the study director and an appropriate note placed in the study file. Deviations must be recorded by the field scientist and then approved by the study director, who may also add written comments.

Standard Operating Procedures

SOPs for routine field studies should be written in a form that is general enough to allow for variations in cultural practices, equipment brands, and personal preferences of the field scientists, as long as these variations will not adversely affect the scientific integrity of the study. The SOPs must also be written to provide specific useful information for the field scientist and for the historical study records. For example, the SOP for a sprayer used to apply test chemicals may allow the operator some freedom to choose the exact method for calibrating output and ground speed, but it should be specific about the frequency of calibration and limits of variation and what the operator must do if the sprayer malfunctions while treating a study test plot.

SOPs for nonroutine studies must often be very specific about how a task should be performed. These SOPs might be written for only one study. Because study personnel might be somewhat inexperienced with the type of study being performed, diagrams or photographs might be helpful. Nonroutine studies are often performed at only one or two study sites, so SOPs for these studies usually need not allow for much variation except, perhaps, that caused by weather. The number of SOPs for a nonroutine study may be greater than that for a routine study because nonroutine studies are often much more complicated. SOPs for a nonroutine study often must be amended after the study has started because of problems that were unforeseen when the SOPs were written. The QAU should carefully study the SOPs before doing an inspection and should be alert for undocumented SOP changes and deviations.

Test Chemicals

Test chemicals used for routine studies are usually from a well-documented batch of product, and the system for field scientists to obtain, use, store, and dispose of any unused quantity is usually well established. Also, there is usually a routine system for archiving test chemicals. These procedures then need only to be monitored by the QAU. For nonroutine studies, the QAU may need to advise the study director of the requirements for clear custody, use, and disposal records for test chemicals. Also, the QAU must ensure that proper, secure storage is available at the study site and that a portion of the test chemical has been submitted to the test-chemical archive.

Study Personnel

GLP regulations require that there be adequate personnel to do the required work, that all scientific personnel be properly trained, and that the qualifications of the personnel be documented. For large nonroutine studies, the amount of labor that will be required may be underestimated. The QAU should immediately report any personnel shortage found during an inspection. Except for entry-level employees, scientists conducting most routine field studies do not need special training for each study. Quite often the field scientists have several years of experience doing these studies. For unique or complex studies, some or all of the personnel may have had little or no experience conducting that particular type of study; therefore, a carefully planned training program will be needed. The best training program will include both lectures and field practice. The QAU should have an opportunity for direct input at all of these sessions. Training sessions that are dedicated solely to GLP compliance procedures are also useful. As the study progresses, the procedures used by study personnel must be monitored by the QAU to ensure that the training was effective.

Data

Documentation of a nonroutine study should receive special attention from the QAU. Often, a large amount of data will be recorded by several people over a long period of time. Data must be secured as well as properly recorded. The QAU should ensure that all data are being recorded as required in the protocol and that any opportunity for data tampering is minimized. For routine studies, study personnel will likely have standardized data forms and will have experience using these forms, and most of the data at a study site will be recorded by only one person. For a nonroutine study, data forms may not exist, forms designed for routine studies may be manually altered for use, or new, unfamiliar data forms may be used. Each of these situations may cause study personnel to record data that are not easily interpreted. Because the field scientist might not understand exactly how the study director wants the data forms to be completed, data might be recorded for later transcription to data forms. Recording of data should be carefully explained in the prestudy training sessions and then closely monitored in the early stages of experimental work. An SOP for recording data, including examples, is desirable.

If a study is to be conducted over a long period of time or if there is to be a large amount of data, an SOP for managing data should be written. A data custodian may also be appointed for these studies. Data should be organized for easy auditing by the QAU or a regulatory agency at all times. The data should be stored so they are reasonably safe from tampering, such as in a locked file cabinet in the field laboratory. As soon as the data are no longer needed in the field, they should be transferred to the study director or to the archive.

The QAU should carefully inspect data from nonroutine studies throughout the progress of the study, whereas such close inspections may not be necessary for all routine studies.

Inspections

Inspections of routine field studies by the QAU tend to become routine as well. For magnitude-of-residues studies, crop-rotation studies, and other routine studies, annual inspections at each field site where studies are being performed may be sufficient. The inspector should be present when a critical phase of one of the routine field studies at the site is being performed (such as application of the test chemical or collection of samples). If the inspected procedure is being performed according to the appropriate SOP, study records are current and properly recorded, the study is progressing according to the protocol, and general GLP compliance is in good order, then other similar studies at the site are also likely to be conducted properly. On the other hand, the QAU should schedule frequent inspections for nonroutine studies.

When inspecting field studies, the QAU inspector should expect to find pertinent SOPs at the test site. For routine studies, procedures become so familiar to field scientists that the presence of the SOP file is of little immediate value; therefore, the QAU inspector often must explain to the field personnel that the presence of the SOPs is a regulatory requirement. Study personnel doing nonroutine studies often may need to refer to the SOP manual and so a much-used SOP manual will usually be found at the test site. For these nonroutine studies, the QAU should ensure that there are sufficient SOPs for work being done and that the SOPs are actually being followed. Any changes in the study procedures must be shown in the study records, and the SOPs must be appropriately amended and approved. The QAU inspector should diligently review the SOPs before any inspection of a nonroutine study.

Field work often must be done during hot weather, when the wind is blowing, or under dirty conditions. There is rarely a convenient place to record data. This creates a situation that makes it difficult to record data neatly and legibly; therefore, the study director and the QAU inspector must ensure that data are being recorded directly on study data forms or notebooks rather than being recorded on a sheet of paper for later transfer to forms or a notebook. Also, data in the field must be secure from loss or tampering.

Conclusion

GLP regulations do not preclude a common-sense approach to developing compliance procedures. These procedures should be appropriate for the study being conducted and may differ for the various types of studies performed by an organization even though the regulations do not differentiate between study types.

Both routine and nonroutine field studies must be inspected by the QAU, and the QAU must keep its "management"[2] properly informed about the compliance of the studies with GLP regulations. For nonroutine studies, the frequency of inspections and interaction of QAU personnel with study personnel will usually need to be much greater than for routine studies.

RECEIVED for review September 17, 1990. ACCEPTED revised manuscript August 17, 1991.

---

[2]Management is defined here as the person or persons responsible for setting policy at a facility. For a description of the duties of testing facility management, see § 160.31 of the GLP Standards.

Nonbiased Field Sampling

Duane D. Ewing

Pan-Agricultural Laboratories, Inc., 32380 Avenue 10, Madera, CA 93638

Developing a study protocol that includes methods for the control of bias is required by 40 Code of Federal Regulations *Part 160, Subpart G, and is a basic scientific requirement for field studies. Generating nonbiased representative field samples is affected by such variables as study type, study design, sampling equipment, and sample size and must be considered while developing the protocol to satisfy the study. The thought process behind generating field samples that are free from bias and representative of the plot is illustrated through a discussion of several types of field studies.*

FIELD SAMPLES, whether collected randomly, partially randomly, or non-randomly, must be free from sampling bias and must provide an accurate representation of the plot. Sampling is an integral part of the experimental design; it is a strategy that considers all aspects of the study design in developing a well-planned, efficient procedure.

Every trialist possesses tendencies and attitudes that affect how samples are collected and that may affect representativeness. Randomization is one means of removing sampling bias, but it does not automatically guarantee a representative sample. Other methods can be used to minimize bias, such as systematic sampling, as long as the sampling design is well thought out and not arbitrary or haphazard. A nonbiased sample must be generated impartially, even when it is not collected randomly.

Sampling should be described in sufficient detail in the field protocol that there can be no misinterpretation of the procedures. Describing the sampling can be a creative effort that considers many details of the study design, including plot size and design, sample size, and test equipment. Designing a sampling protocol may be more enjoyable than implementing one because sampling can be boring, tedious, and uncomfortable in many field situations. Thought should be given to making procedures simple and convenient so that sampling can be accomplished quickly and accurately.

Three common types of field studies will be discussed to illustrate the thought process and the flexibility necessary when developing a sampling strategy: terrestrial-dissipation studies, dislodgable-residue studies, and field-residue trials. My focus is on obtaining nonbiased samples with adequate plot representation, although I will discuss other aspects of the study design as they influence the thought process.

Terrestrial Dissipation Studies

Field dissipation studies are conducted to determine the persistence and leaching potential of pesticides under actual field conditions. Three samples for each depth increment are generated by collecting five cores per sample and compositing by depth. The three samples for each depth increment are used to indicate variations in concentrations of pesticide residue in the soil.

A common plot design is illustrated in Figure 20.1, in which a large plot is divided into cells identified by rows and columns that create a sampling grid. An accepted method for taking soil samples is to collect three cores from each of five predetermined randomly selected cells and to label the cores A, B, and C (*1*). All A cores are composited by depth, as are all B cores and all C cores. Sampling bias has been removed because the sampling locations have been determined randomly, yet the sample may not be representative of the plot. Only five of the 100 cells are sampled during any event, and these five cells may not be well distributed within the plot. The cells could be made larger and the number of cells reduced, but that may allow the trialist excessive freedom in determining where the cores will be taken, possibly biasing the sample.

Another way to optimize representativeness while minimizing sampling bias is to increase the spatial distribution of the 15 individual soil cores. Figure 20.2 illustrates a design where the plot is divided into three subplots labeled A, B, and C. A sampling grid is established in each sub-

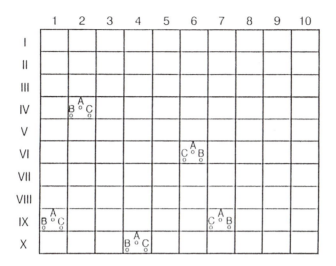

Figure 20.1. Randomized sampling design for terrestrial dissipation studies.

plot, and one core is collected from five randomly selected cells within each subplot. The cores are composited by depth and by subplot. Sample collection is not truly random in this design because restrictions are made on the number and location of cores collected. However, the goal to generate representative, nonbiased samples is achieved.

Both designs are easy to implement using hand-sampling equipment. The trialist can move quickly from one coring location to the next while causing minimal plot disturbance. These designs can also be used with a tractor-mounted soil probe, but the destructive nature of the tractor must be considered. Soil becomes compacted beneath tractor wheels, introducing a new variable. Sampling from within wheel tracks must be avoided, so great care should be taken to minimize travel within the plot. The plot design in Figure 20.2 would be the most cumbersome to implement because 15 cores are collected from 15 different cells, requiring extensive travel throughout the plot area. Travel routes must be well defined within the plot to lessen the negative impact on the study, which can increase sampling time and, ultimately, study costs.

The plot design in Figure 20.3 evolved from using tractor-mounted soil probes in an attempt to maintain plot integrity while allowing for quick and efficient sample collection. Three subplots labeled A, B, and C are established with a buffer area between each subplot, allowing the tractor to turn outside the sample areas. The width of the subplots is the same as the spray swath. Predetermined sampling lines are identified

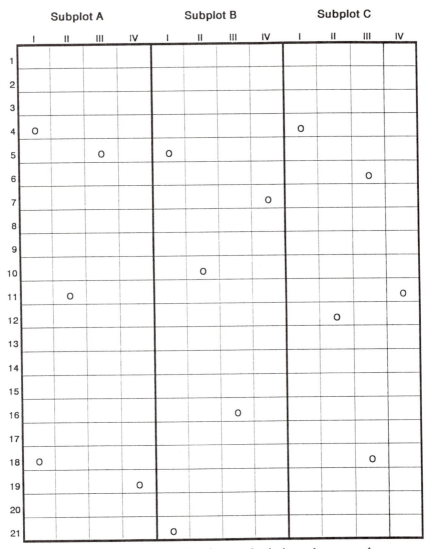

Figure 20.2. Terrestrial dissipation study design when samples are collected randomly within three subplots.

within each subplot (parallel to the spray swath) and are spaced so that each tractor wheel travels exactly in the middle of two adjacent sampling lines during sampling. One line from each subplot is randomly sampled by collecting five cores along the line and compositing by depth incre-ments. Therefore, samples generated from each subplot during a sampling event represent the spray swath.

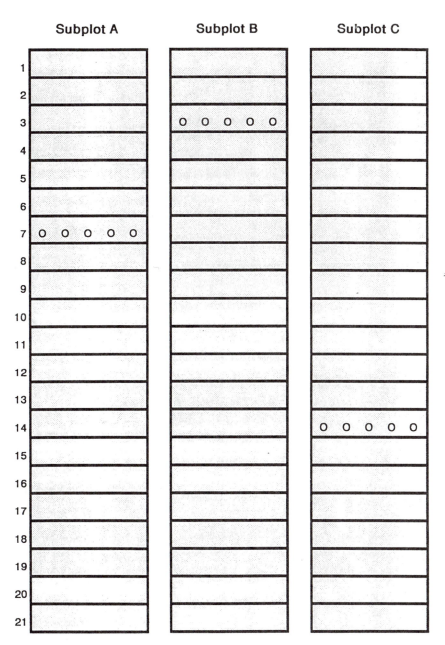

Figure 20.3. A terrestrial dissipation study design for tractor-mounted soil probes.

Two advantages of this design are that the tractor passes over a sampling area once throughout the study, leaving unsampled plot areas intact and that nonbiased samples can be collected in a timely manner. Individual cores may not be distributed as well as those in Figure 20.2, but representativeness is adequately achieved.

Dislodgable Residue Studies

Dislodgable residue studies are conducted to measure potential human exposure to pesticide residue after a pesticide has been applied to a crop. For many studies, a sample consists of 40 leaf disks that are collected from specified areas of the plant. Triplicate samples are often collected. The representativeness of foliar dislodgable residue studies depends on where the samples are collected within the plot and where they are collected from the plant. Iwata and co-workers (2) suggested two ways of collecting nonbiased samples from citrus trees while maintaining representativeness. Figure 20.4 illustrates a method in which five leaf disks are collected at 45° intervals around each of eight trees. The sampling interval is offset by 45° for each successive tree so that each sampling interval is equally represented in the sample. A second method is to sample all eight 45° sampling intervals from each of five trees, as illustrated in Figure 20.5.

Neither approach generates random samples, but both produce nonbiased and representative samples. One variable that dictates which approach to use is the number of trees available.

Field Residue Trials

Field residue trials are conducted to determine the magnitude of pesticide residues on a crop. Trials are conducted under commercial agricultural conditions or closely simulate those conditions, and plot size is large relative to sample size, especially with aerial applications. Field residue protocols specify sample size but rarely address how to obtain a representative sample, a detail often left to the trialist. If the trialist does not develop a plan before sample collection, samples may be collected arbitrarily and for convenience, thereby compromising study integrity. True random sampling is rarely, if ever, performed in field residue trials because of time constraints and cost, as these studies frequently have single sampling events.

A common approach is to collect sample units systematically in a W, X, or Z pattern within the plot until the required sample size is obtained. The sampling locations must be nonbiased. This lack of bias is accom-

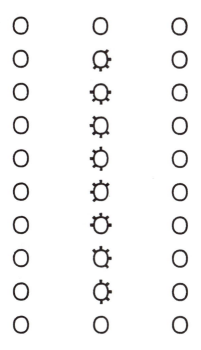

Figure 20.4. Sampling design for a dislodgable residue study on citrus trees using eight trees and offset sampling points.

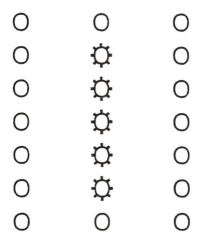

Figure 20.5. Sampling design for a dislodgable residue study on citrus trees using five wholly sampled trees.

plished by collecting a sampling unit at predetermined locations, such as every tenth step or every third tree or by using other, similar means. A major advantage to systematic sampling is that representativeness and freedom from bias can be attained with a smaller sample size than with random sampling because the sample is uniformly collected over the entire plot (3). However, freedom from bias is difficult to obtain with this method unless the trialist uses good judgment and planning prior to sampling.

Conclusion

There are many ways to collect samples that generate reliable study results, allowing much flexibility and creativity in the sampling design. Variables exist in each study design that influence the sampling methodology. However, simple, efficient procedures should be used whenever possible. A well-planned and well-executed sampling strategy will yield non-biased field samples that provide consistent analytical results representative of the test plot.

References

1. "Standard Evaluation Procedure: Terrestrial Field Dissipation", Environmental Fate and Effects Division, U.S. Environmental Protection Agency: Washington, DC, 1989.

2. Iwata, Y.; Knaak, J.B.; Spear, R.C.; Foster R.J. "Worker Reentry into Pesticide-Treated Crops. I. Procedure for the Determination of Dislodgable Pesticide Residues on Foliage"; *Bull. Environ. Contam. Toxicol.*, 18:649–654.

3. Punteven, Werner. *Manual for Field Trials in Plant Protection*; Ciba-Geigy Limited: Basel, Switzerland, 1981; p 28.

RECEIVED for review September 17, 1990. ACCEPTED revised manuscript September 13, 1991.

Processing Studies

Malcolm F. Gerngross

Food Protein Research and Development Center, Texas A&M University System, College Station, TX 77843–2476

The history, growth, and present practices of the processing studies program are related to good laboratory practice (GLP) regulations. The program is responsive to sponsor needs by providing services in research and in support of preparing applications for marketing permits for chemicals regulated by the Environmental Protection Agency. To reduce amendments and deviations, the established processing protocol must complement standard operating procedures and also meet each sponsor's needs. The sponsor is responsible for performing inspections, writing protocols, providing Material Safety Data Sheets, and ensuring that the raw agricultural commodities received by the study facility are in a condition that is appropriate for the study. The responsibilities of the processing facility include observance of GLP regulations and study protocols, realistic simulation of industrial processes, reproducible processing of identical samples, and maintaining confidentiality of the sponsor's interests.

2192-8/92/0249$06.00/0 © 1992 American Chemical Society

Definition and Purpose of Processing Studies

Consumers, regulatory agencies, and chemical registrants benefit from knowing the fate of residues in crops during processing. The objective of processing studies is to process field plot samples in manners simulating industry practices and to provide samples for analyses to determine if the chemical residues are present in the raw materials reaching the processing facility and if such residues are concentrated in product and byproduct streams during processing.

Under the current Federal Insecticide, Fungicide, and Rodenticide Act, processing studies are required to support registration of chemicals intended for use in production or storage of raw agricultural commodities (RACs) intended for later processing into food or feed. Processing studies are classified into the following groups: cereal, legume, forage, root, fiber, tuber, sugar, drug, and oil crops.

Many RACs are consumed as food or feed with little or no preparation, but the majority are subjected to some processing. A processing study is defined as the fractionation of control and treated samples of RACs using an actual or simulated commercial procedure. In commerce, the fractions produced would be used as ingredients for food or feed, but in some cases are the final food or feed. On occasion, although no detectable residues are found in a RAC, residues will be concentrated or will form degradation products in one or more of the fractions. For example, a chemical residue may be detectable in crude corn oil, but not in the whole kernel. This fact may mean that the chemical residue has migrated systemically to the oil-holding structures in the kernel, which are located predominantly in the germ.

History and Growth of Processing Studies

The first processing study at the Food Protein Research and Development Center of Texas A&M University to produce fractions for chemical residue analyses occurred in 1953. At that time, our processing group (known as the Cottonseed Lab) fractionated cottonseed into crude oil and meal. At this stage of processing study development, no documentation was required by the sponsor or then-existing governmental regulatory agencies.

By the mid-1960s, processing studies were required of additional RACs, and sponsors requested that written processing procedures accompany the returned fractions. By the late 1970s, a sample materials balance (an accounting of amount of original sample and amounts of final process fractions), processing flow diagram, and additional fractions were added to the list of requirements. From 1953 to 1985, growth and requirements of processing studies progressed at a slow rate. Chemical company representatives rarely visited the processing facility, and government agency representatives never visited. In 1985, Good Laboratory Practice (GLP) Standards were introduced to the processing facility environment. Word-of-mouth instructions became a thing of the past, and a GLP program with protocols and written standard operating procedures (SOPs) for operation of a specific facility was forthcoming. Although the regulations were not official, we adopted the spirit of GLPs. On October 16, 1989, the finalized GLP regulations became effective, and processing facilities were required to be in compliance.

Sponsor Responsibilities

Ensure That the Facility Operates Under GLP Standards. Representatives of the sponsor's organization should initially visit and periodically revisit the processing facility to ensure that a GLP program is in place and practiced. Because government regulations require the study director to be the focal point of study control, his or her understanding of processing operation is essential for interpreting the final data package and handling situations during the processing study. Personnel from the sponsor's quality assurance unit (QAU) should also satisfy themselves that the facility is in compliance with GLP Standards by performing a data audit and facility inspection. Both the sponsor and processor would benefit if a standardized inspection form were developed and used by all sponsor QAUs for processing facility and data inspections.

Provide a Study Protocol. Government regulations do not specifically state that the sponsor must provide the processing facility with a copy of the study protocol, but a protocol reduces the potential for miscommunication between the sponsor and the processor. A protocol planned by both parties that complements the processor's SOPs should take into account the proposed "experimental start date"[1] and "experimental termination date"[2], special processing conditions, the processing procedure, workable coding and identification systems, the fractions to be returned, and shipping instructions. A copy of the protocol should be received before the samples arrive to allow adequate time for personnel to set the facility's master schedule and to meet other requirements.

---

[1] The EPA defines "experimental start date" as the first date the test substance is applied to the test system.

[2] The EPA defines "experimental termination date" as the last date on which data are collected directly from the study.

Although all chemical companies abide by the same GLP regulations, the use of each sponsor's SOP system in developing the protocol would cause confusion at the processor's location. Processing facility SOPs that have received sponsor approval should be used to develop the processing section of the protocol. For example, the processing facility's packaging SOP should be used to ship processed fractions.

Provide a Typical Raw Agricultural Commodity. Because samples are to be processed as closely to commercial methods as possible, the RAC must be received in a typical and acceptable condition. RACs that are high in moisture content, immature, or of small quantity cannot be completely avoided for reasons such as required harvest dates, prevailing weather conditions, and application of large quantities of chemical. However, samples with broken bags, improper study identification, or no identification should be eliminated. Ideally, one sponsor's representative should be the liaison between the field investigator, the processor, and the analytical laboratory. To avoid receipt and log-in of RACs not intended for processing, field investigators could ship the samples directly to the sponsor for dispersion. Unfortunately, regulatory requirements, geographic location, and sponsor receipt and storage capabilities restrict this practice on occasion.

Provide a Material Safety Data Sheet (MSDS). In reality, the processor is not concerned with the specific test substance because the SOP is followed regardless of the chemical applied. The processor follows safety rules that apply to the physical nature of the process. However, chemical concentrations in the RAC may vary due to the amount applied, elapsed time, and nature of the compound. To reduce potential health hazards of an unknown chemical concentration to the sample handler, a MSDS should accompany each set of samples or each protocol.

Responsibilities of the Facility

GLP Compliance. A complete SOP system suitable for the processing facility, with appropriate GLP situations addressed, must be in place and available for inspection. The major topics to be emphasized in the SOPs are an organizational chart, equipment and building layout, storage procedures, curricula vitae with training records, job descriptions, equipment maintenance logs, master schedule, and processing procedures.

In order for the sponsor to submit the required report for government review, a data package must be submitted to the sponsor after completion of the process. This report shall include a table of contents, the main body of the processing report, a compliance statement, the facility's QAU statement, sample materials balance, original raw data with communication logs, an exact copy of the freezer or cooler temperature records, personnel identification page, sample receipt and shipping records, and a copy of the processing procedure used.

Processing by Typical Commercial Methods. Samples must be processed in a manner typical of commercial methods. In some instances, industrial practice must be simulated because of the great difference in quantities processed and the equipment used. Occasionally, equipment is fabricated to fit the requirements of commercial procedures and still handle small quantities of sample. In many cases, one industrial plant will process a commodity differently from another plant because of equipment variation and proprietary techniques. The GLP processor should know the practice of the commercial majority and develop a similar standard procedure. When different commercial processing parameters exist, the less extreme condition shall be chosen. For example, if a temperature range of 150–165 °F is used in a commercial process, the GLP processing facility shall process at the lower temperature of 150 °F. The less extreme conditions are chosen to reduce the risk of chemical degradation that may be caused by extreme conditions. Periodic review of the industrial procedures is necessary to consider processing changes and the number of facilities making the changes. Normally, facilities within an industry are slow to react to processing changes for various reasons, including equipment cost.

Reproducibility of the Study. The processor must maintain a standard of performance with each study so that data are readily interpreted for future reproduction of the process. Documentation should be completed on standardized forms to minimize the potential for unaccounted time gaps and to enable reasonably informed individuals to follow the processing flow by reviewing recorded raw data. Documentation is not limited to the processing procedure but should encompass all areas from receipt of RACs and protocol to shipment of processed fractions and archiving of data.

Confidentiality. Processing facilities should follow strict standards of confidentiality by not revealing sponsor interests to outside parties. A fine line exists between the business of confidentiality and GLPs. All processing data generated are the property of the sponsor and should not be released to other parties unless permission is granted in writing.

Confidentiality is also practiced when the sponsor is a study management group (an organization, contracted by the chemical company, that essentially performs all duties involved with registration or reregistration of a chemical compound). On numerous occasions, the study management group does not reveal the identity of the corporate sponsor (chemical company) to the processor. The corporate sponsor will find difficulty in obtaining raw data from the processor if the study management group does not complete the study.

Provision of Research and Information Services. The expertise of the study lies with the processor when processing questions arise. The processor should be in a position to retrieve information or provide information to the sponsor and government agencies. The capability to perform special processing assignments and to expand the procedure by providing added fractions or products should be readily available.

GLP Requirements for Processing Facilities

The following procedures are what works for this facility, and these procedures have been agreed upon by the chemical company sponsors.

Processing facilities must abide by GLP requirements:

- The processing protocol must be in order and on hand (one copy for the QAU and one copy for processing personnel).

- The samples are received, and pertinent information (time and date received, number of samples, and identification and condition of samples) must be logged.

- The study must be added to the master schedule by the QAU or some other responsible party. The QAU must maintain a copy of the master schedule.

- A processing notebook containing appropriate forms for raw data entry, a protocol with attached processing SOPs, and any other information pertinent to the study is prepared.

- The in-house QAU must perform random inspections to include at least one critical-phase inspection.

- Processing equipment logs must be used.

- Communication with the sponsor must be documented and incorporated with the raw data.

- Processed fractions or products are shipped to the analytical laboratory.

- A final data package must be prepared and submitted to the sponsor.

- An exact copy of the original submitted data package is archived at the processing facility.

Conclusion

Communication between the sponsors and processors is necessary for GLP Standards to be effective. GLPs should be used as an asset and not as a liability in the quest to register chemical compounds. Front-end costs of GLP implementation and routine maintenance costs will be overridden by long-range benefits. Food and feed safety is an issue requiring more testing, and GLPs will ensure that this work is not in vain. GLPs are relatively new to processing studies and with time shall promote a feeling of security to chemical companies, government agencies, and, most of all, the consumer public. Most of the food consumed in the United States is exposed to some type of processing, and as a result, processing studies will continue to grow.

RECEIVED for review September 17, 1990. ACCEPTED revised manuscript August 17, 1991.

Chapter 22

Livestock Studies

Gino J. Marco[1], Roger A. Novak[2],
and Judy H. Hochman[3]

[1]Marco-Tech, 1904 Tennyson Drive, Greensboro, NC 27410
[2]NPC, Inc., 22636 Glen Drive, Suite 304, Sterling, VA 22170
[3]Compliance Reviews, 8405 Devilbiss Bridge Rd., Walkersville, MD 21793

*If a pesticide is used on a crop that the Subdivision O
Guidelines indicate will provide a feed component, a livestock metabolism study and possibly a feeding study could
be required for that pesticide. Also, a metabolism study
would be required if there is direct treatment with pesticides
of livestock or premises. Key elements for initiating a
metabolism study are specific activity of the radioactive pesticide, dose level, and number of animals. A practical goal
of 70–80% accountability of the radioactive content in
metabolites requires rigorous decisions on chemical strategies. Time-consuming procedures and sophisticated instrumentation are required. Nonradioactive feeding studies can
be required if detectable pesticide residues are found in field
crop studies or if positive results are obtained in metabolism
bioaccumulation studies. Key elements for initiating feeding
studies are the number of animals, dose levels, treatment
length, and sample storage stability. In all the studies, the
concerns for protocols and the role of good laboratory practices are discussed.*

THE INTENT of this chapter is to describe (1) how the regulatory requirements for a study and the Good Laboratory Practice (GLP) regulations interact with and affect the scientific effort, and (2) how the scientific procedures respond to these regulations and guidelines. In the practical sense, this interaction has become one of the more difficult aspects of present day livestock studies.

In the Environmental Protection Agency (EPA) Pesticide Assessment Guidelines, Subdivision O, Residue Chemistry (1), livestock is considered to include cattle, poultry, and swine. For radioactive metabolism studies, the milking goat normally is used to represent ruminants and the laying hen represents poultry. Swine are not usually tested unless metabolites found in poultry or ruminant metabolism studies differ greatly in relative levels or are not seen in rat metabolism studies. For feeding trials, dairy cattle and laying hens are usually used.

Over the years, livestock metabolism studies and feeding trials were called for when detectable residues of concern (usually above 0.01 ppm) were found in feed components from crop field trials. Residues of concern are those that are toxicologically significant and are often determined by structure–activity relationships. The residue levels are detected by an approved crop residue method. Interestingly, improvement of a method's sensitivity could invoke a request for animal studies because residues not previously seen will now be observed. At times, should a potential residue be considered toxicologically significant, animal studies could be requested, even if no measurable residues were found. In the situation in which residues were detected, a multiple of the residue value, sometimes up to 1000×, would arbitrarily be set as the dose level. For radioactive metabolism studies, this procedure often resulted in tissue levels of 10 ppb or slightly higher, permitting some extraction and limited cochromatography. However, these levels rarely were sufficient to isolate, purify, and identify many metabolites. Finally, metabolism studies were, and still are, required for pesticide uses that involve direct treatment of livestock or premises.

The Good Laboratory Practice (GLP) Standards (2) for metabolism studies and field trials contain the same elements as those discussed and established by authors of other chapters in this book. Development of protocols, identification of qualified personnel, calibration of equipment, documentation of effort and procedures, and quality assurance (QA) oversight are needed. The differences result because the facilities, chemistry, procedures, and instrumentation are unique. This chapter addresses these unique situations.

The following components are involved in livestock studies:

1. in-life metabolism studies
2. metabolite characterization and identification
3. residue methods and analysis
4. in-life feeding studies
5. feeding study storage stability and sample analyses

Metabolism Studies

Need for Study. At present, ruminant and poultry metabolism studies could be requested even if very low or nondetectable residues are present in feed. The rationale is to require animal studies for a pesticide used on any crop that provides an animal feed component, as stated in Table II of the Subdivision O Guidelines (*3*). Table 22.I shows the first three crops in this list that have feed components. The totals indicate the number of crops, raw agricultural commodities (RAC), and feeds in Subdivision O Guidelines Table II. If a major plant metabolite is not also found as an animal metabolite, dosing animals with that metabolite could be required (*3*). Thus, animal studies could be required for virtually all pesticides. In this situation, the question of dose level arises, especially if residues are very low or nondetectable. Because the need for animal studies would not be based on a particular residue level in the crop, establishing a specific activity and dose level to produce measurable amounts of metabolites at levels such that major components can be isolated, purified, and identified would be prudent. Later in this chapter, the rationale will be discussed for establishing a low-level (parts per billion) tissue residue by feeding a dose that is low for metabolism studies, yet exaggerated in terms of the crop residue.

Specific Activity. A key decision is specific activity because radioactive labeling is essential in metabolism studies. Radioactivity can be expressed in terms of disintegrations per minute (dpm) or curies (Ci) per mass unit (grams), leading to expressions like dpm/g or Ci/g. The larger the specific activity, the easier it is to find a metabolic component in the biomass. However, very large specific activities can cause problems. First, too high a value can cause possible self-decomposition of the test material, and thus requires constant purity checks (a GLP concern). Metabolites, in turn, would also have high specific activities and cause the same concerns as

Table 22.I. Raw Agricultural Commodities and Feeds Derived from Field Crops

| | | | | Percent of Livestock Diet | | | | | |
| | | | | Cattle | | Poultry | | Swine | |
Crop	Raw Agricultural Commodity	Processed Commodity	Feed	Beef	Dairy	Broiler Turkey	Laying Hen	Boar and Sow	Finish Animal
Alfalfa	forage	meal	forage	50	80	NU[a]	NU	NU	NU
	seed		seed	30	10	20	20	10	10
	hay		hay	25	80	NU	NU	NU	NU
			meal	25	80	5	5	50	5
Almond	nutmeat hulls		hulls	25	25	NU	NU	NU	NU
Apples	fruit	wet pomace dry pomace juice	dehydrated pomace	50	25	5	NU	NU	NU

NOTE: The crops are three of a total of 48; raw agricultural commodities are 6 of a total of 111; and feeds are 6 of a total of 141 in Subdivision O Guidelines, Table II.
[a]NU means nonfood use.
SOURCE: Data are from references 1 and 3.

those for the dose material. Second, costs increase with higher activity. Increased costs would encourage stringent use of the synthesized test material, and study strategies could be minimized in both use of test compound and in the depth of study. For example, high specific activities can help discern low metabolite levels in the tissue, a feature that encourages the use of comparative chromatographic systems rather than isolation techniques using spectral identification. Finally, high specific activities, along with high dose levels, could result in exceeding the animal testing facility license for total radioactivity.

Dose Level. Another key decision is the dose level that will be used because crop residues may not exist to help set a level. Up to some point, animal tissue residue levels normally are proportional to the dose level. At a dose level of 100 ppm or higher in the feed, the residue incorporated could reach a plateau. Very high doses could cause pharmacological effects, altering metabolic pathways and even decreasing tissue residues. One method is to arbitrarily use 10% of the chemical's LD_{50} for the test animal as the high-level dose. At 10%, a 50-ppm feeding level would represent a 500-ppm LD_{50}, or 1.5 mg/kg for 50-kg goats and 4.0 mg/kg for 1.5-kg chickens, as shown in Example 22.I. For chemicals with low LD_{50} values, the use of 10% of the LD_{50} could result in a high-dose level that is too low, making observation and isolation of tissue metabolites difficult, if not impossible. A second method for setting the high-level dose is to establish the chemical's maximum tolerated dose by measuring some pharmacological response, such as equilibrium or weight loss, rather than death. Thus, the use of a low dose, such as 5 ppm or less, could be justified by using as high a specific activity as possible. Also, test animals should not be predosed with nonradioactive test material because the result could be overloading of tissues with nonradioactive compound (or metabolites); if turnover rates are slow, the radioactive dose would not equilibrate and give a true value for radioactive residues in tissues. This may be the reason that the EPA's Phase 3 Technical Guidance document [tracking number 171–4(b)] (3) specifically states that no preconditioning of the animal with nonradioactive test material should be done. In Example 22.I, the result of dosing a test animal with a compound that has a specific activity of 20 μCi/mg at 50 ppm in feed is shown. A reasonable assumption is for 1% of the dose to appear in milk or eggs. This example shows a 0.5-μg/g (500 ppb) residue mass or 22,000-dpm/g radioactive content. These levels permit at least a reasonable attempt to identify major metabolites. The larger the sample extracted (e.g., 50–100 g) the better

Example 22.I. Relationship of Dosage Characteristics to Tissue Residue

- **Dose:** 20 μCi/mg = 4.4 × 10^7 dpm/mg = 4.4 × 10^4 dpm/μg

- **LD$_{50}$:** 500 ppm × 0.1 = 50 ppm

- **Chicken:** weight, 1.5 kg; feed, 120 g/day
  6.0 mg/120 g × 10^3 μg/mg = 50 ppm *or* 4.0 mg/kg/chicken
  6.0 mg × 4.4 × 10^7 dpm/mg = 2.6 × 10^8 dpm/chicken/day

- **Goat:** weight, 50 kg; feed, 1.5 kg/day
  75 mg/1500 g × 10^3 μg/mg = 50 ppm *or* 1.5 mg/kg/goat
  75 mg × 4.4 × 10^7 dpm/mg = 3.3 × 10^9 dpm/goat/day

- **Residue:** Assume a 1% dose in egg or milk residue; 1% = 0.5 ppm (500 ppb) = 0.5 μg/g.
  Radioactive content is 0.5 μg/g × 4.4 × 10^4 dpm/μg = 22,000 dpm/g.
  A low 5-ppm dose in feed equals 0.05 ppm in milk or egg or 2200 dpm/g.

the chances of identifying major metabolites. Often, the detection limit of crop residue analysis is 0.01 ppm, so the low 5-ppm dose in Example 22.I still represents a 500× exaggerated dose, assuming 100% of that feed in the animal diet.

Establishing the specific activity and dose level is an area that quite frequently leads to protocol amendments or deviations. For example, using too low a level for specific activity or dose could require redosing at higher levels than the protocol states and result in additional time and costs. This problem can be prevented by performing preliminary pilot and range-finding studies, which do not necessarily require QA oversight unless the results are to be included in the final report. For small animals such as poultry, administering five doses over a range of 1–50 ppm to one or two animals would be reasonable. Large animals, such as goats, are costly and often difficult to obtain; so the initial goat could be used to establish the need for an altered dose. A protocol amendment should suffice in this case.

Study Size. A critical starting point for study design is the number of animals required for metabolism identification. A dominant consideration is the availability and cost of the radioactive chemical. Depleting supplies with false starts could require additional syntheses and thus cause increased costs and lost time.

Table 22.II shows some typical sample weights and amounts obtained from a 1.5-kg chicken and a 50-kg goat. When many metabolites are present, 70–80%, at best, of the radioactive residue can be identified and characterized. For a reasonable attempt, 200–400 g of tissue containing at least 10,000 dpm/g or 230 ppb of radioactive material is desired. Again, a pilot study run with one or two chickens would be wise as a preliminary screen for the number of metabolites in eggs and liver. Sufficient tissue can be obtained from one goat, and if needed, a second animal at an adjusted dose could be readily added. If only several major metabolites are found in edible tissues, then 10 chickens and one goat would be the minimum number necessary for a study, assuming a 20-μCi/mg, 50-ppm dose. Another concern is the availability of milking goats and laying hens; acceptable animals cannot always be obtained when the study initiation date arrives. Again, changing judgments in this area can lead to protocol amendments or deviations, as can problems with animal health during the study. Poorly performing, unhealthy animals can necessitate reinitiation of a study, possibly requiring a protocol amendment.

This biological step is often referred to as an in-life study and is often

Table 22.II. Typical Animal Sample Amounts

Sample	Sample Weight per Animal	
	1.5-kg Chicken	50-kg Goat
Fat	20	600
Gizzard	20	—
Heart	7	225
Kidney	9	175
Liver	33	850
Skin	15	—
Whole egg (daily)	40	—
Preformed yolk	30	—
Thigh muscle	60	—
Breast muscle	120	—
Leg muscle	—	750
Loin muscle	—	350
Morning milk (daily)	—	1400
Evening milk (daily)	—	1100
Excreta (daily)	150	—
Urine (daily)	—	1000
Feces (daily)	—	1500

NOTE: All values are given in grams, except for milk and urine values, which are in milliliters.

done at a facility separate from that where the chemical analyses will be performed. Thus, a separate protocol with all the GLP and QA requirements applies to in-life studies as well as analytical studies. Early QA involvement is important to verify dose calculations, animal identification and care, and proper administration of the test substance. The QA officer ensures that all standard operating procedure (SOP), protocol, and GLP requirements are met.

At the start of the study, the QA officer verifies that the required number of animals are placed in the study and that they are of the appropriate weight and sex for milk or egg production, are identified in accordance with the study protocol and laboratory SOPs, and are properly housed and receive proper care. A detailed checklist of phases to be audited or inspected is shown in Figure 22.1, and a general list follows.

1. protocol preparation
2. test animal quarantine, randomization, housing, and identification verification
3. test substance preparation
4. test substance administration
5. test system observation and sampling
6. sample extraction and combustion
7. analysis
8. necropsy
9. data review
10. report audit

Sample collection and animal-dosing procedures should be observed at least once by the QA officer. Samples are generally collected before each dosing. At this time, the QA officer verifies sample container labeling, measuring or weighing of the sample, methods used to prevent sample contamination, and sample storage. During dosing, the QA officer verifies that each dose group receives the correct dose level, which is usually administered in the form of a gelatin capsule. The dosing procedure is monitored for compliance with company SOPs. In addition to these verifications, the raw data documentation should also be checked to ensure that all methods have been described and that all information on animal husbandry, observations, dosing, and sampling has been supplied. Finally, the process of tracking the samples from facility to facility is important and must follow GLP guidelines.

Metabolites. The terms *characterization* and *identification* will now be defined. *Characterization* means to establish chemical structure by specific reactions (chemical or enzymatic), physicochemical properties, or cochromatography against reference compounds, and often involves a combination of these methods. *Identification* means to establish chemical structure by isolation, purification, and use of analytical instrumentation,

Dates of Inspection: _____   Performed By: _____

ORGANIZATION AND PERSONNEL	YES	NO
Personnel trained/experienced		
Records of training/ experience maintained		
Sufficient no. of personnel		
ANIMAL CARE		
Facility has animal rooms/ areas to assure:		
o separation of different species, test systems		
o isolation of different projects		
o quarantine of animals		
o routine/specialized housing		
o isolation of studies with biohazardous, volatile aerosols, radioactive or infectious agents		
o diagnostic treatment & control of experimental animal diseases		
Facilities exist for collection/disposal/storage of animal waste		
SOP for animal care/ID		
ANIMAL SUPPLIES		
Specified separated storage areas		
Protected against infest./ contamination		
Refrigeration, if necessary		
TEST AND CONTROL SUBSTANCE HANDLING		
There are separate areas for:		
o receipt/storage		
o mixing with carrier		
o storage of mixture		
Storage area for test/control articles/mixtures separate from test system housing		
Reagents/solutions labeled for ID, storage, expiration, concentration		

TEST AND CONTROL SUBSTANCE HANDLING (CONT'D)	YES	NO
Storage containers properly labeled		
Reserve samples (ea. batch retained		
Proper storage as per SOP/protocol		
Distribution designed to preclude contamination, etc.		
Proper ID throughout distribution		
Receipt & distribution of each batch is documented		
TEST SAMPLE HANDLING		
o Proper labeling		
o Proper storage separate from test substances		
o Chain of custody forms completed		
EQUIPMENT-MAINT. & CALIBRATION		
All equipment used in generating data and equip. used for facility environment control is/was:		
o inspected, cleaned maintained		
o tested, calibrated/ standardized		
o detailed SOPs for above		
o written records for above		
o designated responsible person		

Figure 22.1. Detailed quality assurance unit checklist.

such as IR, NMR, and mass spectroscopy, ideally by comparison with a synthesized standard. The synthesis facility providing the reference metabolite should follow GLP guidelines in metabolite preparation and chain-of-custody transfer. As already indicated, the degree of characterization and identification depends on concentration in the tissue and the amount of tissue available.

Figure 22.2 shows a decision-making scheme for metabolite identification based on practical laboratory techniques and potential technical difficulties (*3*). The gray zone is the 10–50-ppb level. Separation into polar, nonpolar, and nonextractable components is possible, preferably using neutral, cold, and nonoxidative conditions. Some cochromatography could be possible, but the chances of characterizing 70–80% of the radioactive content are limited. Of course, the greater the concentration above 50 ppb, the greater the chances of identifying a major portion of the metabolites and giving a convincing discussion on 70% or higher of the residues. Analyzing the nonextractable fraction, if of a high value, is a step for which inventiveness and sophisticated techniques are required to release and identify the true unaltered metabolites. In writing the protocol for this effort, general statements and flexible approaches should be used to minimize protocol amendments and deviations. As part of the protocol, a sample fractionation scheme (Figure 22.3) with general procedures is a useful approach because it allows the results to dictate specific procedural details as the laboratory work proceeds.

Information on the possible bioaccumulation of the chemical is also desired. To establish bioaccumulation, a common approach is to consecutively feed the chemical for 3 days. A preferred approach is to measure the accumulation of radioactivity in some key tissue, such as milk or eggs, and to stop dosing when a plateau is reached. Simple methods for estimating the plateau are determining when two consecutive values are within 10% of each other or using the Student *t*-test to determine that consecutive values are not significantly different from each other. Normally, feeding chickens for 7 days and goats for 5 days will accomplish a radioactive plateau in tissues. Also, the normal accepted sacrifice time is within 24 hours after the final dose. The total radioactive residue is measured in the edible tissue and milk or eggs. This residue is expressed both as a percent of dose and as parts-per-million equivalent to the compound dosed. These values serve as a guide in the metabolite determination as shown in Figure 22.2.

Tissue sample collection at the time of sacrifice is a critical phase and should be monitored by the QA officer. During the sample collection

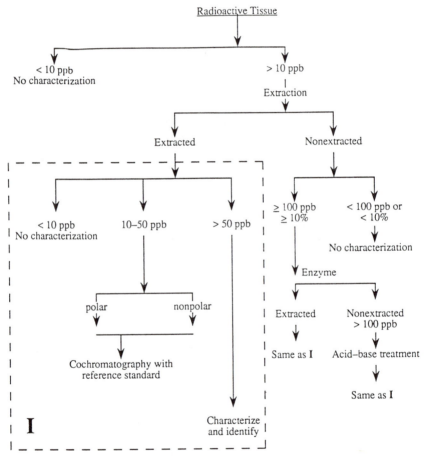

Figure 22.2. Decision-making scheme for metabolite identification.
(Adapted from ref. 3.)

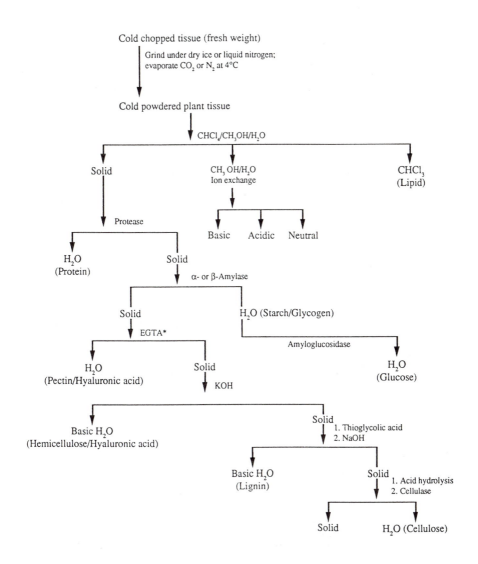

* Ethyleneglycol-bis-2-(aminoethyl)-N,N'-tetraacetic acid at pH 4.5

Figure 22.3. Fractionation scheme for plant and animal material.

phase, the QA officer ensures that the animals are sacrificed in the proper order (normally controls first and high dose last to prevent contamination), that the appropriate tissues and tissues amounts are collected, and that the samples are properly labeled and stored until analysis. If samples are to be shipped to another facility for analysis, sample tracking documents must also be checked. Documentation of necropsy data is also reviewed. This review of data includes time of sacrifice, tissues collected (including amount and order of collection), and gross pathological observations.

Again, changes in length of feeding and sacrifice time are common protocol amendments or deviations. Feeding chickens for 7 days and goats for 5 days with sacrifice no less than 24 h after the last dose should minimize the need for amendments or deviations. Careful attention to sacrifice order and sample collection and labeling can also minimize deviations.

Residue Methods

For a new registration, the definition of metabolites in key tissues serves to guide the residue analyst in developing an appropriate enforcement method for significant metabolites.

For reregistration, the definition of metabolites serves to establish the need for modification of the existing method. Method improvement, by achieving higher accountability, could increase residue values or even require the addition of other compounds to the analysis. A comparison of the accepted residue analysis procedure on the radioactive tissue to the metabolites found in that tissue serves to establish any changes required.

The residue method to analyze tissues in feeding trials must be both validated and sensitive (often ≤ 0.01 ppm). Modification of the procedure may be required for different matrices. The residue analysis area must follow GLP guidelines for such steps as instrument calibration, the use of proper controls, and proper data transfer. Perhaps the analytical phase is the most difficult and complex in terms of QA verification. Not only should the analytical procedures be verified against the analytical method, but also all sample receipt information, calibration and maintenance logs, and test and analytical reference substance receipt and usage logs must be examined. Reagents and solutions must be properly labeled. Analytical procedures must be accurately documented. Calibrations must be performed for instruments such as balances, gas chromatographs, and high-performance liquid chromatographs. Efficiencies must be determined. The residue method must also be used in tissue storage stability studies.

Feeding Trials

Crop field studies, in which the proposed use level of the pesticide is applied, can establish the need for nonradioactive animal feeding studies. If the crop field studies show detectable residues at maximum rate and minimum preharvest intervals, animal feeding studies are needed, as noted in 40 *Code of Federal Regulations* (CFR) §158.240. A common treatment regime is to use part-per-million levels that are 1, 3, and 10 times that of the crop residue value and to include a control group. Another result that could cause feeding studies to be required is finding a significant level of bioaccumulation in the metabolism studies.

Finding very low residues in crop field studies could require evaluation of the metabolism information for decisions on the need for feeding trials. As noted in Example 22.I and Table 22.III, feeding the test substance in the diet at 5 ppm would represent 500 times the low crop resi-

Table 22.III. Feeding Trial Dosing

Crop Residue (ppm)	Feed in Diet (%)	Dietary Burden (ppm)	Dose in Feed (ppm)	Dose/Burden	Tissue Residue
0.01	100	0.01	5	500×	—
0.1	10	0.01	10^a	1000×	0.1 ppm[b]
Feeding Crop Treated at Label Use Rate					
0.1	10	0.01	0.01	1×	0.1 ppb[c]

[a]Data from ref. 3.
[b]Assumed.
[c]Calculated (tissue residue/1000).

due value of 0.01 ppm if the particular crop was used as 100% of an animal's diet. Also shown in Table 22.III is a situation in which a feed that makes up 10% of an animal's diet contains 0.1 ppm of pesticide residue. In this situation, the dietary pesticide burden is 0.01 ppm. If the pesticide is administered in a feeding trial at a level of 10 ppm, the feeding level is equivalent to 1000 times the actual crop residue, and this is the feeding trial dose level proposed by the EPA (*3*). The percentage of pesticide in the diet is calculated on a dry-weight basis. The protocol should state the rationale used in setting the dose level.

If, in a metabolism study, a 10-ppm dose level (representing 1000 times a 0.01-ppm dietary burden) would result in residues in milk, eggs, or meat of 0.1 ppm, an assumed linear relationship would represent a level of 0.1 ppb in tissue samples as a result of feeding crops treated at label use rates. Thus, a tolerance would probably not be required. The metabolism study, in effect, becomes a feeding study, as noted in Category 3 of 40 CFR §180.6(a). Thus, a metabolism study at a 10-ppm dose, as well as at 50 ppm, could be a good investment in time and money, especially if a feeding trial could be eliminated. Also, minor metabolites present in noticeable amounts at the 50-ppm level might not be detectable at 10 ppm, thus lessening structure identification efforts.

For a residue feeding study, normally 3 dairy cows and 10 laying hens are used in each dosage group. The treatments should last for a sufficient length of time to ensure that residue levels reach a plateau in milk or eggs, usually 4 weeks. Again, animals are usually sacrificed within 24 h of the last dose. The validated residue method should be used for measuring residues in edible tissue (muscle, fat, liver, and kidney) and in milk and eggs, as noted in 40 CFR §180.6. Normally, samples must be stored, so storage stability data must be determined for those samples.

The facilities for animal feeding studies have unique GLP requirements, such as quarantine, care, and treatment of animals, and safety and maintenance. Handling large numbers of animals, disease control, sacrifice and sample timing, sample tracking, and study integrity in general require extensive QA oversight. Protocols detailing these procedures must be clearly written.

Study Outline and Report

An outline for the in-life phase of a livestock study that would apply to both ruminants and poultry is shown in Figure 22.4. These topics could also serve as the sections of a protocol. The items under each topic are meant only to highlight key elements to consider in each section; no details are described. The language in each section should define what is to be done but should also be written with flexibility in mind. Overabundant details and specifics results in continual protocol amendments and possible deviations. Also, under item 2 in Figure 22.4, "Diet", the feed should be certified, indicating that the test chemical has not been used on any of the feed ingredients. The test chemical and doses, test samples, and control samples should be stored in separate freezers or at least on separate shelves within the freezer. This requirement applies to both the metabolism and feeding studies. Because these sections of the outline could be used for a protocol, a protocol checklist is given in Figure 22.5.

The contents of a typical metabolism report contain a number of items, and the major issues for each item follow:

1. The cover page shows the title of the study, the sponsor, the facility where the study was done and the dates of the study, as described in Pesticide Registration (PR) Notice 86–5.

2. Confidentiality is based on a "Statement of Data Confidentiality Claim" within the scope of FIFRA S10(d)(1)(A, B, and C).

3. Compliance statements certify that GLPs were followed and include dated signatures per 40 CFR Part 160.

4. The quality assurance statement includes inspection dates signed by QA officers, as set forth in 40 CFR §160.35(b)(6 and 7).

5. The study identification section lists test material, sponsor, study monitor, study director, and study timetable.

6. The certification of authenticity is the certification by the applicant (study sponsor) that the report is a complete and unaltered copy of the report provided by the testing facility.

7. Study personnel are the study director, principal investigator, main researchers, QA officer, and safety officer.

8. The table of contents includes page numbers of the sections of the report.

9. The summary includes the chemical structure and name of the test material with the site of label, specific activity, detection limits, and radiochemical purity; describes diet and dosing procedure; provides

1. Animals (test system)
   - number of animals
   - acclimation to facilities
   - weights and selection routine
   - identification

2. Diet
   - commercial feed description and feeding method
   - water availability and mineral adjuncts
   - feed quality (e.g., noncontaminated)

3. Test chemical
   - source
   - specific activity and purity of test material
   - specific activity adjustments for dosing

4. Animal handling
   - cage design and acclimation
   - dose administration, including dose calculation, dosing schedule, and dose route and means (e.g., oral capsule)

5. Sample collection
   - excreta
   - eggs or milk
   - blood

6. Sacrifice and tissue collection
   - time of sacrifice
   - tissue handling (e.g., dicing or subsamples)
   - tissues taken

Continued on next page.

• sample shipping description

• sample destination

7. Report

• type of report and review process description

• maintenance of records

8. Quality assurance

• quality assurance process

• inspections and signatures

• official protocol and addenda

9. References

• technical

• regulatory (*see* refs. 1–3)

Figure 22.4. Outline of a livestock study.

evidence that the test chemical is not in control feeds; includes method and time (hours after dose) of sacrifice, procedures to quantitate and extract radioactivity, and procedures to identify radiolabeled residues; provides evidence of the ability of the enforcement method to recover residues; and lists the tolerance petition number. These summary items are listed in Phase 3 Technical Guidance Tracking Number 171–4(b) (*3*).

10. The introduction gives the background, registration history, and proposed use of the pesticide, and the study purpose.

11. The materials section names test substances, facilities, and animals.

12. Methods include dosing preparation and procedure; sample collection, handling, and storage stability; and radiotracer techniques, extractions, and characterization procedures.

13. The results and discussion section describes quantitation, distribution, and characterization of metabolites along with accountability of analytical methods and metabolic pathways.

```
Project Number:_____   Date of Protocol Evaluation:_____

Study Director:_____   Inspector:_____
              (Signature/Date)                 (Signature)
```

Reference:	EPA GLP, Section 160.120, 792.185

	Yes	No	PROTOCOL FOR THE ABOVE STUDY CONTAINS THE FOLLOWING
a(1)			1. Descriptive title and statement of purpose of the study.
a(2)			2. Identification of the test, reference, and control substances by name, chemical abstract number (CAS) or code number.
a(3)			3. Name and address of sponsor.
a(3)			4. Name and address of testing facility.
a(4)			5. Proposed experimental start and termination dates.
a(5)			6. Justification for selection of test system.
a(6)			7. Number, body weight range, sex, source, species, strain, substrain, and age of test system, where applicable.
a(7)			8. Procedure for identification of test system.
a(8)			9. Description of the experimental design including the methods for the control of bias.
a(9)			10. Where applicable, description or identification of diet and any material used to solubilize or suspend the test or control substance before mixing with carriers.
a(9)			11. Where applicable, specifications for acceptable levels of contaminants that are reasonably expected to be present in the dietary material and known to be capable of interfering with the purpose or conduct of this study, or a statement that there are no such contaminants.
a(10)			12. Route of administration and the reason for its choice.
a(11)			13. Each dosage level of the test, reference or control substance to be administered, expressed in appropriate units.
a(11)			14. Method and frequency of administration of test, reference or control substance.
a(12)			15. Type and frequency of tests, analyses, and measurements to be made.
a(13)			16. The records to be maintained.
a(14)			17. Date of approval of the protocol by the sponsor.
a(14)			18. Signature of the study director.
a(15)			19. Proposed statistical methods used.

Figure 22.5.   Quality assurance protocol-evaluation form.

14. The conclusion gives a brief discussion of findings, significant metabolites, study problems and deviations, residue levels that must be regulated, and ability of the enforcement method to recover residues.

15. Tables and figures support the discussion and conclusions.

16. References and appendices include technical papers, regulatory sections, protocols and amendments, test material data, calibrations, copies of typical raw data, and other pertinent materials.

Conclusion

The following livestock studies require protocols and must comply with GLP regulations:

1. in-life metabolism studies

2. metabolite characterization and identification

3. residue methods and analyses

4. in-life feeding studies

5. feeding study storage stability and sample analyses

Each study involves unique facilities, procedures, and equipment. The protocols covering each study should contain sufficient detail to guide the researcher and ensure achievement of the study objective. However, overly elaborate details will result in many protocol amendments and probable deviations. For livestock studies, the GLP concerns involving protocols, SOPs, chain of custody, facilities inspections, QAU functions, data auditing, computer validation, and reports are generally similar to other scientific studies.

Four documents from the EPA cover, in greater detail than in this chapter, questions for data reviewers on submittals for analytical methods; storage stability; metabolism in food animals; and residues in meat, milk, poultry and eggs (*4–7*). For the data submitter, the points addressed in these documents will aid in providing the data of interest to the EPA.

References

1. "Pesticide Assessment Guidelines", Subdivision O, Residue Chemistry, U.S. Environmental Protection Agency, Office of Pesticide and Toxic Substances, Washington, DC: EPA 540/9–82–021, October 18, 1982. NTIS PB83–153973.

2. "Pesticide Program; Good Laboratory Practice Standards, Final Rule", *Code of Federal Regulations* Title 40, Pt. 160; *Federal Register* 48:230, November 29, 1983, and 54:158, August 17, 1989; Amended Rule Title 40, Pt. 160, October 16, 1989.

3. "FIFRA Accelerated Reregistration Phase 3 Technical Guidance— Attachments 1, 2, 3, and Tracking Number 171–4 (a–i)", U.S. Environmental Protection Agency, Washington, DC, EPA 540/9–90–078, December 24, 1989.

4. Nelson, M. J.; Griffith, F. D. "Health Effects Division Standard Evaluation Procedures", U.S. Environmental Protection Agency, Washington, DC; Analytical Method(s); Accession No. PB 90–103284; Report No. 540/9–89–062; August 8, 1989.

5. Nelson, M. J.; Griffith, F. D. "Health Effects Division Standard Evaluation Procedures", U.S. Environmental Protection Agency, Washington, DC; Storage Stability Study; Accession No. PB 90–103276; Report No. 540/9–89–063; August 8, 1989.

6. Nelson, M. J.; Griffith, F. D. "Health Effects Division Standard Evaluation Procedures", U.S. Environmental Protection Agency, Wasshington, DC; Metabolism in Feed Animals: Qualitative Nature of the Residue; Accession No. PB 90–103292; Report No. 540/9–89–061; August 8, 1989.

7. Loranger, R. A. "Health Effects Division Standard Evaluation Procedures", U.S. Environmental Protection Agency, Washington, D;. Residues in Meat, Milk, Poultry and Eggs: Feeding Studies/Feed-Throughs; Accession No. PB 90–208943; Report No. 540/9–90–087, April 17, 1990.

Received for review September 17, 1990. Accepted revised manuscript September 13, 1991.

Chapter 23

Specialized Field Testing
Worker Protection

D. Larry Merricks, Donna H. Merricks, and William C. Spare

Agrisearch Inc., 26 Water Street, Frederick, MD 21701

Worker-exposure testing (mixers, loaders, and applicators and clean-up and reentry workers) presents unique data-collection and quality assurance (QA) problems. A major requirement of exposure testing is accurately modeling the correct agricultural work practices. Evaluating each participant's work history and job responsibilities during exposure is critical for obtaining valid samples. Tracking the activities of human subjects and relating those activities to exposure samples and analytical results requires a team effort in field data collection using predetermined sample and personnel numbering systems. Evaluating chemical drift from aerial and air-blast pesticide applications requires the same type of preplanning and sample tracking with preassigned sampling locations. Video and photographic records, with field notes and on-site observations, are discussed within the framework of specialized QA problems for field testing.

MONITORING FIELD WORKERS FOR EXPOSURE to agrochemicals requires special consideration. When working with human subjects, the wide variety of agricultural practices and the large array of commercial formulations must be considered, as well as all of the problems commonly associated with field testing, such as logistics, timing, weather, and equipment. This chapter explains how Agrisearch Inc. designs and conducts studies to comply with the applicable Environmental Protection Agency (EPA) standards (1, 2). We have conducted worker-exposure testing since 1984 for workers handling agrochemicals and for workers reentering previously treated areas. Data have been collected from approximately 800 workers in more than 30 separate testing programs. In each program, Good Laboratory Practice (GLP) Standards (3) have been incorporated to ensure proper data collection and to ensure that the true agricultural practice has been defined in the protocol, performed during the study, and presented in the final report. The discussion in this chapter pertains only to the EPA registration testing and does not necessarily comply with California Department of Food and Agriculture (CDFA) testing requirements. In fact, CDFA does not require field testing to be performed according to GLP guidelines. Other notable differences between the EPA and CDFA requirements are the number of test subjects required and the text of informed-consent forms. Also, CDFA does not require retention of empty test article containers. CDFA requirements pertain only to testing performed in California. Manufacturers generally follow the EPA testing guidelines for studies performed outside of California and seek CDFA approval of the study protocol design.

The quality assurance unit (QAU) is involved in each study program from review of the protocol through study conduct and final reporting. Throughout each study, the QAU reviews documents and documentation; laboratory analyses; field site design, data-collection, data-reduction, and reporting procedures. These areas of QAU involvement are discussed in many places within this book; however, there are unique auditing and review requirements in field-exposure testing that will be discussed in this chapter.

Study Design

Current or Proposed Formulation Label. To meet the general testing requirements of the EPA, the maximum-use rate for a pesticide formulation must be used for human exposure monitoring. The formulation label defines use rates for the test substance for all registered crops, with statements describing frequency and timing for multiple applications.

The label is generally incorporated into study protocols and therefore becomes a document for review by the QAU as it relates to the protocol. QA review is difficult at times because labels are not written for audit but for field use by pesticide applicators. For example, a label may call for multiple applications at 7-day intervals. If rain delays the study for 1 day for an application on day 8, this delay may be considered a deviation from a protocol that defines application according to label directions. However, the intent of the label is 7 days or more; to a pesticide applicator, 8 days would be viewed as normal agricultural practice, but application on a 6-day interval would be considered a deviation. To avoid such confusion, the QAU must understand agricultural practices and become familiar with the formulation label. Also, the protocol must be designed with inherent flexibility to allow for the many changes that may occur in field operations during the life of a study.

The label also defines conditions that are important for chemical formulation receipt and storage at the laboratory and in the field. At the laboratory and field storage site, the temperature and relative humidity should be continuously monitored. This monitoring must be verified by the QAU from the data and from audits conducted in the field. A parallel concern is receipt, storage, and handling of analytical standards (a.i.) for use in the laboratory for storage-stability testing and for analytical standards sent to a field site for use in field fortifications. The QAU must understand distinctions among formulations, standards, and field-collected samples (treated and control) and the storage and handling needs of each.

Often, a label combines special handling requirements and protective-equipment statements, and those requirements must be considered in the protocol design and when defining the agricultural practice. If a label defines particular protective equipment, exposed workers must use that equipment. If the study objective is to help define protective equipment requirements for a new formulation, worker exposure must be measured using a variety of protective equipment. That equipment must meet at least minimal protection standards and be acceptable to the sponsor and to the EPA or the CDFA. The QAU must be aware of the impact that subtle differences in the use of protective equipment may have on the study design. Again, a clear understanding of the agricultural practices involved is essential.

The label defines the acceptable modes of application for the formulation. For example, ground boom or aerial application for many field crops, air-blast or aerial application for tree crops, and high-pressure sprayer or individual hand-held or backpack units for greenhouse use are typical pesticide application modes. Generally, the mode of application and the type of equipment to be used define the potential exposure and must be addressed in the protocol. The QAU must be capable of relating the label-imposed equipment requirements to the protocol. Once the study-specific equipment has been determined, calibration methods can be defined. The QAU should be called upon to help identify the data-collection needs for each study and to help develop associated data-collection forms. A form is used to record application-equipment specifics that make each piece of equipment unique (Figure 23.1).

AGRISEARCH LABORATORIES, INC.
SPRAYER SPECIFICATION SHEET

Project Number: _____ Date: _____ Initials: _____

Site: _____

Method Of Application:
1) Airblast   6) Aerial - Rotary Wing   10) Shank Injection
2) Groundboom Tractor   7) Low Pressure Hand Wand   11) Fumigation
3) Groundboom Truck   8) High Pressure Hand Wand 12) Solid Broadcast Spreader
4) Groundboom Rail Car   (attached to truck)   13) Liquid Broadcast Spreader
5) Aerial - Fixed Wing   9) Backpack   14) Other (describe)

If 14, describe _____

Formulation Added _____ Water Added _____ Tank Size _____

Rate (lb ai/acre)_____ and (gal/acre) _____ Total lb ai applied _____

Total # Acres Treated_____ Final Mix Concentration (lb ai/gal Diluent)_____

Total # Gallons Sprayed_____ Tank Applications Monitored_____

Sprayer Make and Model_____

Pump Make and Model_____

Cab Type:
1) Open Cab(Cockpit)   3) Closed Cab(Cockpit)Window Closed
2) Closed Cab(Cockpit)Window Open   4) Closed Cab (Cockpit)Window Closed/Filtered Air

Ground Speed (mph)_____ Vehicle Make and Model_____

_____

Disc #_____ # Nozzles_____

Nozzle: Type/Manufacturer_____ Model No._____

Nozzle Pressure (psi)_____ Boom Height Above Ground (feet)_____

Swath Width (feet)_____ Length of Boom (feet)_____

Shank/Probe Type: 1) Knives   2) Sweeps   3) Other

Depth of Injection/Incorporation (inches)_____

Describe Application Procedure_____

_____

Figure 23.1. Sprayer specification sheet.

Protocol Approval. Protocol development begins with discussions between the testing laboratory and the sponsor about the purpose, location, timing, current agricultural practices, and QA requirements of a study. Subsequent testing laboratory contact with cooperators assists in determining necessary field functions and conditions such as seasonal timing, growth-stage verification, application modes and frequency, potential worker tasks to be monitored, and specific crop requirements. Therefore, agricultural practices define the key sites for potential worker exposure. The model for testing exposure and data collection must not interfere with worker job performance in actual field practice. At this time, the QAU must review the standard operating procedure (SOP) for each experimental function to determine the appropriateness of the SOP to the field-exposure model. Special attention is required for sample identification, sample media placement, collection of samples from the workers, procedures to prevent cross-contamination, sample handling and storage in the field, sample shipment to the laboratory, and sample analysis.

When the protocol and study design have been developed, the EPA is asked to review the protocol. Often a meeting is held between the EPA, the sponsor, and the testing laboratory to complete the protocol design process. Study timing is determined by the EPA requirements and normal agricultural practices. A quality control checklist ensures that the written protocol is a workable plan that meets the needs of the sponsor, the EPA, and GLP requirements within a valid agricultural model. This protocol review is performed by the QAU, and any discrepancies are discussed with the study director. Finally, the EPA should review and approve the study protocol.

Test Sites. Two important aspects of the test site that must be addressed in the protocol and recorded in the study notebook are location and size. The test site must be in the proper geographic area to model use in the major sales markets for the formulation. That is, a chemical that is used mainly in Iowa and Illinois should not be tested on corn in Florida. The test site must also be of adequate size. Because worker-exposure studies involve human subjects performing a specific task, the size of the treated area (acreage, greenhouse space, or poultry buildings) must be sufficient to provide adequate exposure. For example, if a worker takes 2 hours to spray a 10-acre block of trees and exposure of 15 workers must be measured, then at least 150 acres of trees must be provided for the test site. Worker exposure must also be detectable and quantifiable. That is, if exposure is measured for 2 hours, the theoretical exposure during that time must be above the analytical detection limit, otherwise exposure assessment is not possible. In summary, the test sites must be in areas of commerce for the formulation, must use the correct crop for application, and must be of sufficient size to ensure the validity of analytical exposure data.

Current agricultural practices must also be considered when selecting a test site. The formulation label will specify the mode of application (e.g., aerial or air blast), which in turn will provide guidance for equipment use. Different geographical areas will often dictate a commonly used application procedure. Therefore, a use history for the test site should be obtained. If the normal practice of growers in an area is to use aerial application, exposure of workers using ground-boom or air-blast sprays should not be monitored. Site visits during protocol development and before study initiation will help to ensure that the proper agricultural practices are understood and will be monitored. The QAU must be a part of this understanding in order to evaluate the testing process and perform data audits.

Test Substance. The quantity of test substance that will be used in a field study is determined on the basis of current agricultural practices, label requirements, and site size. A 20–50% overage in the required amount of chemical formulation, depending on study size, will ensure adequate quantities for study performance. Ideally, all test substance used in a particular study should originate from one lot or batch, and each container should be clearly marked with that lot number. In cases in which this procedure is not possible, lot numbers should not be mixed for a single test unit, such as one application event or one worker sample collection. Samples of each formulation lot used on each day must be retained to quantify percent of active ingredient (a.i.) and to meet GLP requirements for retention. Samples may also be returned to the sponsor for verification of other formulation components. Samples of the tank mix should be taken periodically throughout the study for GLP-required homogeneity analyses.

The QAU has a vital role in auditing test substance receipt, storage, and use during the study. The test substance storage-container label must contain the substance name, batch or lot number, Chemical Abstracts Service (CAS) number, expiration date, and storage conditions necessary to maintain product integrity. The study records must document the characteristics of the test substance (chemical or descriptive), the date the test substance was received, and the quantity in each lot or batch. If analytical standards are received, the same procedures should be followed. The temperature and humidity of the storage area must be monitored, and the storage containers must not be changed during the course of the study. The QAU audits all these parameters and ensures their documentation. In many cases, the test substance is shipped directly to a facility near the field test site. The QAU will audit all shipping documents and verify the recording of all necessary information, including records of the on-site conditions from time of receipt through use. During the field-study phase, the QAU audits the sampling, mixing, and application of the test substance.

All empty test formulation containers must be kept until the study is completed. Study completion is defined as the date the final report is accepted by the sponsor. This rule makes performing large field studies more difficult and, in some cases, unsafe. Handling empty pesticide containers can result in chemical spills and leaks. Arranging for a cooperator, who is usually a farmer or grower, to take seriously a request to maintain a pile of empty bags or empty plastic jugs is often difficult. However, because this function must be performed, a general guideline is useful. One procedure is to place all empty bags into one or two heavy-duty plastic bags that are well labeled. These bags can be maintained along with bulk chemical storage until final disposal. Large empty drums should be maintained in a shed or under cover. The EPA will accept requests for an exception from this retention requirement, but the agency also reserves the right of denial.

Preliminary Laboratory Work

Analytical Method Trials. Before entering the field to collect data, the analytical procedure needed to quantitate the exposure must be verified. A trial of the method used for each phase of the analysis must be performed with known amounts of the a.i. to demonstrate recovery, precision, and accuracy.

Breakthrough Trials. The filters used to collect a.i. from the air near each worker by personal sampling pump must be tested. Two filters are placed in series. The first filter is fortified with a known quantity (in micrograms) of the a.i., and air is drawn through both filters for a period of time similar to the time required for each job function. Analysis of each filter will provide a quality control check on the efficiency of the filter system to retain the a.i. Thus the appropriate filter system will be identified before collecting field data.

Extraction Trials. Each sampling matrix used to monitor exposure (e.g., filters, detergent solutions, cellulose patches, clothing, hats, and gloves) must be tested to ensure that any a.i. collected will be extractable and quantifiable. A sample matrix that is a good collector for the a.i. but does not allow for quantitative extraction is a poor sampling matrix. On the other hand, if the extraction process removes components from the sample matrix that interfere with the analysis of the a.i., then the sample matrix is not appropriate. This problem may be overcome by subjecting the sample matrix to a pre-extraction process before use so that any components that might interfere with analysis of the a.i. will be removed before the analysis. All method-validation activities are subject to audit or inspection by the QAU.

Laboratory Storage Stability. After selection of appropriate sampling matrices, multiple samples of each matrix should be fortified with the a.i. to assess stability under storage conditions. Samples are usually stored at −20 °C (±10 °C). Sufficient samples of each matrix are fortified and frozen to allow for analyses throughout the storage period. The production and storage conditions of these samples are defined as critical phases and may be inspected by the QAU.

Preparing Sample Media. The QAU may choose to audit the preparation of sample media such as cellulose patches, clothing that was subjected to the extraction process before being worn in a study, pump filters, and hand-wash detergent solutions. During this period, all sample bags and containers to be used in the field are labeled, organized, and sorted. The QAU will evaluate the labeling at this time as well as in the field during sample collection.

Calibration and Preparation of Standards. Weather stations, including thermometers and sling psychrometers, must be calibrated for use in the field. The QAU will verify use of standardized calibration equipment. Prepared solutions of analytical standards for use in the field must have the weight, volume, solvent, date, storage conditions, and preparer's initials recorded in the data notebook. The QAU will audit these data along with the storage and shipping of all standard solutions to the field. Additionally, the GLP label requirements for standard solutions, including identity, concentration, storage conditions, and expiration date, should be audited.

Field Conduct

The Field Notebook. The study notebook is the on-site recording medium for all field functions. The following items must be recorded in the field notebook:

- study number
- location
- dates

- weather data
- signatures and initials of field team members
- calibration data of the test equipment
- job function of each worker (test subject)
- crop type or variety
- crop maturity
- unique characteristics of the crop
- test chemical lot numbers
- test chemical use data
- application methods (including equipment model and description)
- abnormalities in test subject performance
- any instances of equipment failure (including repairs and replacement)
- description, location, size, and number of sampling media on each test subject

The QAU must verify all of these procedures and subsequent documents in the notebook and must check for protocol deviations.

Each worker (test subject) must sign and initial the notebook under his or her job assignment and worker number. Spaces should be available for documenting start and stop times for each job function. Loss or dislodgement of sampling media, contamination of samples, or any deviation from the normal agricultural practices must be noted during the field phase. The QAU will observe actual field activities and will verify the documentation of any abnormalities, equipment failure, or environmental factors that influence sample collection. The QAU will also check the data entries, frequency of data collection, and field equipment calibration entries into the study notebook for accuracy.

Field Team. The field team usually consists of a minimum of three people plus a QA auditor and the study director. The team assigns each test subject a unique number that is permanently displayed on outer clothing throughout the study. All test subjects are informed of the study's purpose and requirements and the health and safety aspects of the test substance. Adherence to normal agricultural practices is stressed. Interpreters are provided for non-English speaking test subjects. All workers must sign an "informed consent" form (Figure 23.2); by signing, the test subjects acknowledge an understanding of safety information for the test

---

AGRISEARCH INCORPORATED
INFORMED CONSENT

CHEMICAL_____

FORMULATION_____

USAGE_____

    I acknowledge that I have been informed and under-

stand the safety information provided for the above chemical

formulation described to me on _____

by _____,

representative of _____.

_____
Print Name

_____   _____
Signature   Date

Figure 23.2. Informed consent form.

substance and their involvement in the study. The field team begins weather data collection one hour before test subject exposure and continues until study termination. The weather data are recorded in the field notebook or are captured by computer, and are initialed by a team member. Each sample matrix is fortified with known amounts of the test chemical, and control samples are prepared. The QAU audits for correct procedures, fortification solution and sample media identification, correct labeling on all sample containers, and correct number of samples based on the protocol. If personal or high-volume air sampling pumps are being used, they are calibrated before study initiation, and calibration specifics are recorded on a pump calibration form (Figure 23.3). The test subjects are then equipped with sampling media and sampling pumps as outlined in the protocol. All activities are inspected and audited by the QAU for correct sample location, number, and type; recorded pump number; and correct suiting-up procedures to prevent sample contamination. A team member takes a still photograph of each suited worker, including the unique number for future identification. The time at which each worker begins to apply the test substance is recorded.

The agricultural practices of the worker are monitored by a field team member and the QAU. Critical stages or activities of the study are recorded by using both video and still photographs. This recording method provides a permanent record of both agricultural practices and study conduct. The QAU uses an inspection form for worker-exposure studies. This form identifies critical study aspects in a comprehensive checklist format (Figure 23.4).

While test workers are in the field, field team personnel should organize the labeled sample bags, jars, and hand-rinse solutions. This organization speeds the sample-collection process following completion of the exposure period. Part of this organization is a precise sample-collection order and procedure that is described in an SOP and inspected by the QAU. Preventing cross-contamination between samples is just as important as collecting valid samples. Therefore, part of the job of the QAU inspector is to continuously observe the field research team for organization, compliance with procedures, and cleanliness of sampling equipment. Particular attention must be paid to cleaning forceps and changing gloves between collection of samples from each test subject.

After samples have been collected, each personal sampling pump is recalibrated, and the pump flow rate is noted in the field notebook. This recalibration is critical to calculating the actual pump rate during use, which is then used to extrapolate to the human breathing rate. The QAU inspects the procedures and audits the data-collection and -reduction phases of this operation.

Following sample collection, all sample bags are heat sealed and placed on dry ice for initial freezing. Chain-of-custody forms are filled out for shipment of the collected samples to the analytical laboratory. The QAU inspects sample bag sealing, labeling, and packing procedures and audits the chain-of-custody forms. Figures 23.5 and 23.6 are examples of chain-of-custody forms. All samples should be shipped on dry ice by over-night carriers.

AGRISEARCH INCORPORATED
HIGH VOLUME AIR SAMPLER CALIBRATION WORKSHEET

Disc Number	H_2O $H(P_a/P_o)(T_o/T_a)$	Q True Air Flow	I	$I-(P_a/P_o)(T_o/T_a)$ Observed Flow Meter Readings
5				
7				
10				
13				
18				

Project: _____   Initials: _____   Date: _____

Comments: _____
_____

High Volume Air Sampler was calibrated at the following conditions:

Barometric Pressure: _____ "Hg; convert "Hg x 2.54 cm/in. x 10 mm/cm = mmHg

_____ = _____ mmHg = P_a

Ambient Temperature: _____ °F; convert °F - 32 x 5/9 = °C + 273.16 = °K

_____ = _____ °K = T_a

Orifice Calibrator Serial No. S194D

Hi-Vol Sampler No. _____

Figure 23.3. Pump calibration form.

QUALITY ASSURANCE INSPECTION FORM
WORKER EXPOSURE, FIELD STUDIES

	Yes	No	NA
General			
Final study protocol - including any revisions	—	—	—
Test Substance			
Record of the amount of test chemical received, used, balance remaining, storage conditions	—	—	—
Recordkeeping - black ink, dated, signed, used study number, and trial ID number, etc.	—	—	—
Test substance formulation, labelling, lot #, batch #, expiration date.	—	—	—
Site Selection			
Plot size	—	—	—
Layout/plot map	—	—	—
Marked properly	—	—	—
Good representative area field	—	—	—
Site Preparation			
Documentation	—	—	—
Sample collection materials placed according to protocol	—	—	—
Sprayer, Granular, and Field Equipment			
Documentation of calibration of application equipment	—	—	—
Calibration of air samplers (personal, hi-vol)	—	—	—
Calibration of thermometers, weather station	—	—	—
Record of maintenance of sprayer and equipment	—	—	—

Figure 23.4. QA inspection checklist for worker-exposure field studies.

Continued on next page.

	Yes	No	NA

Worker Protection Procedures

| Workers protected according to label/protocol | ___ | ___ | ___ |
| Workers informed of possible health risks/sign consent forms | ___ | ___ | ___ |

Pesticide Application

| Check calculation - rate, amount compound weighed out | ___ | ___ | ___ |

QC Procedures

| Tank mix samples | ___ | ___ | ___ |
| Field Spikes - positive control worker
            - negative controls | ___ | ___ | ___ |

Sample Collection/Storage

Sample collected according to SOP/Protocol	___	___	___
Steps taken to avoid loss - contamination of samples	___	___	___
Sample bags properly labeled	___	___	___
Condition of storage/shipment	___	___	___
Sample shipment paperwork, Federal Express, etc.	___	___	___
Chain of custody	___	___	___

Clean Up

| Waste disposal | ___ | ___ | ___ |
| Equipment decontamination | ___ | ___ | ___ |

Figure 23.4. (Continued) QA inspection checklist for worker-exposure field studies.

AGRISEARCH INCORPORATED
Sample/Specimen Transfer Sheet

Study Title:_____   Project Dir:_____

Sponsor:_____   Shipper:_____
                                          Signature

Project No.:_____   Shipping Date:_____

   The sample specimens listed below have been transferred from Agrisearch Incorporated to the custody of the project sponsor.

Description	ID Code Number	Agrisearch No.

   Upon verification of sample/specimen identification, please return a signed copy of this sheet to:

          Agrisearch Incorporated
          26 Water Street
          Frederick, MD 21701

_____
Sponsor   Study Director   Date

Figure 23.5. Sample or specimen transfer form.

AGRISEARCH INCORPORATED
Chain Of Custody

Project No.: _____   Sample Type: _____

Time Sampled: _____   By: _____   Date:_____

Date Stored: _____   By: _____ Condition: _____

Date Shipped: _____   No. Samples Shipped: _____

Shipped To: (Name & Address): _____

_____

_____

Shipped By: _____
            Signature   Type Name

            Company Name: _____

            Address: _____

Amt. Dry Ice/Container: _____   Commercial Shipper: _____

                                   Waybill No.: _____

_____

Received By: _____
             Signature   Type Name

             Company Name: _____

             Address: _____

Date Received: _____   Condition: _____

    Please maintain a copy of this form through all phases of the study.
Return completed form to sender.

Figure 23.6. Chain-of-custody form.

Conclusion

The QAU is involved in a field study both in the laboratory before going to the field and at the field site. The QAU has the responsibility not only to help build and maintain quality standards but also to help ensure that a data trail is generated that enables a reviewer to understand the study and that complies with the GLP Standards from protocol development to signing of the final report. The QAU must perform the following study-specific tasks:

• audit protocol for regulatory requirements

• ensure agricultural practices and test-site requirements

• audit preliminary laboratory work

• inspect field procedures

• inspect field-data collection

• ensure proper test formulation, analytical standard, and sample storage and shipping

• audit laboratory analyses

• inspect laboratory data reduction and draft report

• audit and sign final report

References

1. "Pesticide Assessment Guidelines", Subdivision K, Exposure: Reentry Protection, U.S. Environmental Protection Agency, Office of Pesticide Programs, Washington, DC 20460; EPA 540/9–84–001, October 5, 1984.

2. "Pesticide Assessment Guidelines", Subdivision U, Applicator Exposure Monitoring, *Code of Federal Regulations* Title 40, Pt. 158.140; U.S. Environmental Protection Agency, Exposure Assessment Branch, Hazard Evaluation Division, Office of Pesticide Programs, Washington, DC 20460; NTIS #PB87-133286, October 1986.

3. "Federal Insecticide Fungicide and Rodenticide Act (FIFRA) Good Laboratory Practice Standards"; Final Rule; *Code of Federal Regulations* Title 40, Pt. 160; *Federal Register* 54:158 (August 1989) pp 34052–34074.

RECEIVED for review November 5, 1990. ACCEPTED revised manuscript September 30, 1991.

Mesocosm Studies and Other Aquatic Field Studies with Pesticides

Jeffrey M. Giddings

Environmental Sciences Division, Springborn Laboratories, Inc., 790 Main Street, Wareham, MA 02571

Aquatic field studies for ecological effects and environmental fate (particularly mesocosms) are now required for registration of many agricultural chemicals under the Federal Insecticide, Fungicide, and Rodenticide Act. These studies must comply with recently issued Good Laboratory Practice (GLP) Standards, which were written primarily for laboratory situations. Conducting a field study under GLP Standards presents a combination of challenges related to study complexity (volume and diversity of data), logistics (remote locations and subcontractors), and technical uncertainties (new techniques, changing regulations, and uncontrolled events). These challenges are not insurmountable. Careful planning, preparation for contingencies, and attention to the principles of quality assurance can result in field studies that meet scientific and regulatory objectives while maintaining compliance with GLP Standards.

As THE ENVIRONMENTAL ASSESSMENT of a pesticide proceeds under the Federal Insecticide, Fungicide, and Rodenticide Act (FIFRA), the product is subjected to tests of increasing complexity and sensitivity. Information

2192-8/92/0297$06.00/0 © 1992 American Chemical Society

is obtained about toxicity to terrestrial and aquatic organisms. Also, concentrations of the product in the environment are either estimated by using computer models or are measured in the field. At major decision points in the assessment process, toxicity is compared with environmental concentrations, and the need for further testing is determined (1). If this comparison indicates that use of the product could result in environmental concentrations that are toxic to some organisms, the Environmental Protection Agency (EPA) may require the party seeking to register the product to provide data demonstrating that adverse effects will not occur under actual or simulated field conditions. When aquatic environments are of concern, the currently accepted procedure for rebutting the presumption of adverse effects involves the use of small constructed ponds called mesocosms (2).

Mesocosm studies are a relatively new addition to the environmental assessment process for pesticides. The first mesocosm studies for pesticide registration were conducted in 1987. To date, fewer than a dozen mesocosm studies have been completed and reported to the EPA. Mesocosm techniques are still evolving, and guidelines for conducting and evaluating mesocosm studies still lack the specificity and definition that exist for simpler, more conventional laboratory tests. Nevertheless, mesocosm studies must adhere to Good Laboratory Practice (GLP) Standards as finalized for FIFRA (3), even when this adherence means stretching the GLPs to applications much different from those for which they were originally designed. Indeed, the considerable technical and scientific complexity of a mesocosm study may present less of a challenge than interpretation of and compliance with the FIFRA GLPs.

The purpose of this chapter is to describe how a mesocosm study is typically conducted, with particular attention to quality assurance (QA) procedures and GLP issues. This information is presented from the viewpoint of a study director endeavoring to satisfy data requirements, overcome logistical obstacles, and adapt to the unpredictable behavior of the natural world—all while maintaining GLP compliance.

The Mesocosm Approach

A mesocosm study is conducted by applying the test material (a pesticide end-use product) to a set of uniform experimental ponds and monitoring the ponds' ecological response. The treatment regime is designed to simulate the routes of entry of the pesticide into aquatic habitats under conditions of actual use. The ecological monitoring program typically includes measurement of water quality, algae and rooted aquatic plants, planktonic animals, insects and other invertebrates, and fish. Monitoring usually starts at least several months before treatment begins and continues for several months after treatment (sometimes into subsequent years). Concentrations of the pesticide (and sometimes of its degradation products) are measured in water, bottom sediments, and fish tissue. From 12 to 20 or more individual mesocosms may be included in a study, and usually three or more levels of pesticide treatment are used. As in conventional toxicity studies, treated mesocosms are compared statistically with untreated controls to aid in the evaluation of effects.

Physical Design

Mesocosms for pesticide registration have been constructed in Kansas, Alabama, Texas, North Carolina, Ohio, Virginia, and Mississippi. To satisfy the EPA guidelines (2), each mesocosm pond should be at least one-tenth acre in area, and at several of the sites mesocosms of one-quarter acre have been constructed. The ponds have maximum depths of approximately 2 m and contain approximately $400-1000$ m^3 of water. To ensure that the ponds retain water and especially to prevent contamination of groundwater with the test material, mesocosms are usually lined with packed clay or with impermeable plastic sheets. A layer of topsoil or natural aquatic sediment is placed on the bottom and sides, and the ponds are then filled with water. Before the mesocosms are treated with the test material, fish (usually adult bluegill sunfish) are stocked in each pond.

Biological Development

When water and sediment from an existing, well-colonized source (such as a farm pond) are used to initiate a mesocosm, an ecologically diverse and productive biological community can develop within a few months. If the mesocosms are constructed instead with topsoil and well water (or other uncolonized materials), the EPA guidelines require at least one year of maturation before the mesocosms are considered ecologically representative and suitable for a pesticide registration study.

Unlike laboratory toxicity studies, the test systems in a mesocosm study are not fully characterized and are only partially under the control of the scientist. Many of the specific requirements in § 160.90 of the FIFRA GLPs (animal and other test system care) are clearly relevant to mesocosm studies in their intent, yet are difficult to interpret in the present context.

For example, "At the initiation of a study, test systems shall be free of any disease or condition that might interfere with the purpose or conduct of the study" (§ 160.90(c)). This requirement is applicable to stocked fish (one component of the test system), but not readily applied to the test systems in their ecological entirety. Before a mesocosm study begins, the EPA, the sponsor, and the testing laboratory often discuss whether the mesocosms are healthy, mature, representative, and generally suitable for the proposed study. But no specific, generally accepted criteria currently exist for evaluating the condition of a complex ecological system like a mesocosm, nor is it clear what conditions, if any, might interfere with the purpose or conduct of the study. Most mesocosm studies include several months or even years of pretreatment monitoring, with the partial objective of demonstrating that the test systems are not behaving abnormally (although normal is not clearly defined in this context). Neither the mesocosm testing guidelines (2), nor the standard evaluation procedures (1), nor the GLPs (3) describe how the pretreatment data are to be interpreted; the matter, like many issues related to mesocosm studies, must be left to professional judgment.

Ecological Monitoring

The ecological monitoring program in a mesocosm study is essentially a broad limnological survey. The mesocosms are sampled at regular intervals—usually once a month during the pretreatment year, accelerating to every other week beginning early in the spring of the treatment year, continuing biweekly through the treatment and posttreatment phases, and possibly dropping back to monthly in the fall. In the final sampling event, the mesocosms are drained so that all remaining fish can be collected, measured, and examined.

The measured parameters typically include the following:

- *Water chemistry:* dissolved oxygen, pH, conductivity, hardness, alkalinity, suspended solids, dissolved organic carbon, total organic carbon, total nitrogen, and total phosphorus

- *Plants:* phytoplankton chlorophyll and species abundance (phytoplankton are free-floating microscopic algae); periphyton chlorophyll, dry weight, and species abundance (periphyton are algae, fungi, and other microscopic organisms that grow attached to surfaces such as rocks and plant stems); and macrophyte biomass and spatial distribution (macrophytes are large plants, usually rooted vascular plants like pondweed and cattails)

- *Invertebrates:* zooplankton species abundance (zooplankton are small, free-floating invertebrates, mainly crustaceans); benthic macroinvertebrates (larger invertebrates, mainly insects and molluscs, living in or on the sediment); and emergent insects (adult insects that emerge from the water after completing the aquatic immature stages of their life cycles)

- *Fish:* at the final harvest, the number, length, weight, reproductive condition, and gross appearance of adults and the number, length, weight, and appearance of juveniles

- *Ecosystem metabolism:* rates of primary production (photosynthesis) and respiration measured in the ecosystem as a whole, calculated from changes in dissolved oxygen concentration in the light and the dark

From these measurements, many other parameters are calculated, such as taxonomic richness (the number of different taxa of phytoplankton, zooplankton, benthic invertebrates, or emergent insects in a pond), community similarity (similarity of types and numbers of organisms between ponds), and the production–respiration ratio.

This great variety of measured parameters is perhaps the most significant technical challenge of a mesocosm study. Some types of measurements can be performed by trained technicians, but others (especially taxonomic identification and enumeration) are best done by specialists. Coordinating the team of field workers needed for efficient and consistent execution of such a comprehensive ecological monitoring program requires advanced planning, logistical support, managerial skill, and scientific competence.

The huge array of monitoring points also presents a large number of QA and GLP requirements. For each type of measurement, there must be standard operating procedures (SOPs) for sample collection (or in situ measurement); use, calibration and maintenance of sampling equipment (and analytical instrumentation); and sample processing. There must be data forms for documentation of sample collection (e.g., date, time, mesocosm identification, location within the mesocosm, depth, sampling apparatus, volume sampled, subsampling, and preservation), chain of custody, sample processing (e.g., sorting and separating invertebrates from debris in a sediment sample, extracting chlorophyll from a phytoplankton sample, or dissecting and examining a fish), and sample disposition. Systems for recording, verifying, keypunching, organizing, analyzing, summarizing, and reporting each type of data pertaining to the study must be available. Many of these functions are performed with computers, and each computer program must be validated and adequately documented.

Treatment with Test Material

The mesocosms are treated with test material in a manner intended to simulate the routes of entry of the pesticide into aquatic systems under actual use. The most likely routes are spray drift and surface runoff. Spray drift occurs when pesticide droplets are carried away from their intended target by the wind during application and are deposited on the water surface. Surface runoff occurs when pesticide residues on soil or foliage are washed off the field by rain or irrigation. The pesticide may be dissolved in the water, or it may be associated with soil particles carried by the water.

Spray drift is usually simulated in a mesocosm study simply by spraying the pesticide onto the pond surface at some fraction of the application rate specified on the product label. A backpack sprayer with pressure provided by a CO_2 cartridge works nicely; one person sits in the back of a boat holding the spraying device close to the water while another person guides the boat around the pond. More sophisticated spraying systems have also been developed with nozzles spaced out across a boom that is moved slowly across the pond surface while the test substance is sprayed from a pressurized reservoir. Whichever method is used, one of the major concerns is to avoid accidentally contaminating adjacent mesocosms. The sprayer must be held close to the water surface, and special baffles may be provided to prevent the spray from blowing off-target.

Runoff simulation is more complicated. Runoff typically enters ponds at a few discrete points around the shoreline and typically carries a load of suspended soil particles. We have simulated runoff events by using recirculation systems: Water is pumped from one or more points in the pond into a mixing tank, where the pesticide is added in either aqueous or soil-bound form. The mixture then flows back into the pond through several pipes that discharge just below the surface. Water is recirculated for 6 hours—a reasonable duration for a runoff event—during which time about one-sixth of the pond volume is removed and returned. In other studies, runoff has been simulated by using the same kind of boom system previously described for spray drift, with a pesticide-treated soil–water slurry or water alone spread evenly across the pond surface.

For both spray-drift and runoff simulation, the frequency of treatment and the amount of pesticide applied are usually selected to mimic some hypothetical exposure scenario. For example, spray drift might be applied three times at biweekly intervals, approximating the maximum usage rate of the product on a crop, and simulated runoff might be applied six times at 2-week intervals, approximating the worst-case runoff patterns predicted for a growing season in a particular region. The two types of mesocosm treatment are sometimes superimposed within a single study, creating a very irregular sequence of exposure events. This sequence is in great contrast with a typical aquatic laboratory study, in which exposures are defined in terms of concentrations rather than loading rates and an effort is made to ensure that exposure concentrations remain nearly constant throughout the study. Interpreting the exposure patterns in a mesocosm study, with concentrations varying continuously over time and often varying with location within each pond, can be very difficult.

The GLP issues relating to test substance are readily applied to mesocosm studies, although some GLP requirements can be cumbersome under field conditions. Establishment of separate areas for test material receipt, storage, and mixing (§ 160.47) is straightforward, as is documentation of test material usage (§ 160.107). Determining the homogeneity of mixtures and the uniformity of application rates within a test system (§ 160.113), on the other hand, may be next to impossible (or irrelevant) in a mesocosm study, depending on the methods used to treat the ponds.

Statistical Analysis and Interpretation

Statistical analysis of mesocosm results has been a controversial topic since the inception of the FIFRA mesocosm guidelines. This controversy stems in part from the inherent complexity of mesocosm data, with observations on dozens of interrelated variables over time. Mountains of data must be sifted to reveal patterns of response to the experimental treatment. A more troublesome source of disagreement over statistical approaches, however, is an underlying uncertainty about the objectives of the study.

One view of the purpose of a mesocosm study—a view generally taken by the EPA—is that it should answer the question of whether or not a given exposure pattern is likely to cause adverse ecological effects. The study is then designed to test a hypothesis (such as the hypothesis that the treatment causes no effects), and hypothesis-testing analysis of variance techniques is appropriate. An alternate viewpoint is that the results of the study should allow us to describe the relationship between exposure and response over a range of exposure levels. In the latter case, regression techniques are most appropriate.

An evaluation of these two alternatives is beyond the scope of this chapter. However, the divergence of opinion has many implications for the conduct of a mesocosm study. Because study design should reflect the planned statistical approach, such fundamental decisions as the number of treatment levels and the number of replicates at each level become the subject of heated discussion. Conventions for statistical analysis and interpretation (like probit analysis for LC_{50} determination in conventional acute laboratory tests) have not yet emerged for mesocosm studies, and the GLP requirement that the protocol include a statement of the proposed statistical method (§ 160.120) is cause for considerable uneasiness among the parties involved. No matter what method is proposed, it is likely to be disputed after the fact.

Regardless of the statistical approach used, interpretation of a mesocosm study will almost invariably come down to a matter of professional judgment based on the weight of the evidence. This, too, can be an unsettling prospect for the study sponsor, who might prefer explicit pass–fail criteria to be established before the study begins. For the study director, the path is very clear: Statistical methods must be rigorously documented, and care must be taken to avoid overreaching the evidence when drawing conclusions.

Program Management

One section of the FIFRA GLPs that has been particularly difficult to adapt to mesocosm studies is Subpart B, which deals with organization and personnel. The difficulties arise because a mesocosm study is often, by necessity, managed differently from a conventional laboratory test, and the definitions and specifications of the GLPs do not directly apply.

For example, the mesocosm site is often not the main headquarters of the organization responsible for conducting the test. In most cases, neither is it simply a field site without facilities or permanent staff, but more like a branch or field station of the parent organization. Is the testing facility the parent organization or the remote field station? Where is the distinction drawn between an organization that is simply a branch of another organization and an organization that functions as a testing facility in its own right? Who—the parent organization or the field station manager—is responsible for designating a study director; maintaining a quality assurance unit (QAU); keeping organizational charts, job descriptions, training records, and master schedule; and fulfilling the other requirements listed in the GLPs?

Coordinating the day-to-day activities of a mesocosm study is usually the job of an experienced field supervisor, typically a biologist or ecologist. Overseeing a team of technicians who are taking samples and measurements, applying the test substance, recording observations and procedures, and maintaining the mesocosm site are tasks that keep the field supervisor busy. Processing and analyzing samples in the laboratory, compiling and managing the data, negotiating protocol issues with the sponsor and the EPA, and providing logistical support for the field team are tasks typically carried out by one or more individuals other than the field supervisor. Several organizations or subgroups may be involved in different aspects of sample analysis (e.g., insect identification, water quality analysis, or residue analysis), and each organization or subgroup will have its own manager or principal investigator.

No individual, then, is directly responsible for all aspects of study conduct, the GLP definition of a study director. In this situation, what constitutes overall responsibility and who is truly the single point of study control (§ 160.33)? Certainly there is a level of "management"[1] that encompasses all the facets of a mesocosm study, but the individual at that level is generally several steps removed from the actual conduct of the study. If this individual is designated the study director, how do we interpret the requirement that this person ensure that the study is conducted according to the protocol and in conformity with GLP Standards?

Many organizations, including test sponsors and contract laboratories, have tried to resolve the dilemma of the over-extended study director by subdividing each project into several component studies. The EPA's Office of Compliance Monitoring (OCM) has rejected this approach, maintaining that all activities associated with an experiment "in which a test substance is studied in a test system" (§ 160.3) together constitute a single, indivisible study that must consequently have one study director.

Even with that stipulation, the boundaries of a mesocosm study are not always clear, particularly if the mesocosm ponds are already in existence and have been subjected to routine ecological monitoring for months or years before they are designated for use in a particular project. If the experimental start date is "the first date the test substance is applied to the test system" (§ 160.3), then pretreatment monitoring is not part of the experiment. At some point in the pre-experimental monitoring program, the study director must sign the protocol (thereby marking the study initiation date), and data collected after that time are considered part of the study. Because the objectives of the pretreatment monitoring phase of the mesocosm study are not clear and pretreatment data are not explicitly included in the analysis and interpretation of the results, the starting point of the study may have to be determined arbitrarily.

---

[1]Management is defined here as the person or persons responsible for setting policy at a facility. For a description of the duties of testing facility management, see § 160.31 of the GLP Standards.

These issues are still in need of resolution. The OCM has issued advisories that present the EPA's judgments on and interpretations of some of the issues but uncertainties and inconsistencies remain. The study director, sponsor, and QAU must work together to devise GLP-compliant policies and procedures on a case-by-case basis.

Facing the Unexpected

A mesocosm study that involves measurement of 30 parameters in 12 ponds over 20 sample events will produce 7200 samples or data points. Some types of samples—phytoplankton, periphyton, zooplankton, and macroinvertebrates, for example—generate not individual data points but lists of taxonomic abundances; the total number of observations will probably exceed 100,000. The raw data from one study will easily fill an office filing cabinet. The archived specimens will fill a small room. The final report may exceed 2000 pages. The study will take at least 18 months, and more likely twice that. It will involve 30 or more scientists and technicians, and it will cost more than $1 million.

Successful execution of a project of this magnitude takes careful planning. The protocol, SOPs, data forms, and data management systems must be created; staff must be assigned their individual program responsibilities and trained to perform them; equipment and supplies must be obtained and transported to the site; and sampling schedules and treatment regimes must be established. With sufficient advance preparation, the routine aspects of the study can be mastered.

The nonroutine aspects—the unpredictable situations and events—can also play a major role in the progress of a mesocosm study. The behavior of a complex ecological system (a small pond) is to a large extent outside of experimental control, more so than the agricultural systems that are the subject of other pesticide field studies, and much more so than any laboratory system. A mesocosm study is subject to uncontrollable biological interactions, to unexpected immigration and emigration of wild populations, and, like any field study, to the weather.

As an example of biological interactions, it is not uncommon for the fish in the mesocosms to reduce populations of some invertebrates to near-extinction, making observation of those populations (which is one of the stated objectives of the study) difficult or impossible. Immigration of some wild populations, such as frogs, can be impossible to control and can lead to serious divergences among replicate test systems. Other animals, such as geese, muskrats, or crayfish, can physically destroy a mesocosm if not removed promptly. And the weather—particularly heavy storms around the time of pesticide application—can necessitate radical changes in study schedules, techniques, and interpretation.

The vagaries of nature are not the only unpredictable elements of the study. In a project of the magnitude of a mesocosm study, equipment failures are bound to occur, and backup equipment or contingency procedures must be in place. Samples or data may occasionally be lost through accidents or schedule changes. Alterations in field methods may occur in response to technical developments during the course of the study. These and other unforeseeable circumstances must be understood and adapted to by all concerned, including the study director and his staff, the sponsor, the agency, and the QAU.

Despite technical difficulties, logistic challenges, and inherent unpredictability, mesocosm studies can and do succeed in answering the environmental assessment questions for which they were intended. Attention to the principles of quality assurance is essential to the scientific integrity and regulatory acceptability of these studies. GLP Standards should be adapted to ensure the validity of the studies without conflicting with the original purposes of the studies.

References

1. "Hazard Evaluation Division Standard Evaluation Procedure. Ecological Risk Assessment", Office of Pesticides Programs, U.S. Environmental Protection Agency, Washington, DC, EPA-540/9-85-001, June 1986.

2. "Hazard Evaluation Division Technical Guidance Document. Aquatic Mesocosm Tests to Support Pesticide Registrations", Office of Pesticides Programs, U.S. Environmental Protection Agency, Washington, DC, EPA-540/09-88-035, March 1988.

3. "Federal Insecticide, Fungicide, and Rodenticide Act (FIFRA); Good Laboratory Practice Standards; Final Rule", *Code of Federal Regulations* Title 40, Pt. 160; *Federal Register* (August 17, 1989).

RECEIVED for review September 17, 1990. ACCEPTED revised manuscript September 13, 1991.

Chapter
25

Terrestrial Field Studies

Mary E. Johnson and Mark Jaber

Wildlife International Ltd., 8598 Commerce Drive, Easton, MD 21601

As of October 16, 1989, terrestrial field studies were required to be conducted in compliance with the Federal Insecticide, Fungicide, and Rodenticide Act Good Laboratory Practice (GLP) Standards. Terrestrial field studies are conducted to assess the effects of pesticides on wildlife. Primary objectives of these studies are to determine if pesticides applied under actual use conditions result in wildlife mortality or other adverse effects and to measure potential routes of exposure to pesticides. Techniques used to accomplish these objectives include determining pesticide residue concentrations in wildlife food sources, evaluating nesting birds, conducting bird surveys, searching for casualties, livetrapping small mammals, and conducting radioradiotelemetry. Although most basic GLP principles apply to laboratory and field situations alike, there are differences, both logistical and technical, that must be considered when applying portions of the GLP regulations. This chapter describes methods that have been used by Wildlife International Ltd. to bring terrestrial field studies into compliance with GLP regulations that were developed for the toxicology laboratory.

2192-8/92/0309$06.00/0 © 1992 American Chemical Society

Wᴵᴛʜ ᴛʜᴇ ᴜᴘᴅᴀᴛᴇ of the Federal Insecticide, Fungicide, and Rodenticide Act (FIFRA) Good Laboratory Practice (GLP) Standards in 1989, a series of studies not previously affected (i.e., ecological effects, chemical fate, and residue chemistry) were required to be conducted in compliance with GLPs. Field studies conducted in support of registration or reregistration of a pesticide are one group of studies affected by the updated regulations. One type of field study affected is terrestrial field studies. These studies are conducted to determine the effects of pesticides on wildlife when the pesticide is used under conditions of actual use. More specifically, the primary objectives of these studies are to determine if pesticides applied under actual use conditions result in wildlife mortality or other adverse effects, and to measure potential wildlife exposure to pesticides. Techniques used to accomplish these objectives include determining pesticide residue concentrations in wildlife food, evaluating nesting birds, conducting bird surveys, searching for casualties, livetrapping small mammals, and conducting radioradiotelemetry.

Because the FIFRA GLP Standards were originally designed for the toxicology laboratory, difficult problems may arise when implementing these regulations for terrestrial field studies. Although most basic GLP principles apply to both laboratory and field situations, some differences, both logistical and technical, must be considered when applying portions of the FIFRA GLP regulations.

Conducting Terrestrial Field Studies

Terrestrial field studies can be conducted in multiple geographic regions on many different crops to assess hazards to wildlife for a variety of pesticide uses. Test sites are selected in areas of known wildlife habitat. Most studies use 8–14 treated test replicates (plus any necessary control replicates) that are owned or leased by growers. Each test replicate consists of a crop field, generally ranging in size from 20 to 200 acres, and its associated wildlife habitat. Crops commonly used include potatoes, corn, cotton, tobacco, cabbage, and apples. To maximize wildlife exposure, the test substance is applied at the maximum use rate and the shortest recommended application interval, as specified on the product label, by the grower or other certified pesticide applicator. All applications are made by using practices and equipment typical for the crop and study area.

Effects of the pesticide on wildlife are monitored, and potential routes of exposure to the animals are determined. Effects noted include mortality, abnormal behavior, and effects on nesting success, reproduction, and population size.

Terrestrial field studies are conditionally required by the Environmental Protection Agency for support of registration or reregistration of a pesticide. According to the Guidance Document for Conducting Terrestrial Field Studies (*1*), the purpose of these studies is either to refute the assumption that risks to wildlife occur under conditions of actual use of the pesticide or to provide some quantification of the effects.

Whether terrestrial field studies are required depends on results of lower tier ecotoxicology studies (e.g., laboratory studies), the intended use of the pesticide to be tested, and environmental-fate characteristics of the pesticide. If the results of the laboratory studies along with intended use patterns and pertinent environmental fate characteristics suggest that the test compound may adversely affect wildlife, further tests, such as terrestrial field studies, are required.

Types of Terrestrial Field Studies

Two types of terrestrial field studies are discussed in the guidance document: screening studies and definitive studies (*1*). According to the guidance document, a *screening study* is designed primarily to demonstrate that a hazard suggested by lower tier laboratory studies does not exist under actual use conditions. The *definitive study* is a relatively detailed study designed to quantify the magnitude of impacts identified in a screening study or from other information. If a screening study indicates that the pesticide has caused little or no detectable adverse effect, concluding that potential adverse effects are minor may be reasonable. However, when effects are demonstrated or if a product has caused adverse effects under actual use conditions, the magnitude of the effects may need to be determined, and a definitive study may be required.

Components of Terrestrial Field Studies

Pesticide applications are made at the time of year when the test substance would typically be used to control a pest outbreak. All applications are monitored by study personnel. Weather conditions, number of acres treated, and amount of test product used are recorded.

One primary technique used to evaluate acute mortality of wildlife exposed to pesticides is casualty searching. Casualty searches are conducted by walking slowly along predetermined routes in an attempt to locate dead or affected animals. Dead animals are recovered and sent to an analytical laboratory for residue analysis. Sometimes only partial remains are found because of scavenging. In this case, the condition of the remains is recorded prior to collection. When animals are found displaying abnormal behavior, any clinical signs that may indicate whether the behavior is related to the treatment are recorded.

Other techniques often used to evaluate potential effects include conducting bird surveys and sampling wildlife food and water sources and soil for residue analysis of the test substance. Bird surveys are conducted in an attempt to identify avian species composition, relative abundance, and use of treated areas. To conduct the survey, bird survey plots are established at several points around the test field, and biologists record all birds seen or heard within a given period of time at each plot. Wildlife food and water sources and soil are sampled and analyzed to determine concentrations of the pesticide to which wildlife may be exposed. Wildlife food includes, among other things, insects, seeds, and vegetation. Samples are collected over a period of time from the test field and from adjacent habitats to account for potential drift or overspray of the test substance during application. Residue concentrations in these samples can then be compared with dietary toxicity data from the laboratory to determine if residues are high enough to pose a hazard to wildlife.

Additional techniques that may be used during terrestrial field studies include evaluating nesting birds, livetrapping small mammals, and conducting radioradiotelemetry. These techniques are used not only to determine if an effect has occurred but also to provide useful information on the extent of any effects noted. Small mammals are livetrapped to monitor changes in species composition and relative abundance following pesticide applications. Small mammals are captured in baited traps placed in and around the test field in the afternoon or evening. The traps are checked the following morning, and all animals captured are identified and released.

Radioradiotelemetry is conducted to monitor habitat use and survival of individual animals carrying radio transmitters. Species tracked can include quail and passerines (e.g., robins, blue jays, and brown thrashers). Birds are captured using either ground traps or mist nets. A transmitter is then attached to each bird by glue or harness. The transmitters emit unique signals that are picked up by a receiver. The location and status (i.e., dead or alive) are determined for each radio-equipped bird daily throughout the study.

The monitoring of nesting birds involves placing nest boxes around test fields in locations where birds are likely to nest or locating naturally occurring nests or both. The boxes or natural nests are periodically checked to determine occupancy and nesting success. Nesting success is determined by how many of the young hatch and live to fledging (the point at which the feathers necessary for flight are grown).

Impact of GLPs

As previously mentioned, the FIFRA GLPs were originally designed for the toxicology laboratory. Terrestrial field studies are seasonal and normally have a duration of less than 3 or 4 months. Several studies may occur simultaneously, each requiring as many as 25 biologists. Such size and timing create logistical problems that stem from obtaining an adequate number of trained personnel and may affect the audit schedule of the quality assurance unit (QAU). Therefore, careful thought and planning need to precede each of these studies to ensure compliance with GLPs. Furthermore, applying these regulations with respect to terrestrial field studies can present some other unique problems. For example, the facility may be a cornfield and its adjacent wildlife habitat, and the test system may be one or more species of birds. Additional aspects of terrestrial field studies that require special attention include maintenance and calibration of equipment, accountability of the test substance, and data collection.

Standard Operating Procedures

Standard operating procedures (SOPs) are necessary for all routine activities "to ensure the quality and integrity of the data generated in the course of a study" (*2*). A well-defined SOP is necessary to provide consistency in data collection. Minor differences in data collection and recording can cause problems with data analysis and report writing. Therefore, SOPs for field procedures, such as small mammal trapping or calibration of application equipment, should be written by experienced biologists. SOPs should be reviewed yearly and updated as needed. Because field studies are influenced by many factors, accounting in the SOPs for all circumstances that may affect a study can be difficult. Therefore, whereas it is necessary to allow some flexibility in the SOP, care must be taken so as not to compromise the intent of the SOP. If written broadly enough to account for all potential problems, the SOP may be too general to have any practical use.

In addition to SOPs, well-written data collection forms will help provide consistency in data collection. Data collection forms are important for proper documentation of the techniques used and observations made throughout a study. Our data recording procedures have evolved so that there are forms for almost everything. The forms are designed so that all the proper questions are asked; therefore, completion of the forms means that all required information is obtained.

Maintenance and Calibration of Equipment

The GLPs state that equipment used in studies must be adequately calibrated and maintained. Because the application equipment is owned by the farmer or certified commercial applicator, past maintenance or calibration records are not likely to be available. However, calibration of each piece of application equipment can be checked to ensure that the equipment is functioning properly and that the rate of delivery of the test substance is in accordance with the protocol. Calibration may be conducted in several ways, depending on the type of equipment and on the delivery mechanism. The objective is basically the same for all equipment types, that is, to keep track of the weight or volume of product put out over a given area of land.

Maintenance is performed by the owner of the application equipment. The time period involved for documenting maintenance procedures is from calibration through application of the test substance. This requirement involves communicating closely with the equipment owner to determine if any maintenance performed will necessitate recalibration of the application equipment.

The Quality Assurance Unit

Field studies represent a unique situation for the QAU. These studies take place throughout the United States, and the duration may vary from several weeks to several months. Each terrestrial field study must be inspected by the QAU at least once while the study is in progress to ensure compliance with GLPs, SOPs, and the approved study protocol.

Our study inspections occur as early in the study as possible and usually coincide with application of the test substance. While the quality assurance officer is on site, as many study phases as possible are inspected. The data also are inspected to ensure that information is recorded properly and that procedures are fully documented. In addition, the facilities, personnel, and equipment are evaluated to ensure compliance with GLPs.

The phase inspection in the field is similar to that conducted in a laboratory. For example, the quality assurance officer will accompany a biologist on a bird survey to ensure proper conduct of the procedure in accord with the protocol and the SOP.

In addition to the site inspection, each study protocol and final report is audited. Each protocol, unique to a specific study and location, is reviewed for compliance by the QAU. The protocol will be reviewed to be certain that the methodology and sampling schedule are clearly described and that the protocol and associated SOPs clearly describe the procedures being used. If further clarification is needed, the protocol or the SOP may be amended or a new SOP may be prepared. The final audit by the QAU on a terrestrial field study is conducted to ensure that the final report accurately reflects the raw data.

Test Substance

Responsibility for the test substance, such as proper storage, distribution, identification, and disposal, are shared between the sponsor and the laboratory conducting the work. Frequently, the test substances being studied are hazardous and require special licenses, equipment, storage conditions, and training in handling. Large amounts of test substance are generally required for these studies, sometimes as much as several tons. Therefore, the safest and best method to handle the product on site is to have the sponsor ship the test substance directly to a chemical distributor. The farmer or commercial applicator then takes delivery of the test substance from the chemical distributor. Chain of custody of the test substance, including date of receipt, any transfers, location of storage, lot numbers, amount used for the study, and disposition of the remaining test substance must be well documented. Empty test substance containers must be retained until study completion. A conditional exception for disposing of the containers may be obtained, on a case-by-case basis, by writing to the Office of Compliance Monitoring at the EPA. If the exception is granted, the EPA will specify certain record-keeping steps that are required to account for the test substance containers in lieu of actual storage of the containers.

Conclusion

Although terrestrial field studies present unique problems when implementing the GLPs, some basic principles apply. Consistent data recording and proper documentation of all observations and techniques throughout a study must be maintained. Problems that may arise when applying the GLPs to terrestrial field studies can usually be solved by remembering the basic principles of the GLP regulations.

References

1. Fite, E. C.; Turner, L. W.; Cook, N. J.; Stunkard, C. "Guidance Document for Conducting Terrestrial Field Studies", U.S. Environmental Protection Agency: Washington, DC, 1988; p 67.

2. "Final Rule for Good Laboratory Practice Standards Under the Federal Insecticide, Fungicide, and Rodenticide Act (FIFRA)", *Code of Federal Regulations* Title 40, Pt. 160; *Federal Register* 54 (August 17, 1989) pp 34052–34074.

RECEIVED for review September 17, 1990. ACCEPTED revised manuscript September 13, 1991.

Problems with Ecotoxicological Field Studies

John A. McCann

U.S. Environmental Protection Agency, EN–342, 401 M Street, S.W., Washington, DC 20460

Conducting aquatic mesocosm and terrestrial field studies in compliance with existing Good Laboratory Practice (GLP) Standards requires not only meeting the requirements of typical laboratory studies but also allowing for special considerations created by such factors as the characteristics of wild test organisms, varying environmental and ecological conditions, and difficulties in recording data. Relatively few ecotoxicological field studies have been completed to date, and even fewer have been evaluated as to their conformance with GLPs. This chapter discusses several areas that scientists and sponsors have neglected while conducting ecotoxicological field studies.

IN 1989, THE ENVIRONMENTAL PROTECTION AGENCY (EPA) expanded the Federal Insecticide, Fungicide, and Rodenticide Act (FIFRA) and the Toxic Substances Control Act (TSCA) good laboratory practice (GLP) regulations to include ecotoxicological and residue chemistry studies, some efficacy studies, and field studies. Many of the laboratories affected by these regulations were aware of the requirements of the GLPs because they had been conducting ecotoxicological studies under TSCA GLPs for several years. Because most FIFRA studies would have to be conducted according to the GLPs once the GLPs became effective, the EPA inspectors began conducting inspections and data audits at pesticide testing facilities in the early 1980s. Most of these laboratories were aware of the regulations when the GLPs went into effect in August 1989. However, not until 1989 did the EPA inspectors begin conducting FIFRA inspections at laboratories and field sites where ecotoxicological, residue chemistry, efficacy, and field studies were being conducted. To date, few compliance inspections have been conducted on field studies. No compliance inspections have been conducted on mesocosm or ecotoxicological terrestrial field studies.

I believe the same type of problems found in laboratories conducting ecotoxicological studies will be found in those facilities conducting field studies. Initially, the problems may appear more frequently in field studies because of the interpretation of the regulations by the regulated community, the inexperience of the laboratories in conducting these types of studies, and, in many cases, the small size of the groups conducting these studies. The administrative requirements for conducting some of these studies, particularly when conducting the studies at multiple sites, are more than many small facilities can handle conveniently. Field studies are labor intensive and can put a heavy strain on personnel and equipment being used to conduct or monitor the studies. For instance, quality assurance (QA) must be conducted by a trained individual who is not involved in the conduct of the study and who reports to someone supervising the study director. The individual to whom the QA officer reports may also be the owner or manager of the company doing the study.

Specific findings from GLP inspections will not be cited in this chapter. For illustrative purposes, I have combined problems found at several sites, so the examples may not be an accurate description of conditions found at any one laboratory.

Inspectors are encountering or expect to encounter problems at some laboratories and field sites. With adequate planning, scientists should be able to avoid problem areas rather than needing to justify a poorly documented study or practice. Scientists or workers who have questions concerning the correct procedures should contact the EPA before initiating a study. Technical problems should be directed to the appropriate branch in the Office of Pesticide Programs or in the Office of Toxic Substances. GLP compliance questions should be directed to the policy branch of the Office of Compliance Monitoring.

During the inspection of testing facilities, numerous problems have been identified. Because of the complexity of ecotoxicological field studies and the frequent strain on resources available to a testing facility, the potential for GLP deficiencies or protocol deviations during a field study is much greater than during a laboratory study. This chapter discusses some of the more serious problems that might occur during an ecotoxicological field study.

Study Director

Problems with the interpretation of the GLP regulations concerning the study director have always occurred. With the expansion of the regulations to field and residue studies, interest in the EPA's interpretation of the regulations in this area has been renewed. Before the 1989 enactment of the regulations, study directors were required to have technical knowledge of the study and were responsible for the conduct of the study. In many cases, the study directors believed they were responsible only for the phase of the study conducted under their direct supervision. They were not required to sign a statement that the entire study was conducted according to GLPs. Many studies did not require the involvement of other individuals who were not under the study director's control when analyzing samples, evaluating data, or archiving samples. Frequently, field studies require activity in several areas not easily supervised by one individual. Because it is difficult for all the activities in a field study to be under the direct control of one individual, the original concept of the study director being present and technically knowledgeable in all aspects of the study was redefined by the EPA.

Under the GLPs, the study director is responsible for the conduct of the entire study. Because the study director is frequently unable to have firsthand knowledge of all phases of the study, he or she must rely on documentation, observations, and records made by others to ensure that the studies were conducted according to the GLPs.

Some contractors or individuals at laboratories have been unwilling to assume the responsibilities of the study director, so the study director is being assigned at the sponsor level. The EPA expects the study director to take the appropriate steps and precautions necessary to ensure that the study is being conducted according to the regulations before completing the compliance statement and signing the report.

In one recent field study, there were three study directors, one at each field site and one at the chemistry laboratory analyzing the samples. A study director treated the fields, collected the samples and data at the field site, or analyzed the samples. The biological data or sample analyses were sent by the study directors to a contractor for evaluation and report writing. The signoff on the report was by each of the study directors for their contribution to the study and by the sponsor. The contractor did not sign anything in order to avoid responsibility for any failures to meet the regulations. Other than the sponsor, no one individual was responsible for the conduct of the study. This study will not meet GLP requirements.

In the example just given, the contractor directed the scheduling of events, controlled the conduct of the study, evaluated the data, wrote the report, and controlled the archiving of the raw data. The contractor actually performed the duties of the study director and should have assumed that responsibility. The designated study directors were actually principal investigators at each site. As study director, the contractor could have delegated the responsibility of conducting a portion of the study according to the GLPs to each of the principal investigators. Because a contractor cannot be present at every testing site, the contractor may have to spend more effort ensuring the quality of the work at each site rather than monitoring the daily activities at the field sites and analytical laboratories.

Currently, study directors and sponsors are apparently unsure of the extent to which they must be able to document the quality of the data and compliance with the regulations at the test sites. Because there are many combinations of factors that can affect the quality of the data at a field study, there is no simple solution. The study director or the sponsor is responsible for collecting enough data to document that all phases of the study proceed according to the protocol and the standard operating procedures (SOPs) of the organizations involved. If the study director does not have enough data from the laboratories or field sites to document compliance with the GLPs for a particular phase of a study, the compliance statement should address those phases of the study not conducted in compliance with the GLPs. The sponsor should either permit the study director to state in the compliance statement that parts of the study may not have been conducted in compliance with the GLPs, or the sponsors should provide data (proof) to the study director that parts of the study were conducted according to the GLPs. This arrangement is particularly effective when chemical analyses are conducted by a laboratory not under the control of the study director. Before signing a compliance statement covering work done at a facility not under his or her control, the study director should have good documentation of the quality of the work. The study director should not sign the statement indicating compliance with the regulations unless there is adequate documentation. The sponsor should also require full GLP documentation for all phases of the study, including any work conducted by sponsor-owned laboratories.

The study director is responsible only for the study-related activities. Test facility "management,"[1] whether at the sponsor, contractor, subcontractor, laboratory, or field site, are responsible for the administrative activities involved in the study (e.g., QA activities, drafting SOPs, archiving, and maintenance of equipment). When the principal investigator or study director is also the owner of the company, an interesting scenario is created. The QA activities for a study might be performed by a contractor, whereas the activities for the facility (e.g., monitoring equipment maintenance, drafting SOPs, or maintaining the master schedule) might be performed by laboratory staff. The non-study-related activities could be handled by the QA staff under the owner, and the study-related QA could be handled by another quality assurance unit (QAU). To prevent conflicts of interest, the QAU must report to someone supervising the study director.

Protocols

The protocol should provide a detailed description of the study. It should contain enough detail that the study could be conducted by using only the protocol and SOPs.

When options are built into phases of the study or protocol, the criteria for making the selection should be stated in the appropriate documents. If there are deviations from or changes in the protocol or SOPs during the conduct of the study, the deviations or changes should be described and addressed in the raw data and the final report. When protocol changes are planned, the reason for the change and the approval of the change by the study director and sponsor should be documented in a protocol *amendment*. Unplanned changes, such as schedule changes due to adverse weather, are *deviations* and should be adequately documented in the raw data and addressed in the final report with a statement as to the overall effect on the test results.

Because of the conditions under which ecotoxicological field studies are conducted, there are ample reasons for deviations due to weather conditions that can affect the conditions of the studies, plant growth, animal movements, and the characteristics of the wild organisms used in the study. Experienced scientists can frequently anticipate many of these variables and can write protocols and SOPs that allow for adjustments in procedures for unexpected conditions. As long as the criteria for the accept-

---

[1]Management is defined here as the person or persons responsible for setting policy at a facility. For a description of the duties of testing facility management, see § 160.31 of the GLP Standards.

able deviations are described in the protocol or appropriate SOP, the deviations can be initiated and then discussed and evaluated in the final report.

Whether because of the state of the art for ecotoxicological field studies, the wide array of possible deviations, or the lack of experience of the biologists conducting these studies, many protocol deviations occur that apparently are not covered by the protocol, protocol amendments, or SOPs. These deviations are frequently not mentioned in the final report either because of neglect of the researcher or failure to properly analyze the data or protocol requirements. Deviations from the agreed upon procedures must be appropriately addressed.

Failure to take a series of samples or to record data is a serious deficiency. Following is a list of serious failures:

• Test concentrations may not be measured at the beginning or end of a study when these measurements are required.

• The location of the water line in a pond may vary unexpectedly because of weather conditions, and thus researchers may fail to make proper adjustments in the time and place that samples will be taken.

• Treatment of the test site or followup sampling may be delayed because of weather conditions.

• The growth stage of the crop to be treated, or the movement or lack of movement of test systems into the area, may be delayed because of the weather.

• The biologists may fail to make appropriate changes in the schedule, resulting in critical observations not being made at the appropriate time.

• When chemicals are being tested under TSCA, all deviations from protocols, SOPs, and test rules standards must be addressed as soon as possible in protocol amendments or deviations. When test rules require the collection of data and the reporting of findings, the information must be included in the final report, or the scientists are in violation of the test rule.

In many instances, extensive changes have been made in the conduct of the study because of either inadequate prestudy information or of poor planning. More than one laboratory has started a study with only a few test organisms because little or no information was available on the toxicity of the chemical to the organisms being used. All the organisms were not exposed at first to allow for last-minute changes in the number of organisms per test container or to make adjustments in test concentrations to be used, based on adverse effects noted in the first organisms exposed.

These changes in procedures may affect such parameters as sampling schedules, stability of the test compound, and comparability of control organisms. These protocol deviations may not be adequately addressed as protocol amendments or deviations in the records or final report. Frequently, the significant deviations are not discussed or evaluated in the report. Sometimes only the starting date is changed.

Whenever a change occurs in the schedule, appropriate changes should be made in the entire schedule to be sure important data are not missed. Changes in environmental conditions can have a dramatic effect on the conduct of a study and the changing of the proposed schedule. The protocol should address the potential problem areas and allow for adjustments in schedules and alternative courses of action whenever possible. Additional adjustments in the conduct of the studies should be addressed in protocol amendments or protocol deviations.

Numerous problems occurred in a field study in a saltwater marsh in the southeastern United States because of inadequate planning. Several ponds had dissolved oxygen levels below that required to keep the test animals alive before the treatment was applied. In other areas, the bow wave from large boats passing by the entrances to several ponds either flooded the area, destroying live boxes, or drew so much water out of the pond that not enough was left to support the fish populations in the pond.

By changing conditions at the treatment area or affecting the movement of organisms into or out of the test area, weather conditions can have a dramatic effect on the scheduling of a study. Weather conditions can delay the start of the study, can change the location and timing of taking samples or making observations, and can adversely affect the availability of personnel by changing the schedule when observations should be taken.

Personnel Records

When the EPA plans to conduct a GLP inspection or data audit at a field site, the appropriate representative of the laboratory conducting the study is notified of the upcoming inspection. The EPA will request that the necessary documents to verify the conduct of the study be available on-site. Included in this list are the qualifications, job descriptions, and training records for the personnel involved in the study. This information is needed by the EPA to determine who is responsible for certain aspects of the study and who has the credentials to conduct the various aspects of the study. These documents should indicate the individuals who are qualified to do the work.

The inspectors are finding inadequate documentation of the qualifications of the staff performing the work. Sometimes the individuals obvi-

ously have the credentials to do the work, but no written records are available. In other cases, the few available records document the fact that the
individuals are not qualified to do the work. There appears to be relatively high turnover of workers in laboratories, particularly at the lower
skill levels. Frequently, these individuals are called on to clean glassware,
set up studies, feed animals, take samples, and even record data, and if any
of these tasks are performed incorrectly, the effect on a study can be serious. On several occasions, individuals with no previous experience and
with less than 2 days of training were observed taking water samples and
recording critical data on a chronic study involving fish and a radioactive
substance. Two years later, when asked about the study, one individual
stated that he now knows what should have been done the first day. This
laboratory is repeating this study for the fifth time because high concentrations of test material were measured in the water at all test concentrations and in the controls. Until recently, no SOPs in the laboratory
addressed collection of samples or the handling of radioactive material.

When management personnel or senior scientists leave the laboratory, positions are frequently filled from within by scientists or technicians
with less experience. The technicians are promoted to fill the vacancies
frequently without the education or the experience of the individual leaving the position. Eventually, technicians may reach management and
study director levels without formal education and with no documented
training or experience. In some cases, management will update the training records and document job experiences the day before designating the
individual as a study director. Technically, these individuals have one day
of documented experience before becoming a study director. When problems are found with a study and the qualifications are basically undocumented for one or more of the staff involved, it reflects unfavorably on the
QAU, the study director, and facility management.

Training and Experience

The GLPs require that the laboratory maintain experience, training, and
job descriptions for all employees involved in the studies. The records
should document the qualifications of the individuals at the time they participated in the regulated study.

In a recent field study, the study director was using his students as
field investigators. The study director could document the training of only
the students in his classes whom he planned to use on the study. On
several occasions, the number of his available students was not sufficient
to make the observations. He then used other students with no docu-

mented experience or training in the area in which they were working (i.e., counting population, carcass searches, and sampling).

Recording Data

All data generated during the conduct of a study, except those collected on automated collection systems, must be recorded in ink. All entries should be dated and signed on the day of data entry by the person recording the data to facilitate reconstruction of the study.

Inspectors have found observations being recorded in pencil by technicians, biologists, and even study directors. Erasers and correction fluid were being used to correct mistakes. Apparently, no one at the laboratories was trained or had experience in recording data by procedures stated in the GLPs. An error in recording data should be lined out with a single line so that it is possible to see what error has been corrected. The corrections should be initialed and dated, and the reason for the correction should be stated. Some common problems related to recording data are detailed in the following list:

- In one laboratory, a technician was recording the daily observations on scraps of paper in pencil and was later recopying the data onto data sheets in ink. The scraps of paper were then thrown out.

- In another case, work sheets for raw data were being set up in bound notebooks in pencil by the study director. Sometimes important data were missed because the space for the data had been left out when the handwritten forms were set up. The laboratory now uses preprinted data sheets for collecting appropriate data.

- About two years ago, a major testing laboratory was recording about 95% of its raw data in pencil on multicopy forms. By examining the carbon copies, it was possible to determine where changes in data were made, but the eraser marks so smudged the carbon copy that neither the original nor the corrected data could be read.

- In another situation, the data for a field study were recorded in pencil. When the study director, who was also the owner, was notified that the studies were to be audited, he learned that the studies should have been conducted according to the GLPs. He neatly recopied all the data in ink and went to considerable effort to document the conduct of the study and to locate missing data. He acknowledged his efforts as he handed his records to the auditors. He had also cleaned up the data base by throwing out the original raw data that would have permitted the EPA to verify the data in his report.

- Until recently, several ecotoxicological laboratories were first recording their raw data in field notebooks and later recopying the data onto preprinted forms at the laboratory. When the EPA tried to verify the reported data using the raw data, it found numerous recording mistakes. Several study directors had never referred back to the field notebooks to verify the accuracy of the data with which they were working.

- When original data are entered into a computer data base for processing, some study directors and QA personnel verify only the accuracy of the computer analyses of the data and not the transfer of data from the original records to the computer data base.

- Several laboratories fail to describe in an SOP the use of field notebooks, the types of data that should be recorded in them, and what should be done with the notebooks when they are full or when the employee leaves the laboratory. Some individuals have taken the notebooks with them because they wanted access to the raw data. Others wanted the books because of personal data recorded in them. Improper processing and archiving of the notebooks, resulting in the destruction of raw data, can be a serious GLP violation. The data recorded in field notebooks must be study specific.

Master Schedule

The regulations clearly state the minimum data that should be listed in the master schedule. The master schedule assists the laboratory staff, the QAU, and inspectors and auditors in determining the status of any study being conducted by the facility. It allows the laboratory to make effective use of their facilities and equipment, and it allows the QAU to schedule inspections and data audits at appropriate points in the studies. It also gives inspectors and auditors an opportunity to evaluate the use of facilities and staff and to select phases of studies to be audited. To be in compliance with the GLPs, the master schedule should list the following information: the test chemical, sponsor, study director, type of study, test system, study initiation date, and status of each study.

The management and QAUs of several small facilities have failed to identify one or more of these criteria. In some cases, the master schedule is maintained by the study director, particularly if the laboratory is making use of a contract QAU. Laboratory management want to limit such information to employees of their facilities. The master schedules in many small laboratories are indexed by sponsor or initiation date because they are used primarily for billing purposes and not for the reasons required by the EPA. The master schedule is required by the EPA to identify the scheduling of work on particular chemicals, to ensure that the laboratory has or had adequate facilities and staff to conduct the studies, and to aid the laboratory with the location of all testing data on a particular chemical.

Quality Assurance

When the EPA begins inspecting small field stations or sites, the inspectors usually find problems with the placement and duties of the QAU. Some of the duties of the QAU are study specific. Study-related problems should be reported to the test facility management, who supervise the study director, so that corrective action can be taken if required. This action is acceptable as long as the study director is not the facility management, which could be the case in small field stations or in cases where cooperators are used. The QAU performing a study audit should not be supervised by the study director.

Some of the QAU's administrative functions, such as maintaining the master schedule, monitoring use and maintenance logs of equipment and test systems, and maintaining such records as personnel records and SOPs, are facility requirements and should be reported to the local test facility management (not the contractors of the study).

When the QA for the study is provided by the contractor or sponsor, the QAU reports should go directly to the study director and the sponsor or contractor. Some field stations or laboratories do not want contracted QAUs examining their master schedule if the QAU also works for or is a competitor. There are also problems if the study director is the owner of the field station or laboratory. The sponsors do not usually provide the independent QA for the administrative responsibilities at the field station office.

In an effort to have an independent QAU at the testing facility and to avoid the appearance of a conflict of interest, some managers are assigning the title of study director to individuals with a minimum of qualifications. The manager still closely monitors the conduct of the study. The QAU would now report to the new study director and manager of the facility.

Field stations and cooperators also experience many of the same problems as laboratories that find it necessary to rely on other facilities to complete parts of the study, such as chemical analyses of water or feed samples, particularly if the study director or QAU is located at one location and the other phase of the study is being done at another facility under different management.

A recent interpretation of the regulations by the policy branch of the EPA's Office of Compliance Monitoring indicates that the study director at one location can delegate the authority for the conduct of a phase of a study to a principal investigator at another location. Test facility management can also delegate the responsibility of conducting QA inspections to another QAU at another location (contract laboratory) as long as the personnel were not involved in conducting the study. Documentation of QAU phase inspections and compliance statements is required from both locations. The study director, the QAU, and the sponsor are responsible for assessing and documenting the adequacy of compliance with the GLPs at all locations involved in the study before signing the report and compliance statements.

Biological testing facilities are having particular problems when the sponsor either performs the chemical analyses or contracts them out to a facility that either is in competition with the biological facility or that has never been inspected by the biological laboratory doing the exposure part of the study. The study director, test facility management for the study, and sponsor should have documentable evidence that all phases of the studies were conducted according to GLPs before signing the report and the compliance statement. In cases where documentable evidence is not available, that phase of the study should be listed as an exception in the compliance statement. When data requirements for TSCA studies and FIFRA studies are specifically identified in the *Federal Register* for particular chemicals or uses, these data must be provided with the study report unless the sponsor or laboratory has received a special exception. Failure to submit the data as requested with or in the final report can be a violation of the appropriate TSCA or FIFRA regulation and the GLP regulation. The protocol and appropriate *Code of Federal Regulations* require submitting data and conducting the study using GLPs.

Archives

The EPA requires that all raw data, protocols, test reports, SOPs, and personnel records be archived in adequate facilities to ensure their availability and usability at a future date. Adequate protection from fire, adverse weather conditions, and unauthorized personnel, which might destroy or misuse the data, should be provided.

Ideally, the data should be properly indexed in a fireproof, locked cabinet in a locked room and stored under adequate environmental conditions. Access to the room should be limited to authorized personnel, and the keys should be handled by a designated archivist, who oversees the use and maintenance of the archives. Sprinkler systems should be used for additional fire protection, depending on the type of storage cabinets being used.

Several laboratories have been unwilling to allocate the space, the equipment, or the designated staff to maintain an adequate archive. Inadequate archiving of data has contributed to the loss of data, payment of fines by the sponsor, rejection of studies, and unfavorable publicity. If sponsors are unable to provide raw data to support study reports, they can be subjected to one or more of these adverse actions. Destruction of raw data for a required study is a serious offense and a violation of GLP, TSCA, and/or FIFRA regulations.

Improper storage of raw data has resulted in data being destroyed. In one case, some raw data were lost when the building in which they were temporarily stored was destroyed. In another instance, an employee destroyed raw data sheets because no provisions were made at the laboratory to archive the original data. All original data should be archived in adequate facilities as long as they are needed in accordance with the EPA's data retention schedules.

When it is possible, field data should be recorded directly onto prepared data sheets and archived in a central file. When data are being recorded for a number of studies or a historical record, a true copy of the appropriate data pages from the logbooks should be filed with the raw data to ensure that it will be available in the future. A copy also provides a convenient reference when the study director prepares the report or when the accuracy of the data is verified by the QAU or an inspector or auditor.

Documentation of GLPs

Scientists and managers must develop ways to ensure that adequate data are being collected and saved. Procedures must be set up to ensure that GLPs are being followed at sample mixing sites, application sites, chemical analysis sites, and data processing areas. If the QAU or study director or a qualified representative cannot be on-site when critical phases of a study are being conducted, two individuals could be used to document and verify that correct procedures were used to collect the data. The data will have to be complete enough to permit a detailed reconstruction of the study. Some laboratories use video cameras to document the conduct of various phases of a study. Others use tapes or movies to train staff members before sending them on-site to conduct a particular phase of a study.

Test Substance

Adequate characterization, handling, and storage of the test substance are essential to conducting any meaningful study to be submitted to the EPA. The investigator at the field site must be able to provide adequate documentation on the identity of a test substance. Without this documentation the study is meaningless.

Records at the test site should document the use of the test substance as called for in the protocol. There should also be documentation that the test substance was stored under acceptable conditions to ensure the quality of the sample, that no adverse environmental conditions affected the test substance, and that inappropriate use or use by unauthorized personnel did not occur.

In one field study in the southeastern United States, the test substance was maintained by local pesticide operators in metal sheds in the field. There were no analyses of the test substance used, no analyses of the tank mixes, and no assurance that the test substance had not been destroyed by environmental conditions. There should be documentation that the test substance has not changed during storage, particularly if there is any question concerning storage conditions.

Another concern of the inspector and the study director should be the handling of the test substance at the mixing site while the inspector and the study director are at the application site. If the activities at either the mixing site or the application site are not monitored by the QAU or the study director, another individual should be on-site to verify what occurred at each site. Some laboratories have gone as far as using video cameras to monitor activities at both locations. Paper trail documentation and clear photographs would constitute a good record of the conduct of a study.

The security of the test substance can be ensured by storing it in limited-access areas. If the test substance is stored in less than adequate conditions, documentation verifying the quality and tracking the use of the test substance should be available.

With respect to disposal of empty test-substance containers, authorization to dispose of the containers may be obtained on a case-by-case basis by contacting the EPA at the following address:

Jack J. Neylan, III
Policy and Grants Division (EN342)
Office of Compliance Monitoring
U.S. Environmental Protection Agency
401 M Street, SW
Washington, DC 20460

Conclusion

To date, the EPA has not been involved in an on-site GLP inspection of a laboratory involved in a mesocosm or terrestrial field study, so it is difficult to say exactly what deviations from the GLPs will actually be found when an inspection is conducted. Because of the complex nature of these studies, many problem areas may not be addressed by this chapter. Study directors, QA personnel, and laboratory management should not limit their concerns during a field study to only the problem areas discussed in this chapter. I hope that when study directors and managers are aware of some of the problems anticipated by the EPA when auditing field studies under GLPs, they will consider these problems when planning a study and thus avoid them.

RECEIVED for review September 17, 1990. ACCEPTED revised manuscript August 17, 1991.

Agrochemical Groundwater Studies

Sandra C. Cooper and James M. DeMartinis

Blasland & Bouck Engineers, 6800 Jericho Turnpike, Suite 210W, Syosset, NY 11791

Application of Good Laboratory Practice Standards presents a challenge to scientists conducting agrochemical groundwater field studies because of unexpected changes in weather patterns, dynamic field conditions, and variable geologic conditions. Standard operating procedures (SOPs) for field practices should be written to provide sufficient flexibility to adapt procedures to site-specific conditions without compromising the study protocol. Corrective changes to SOPs and protocol deviations that could affect the quality of the study can be minimized by involving personnel with field research experience in writing SOPs and study protocols. Problems in recording field data or in modifying SOPs on-site often can be attributed to adverse field conditions, time constraints, lack of proper training, or lack of communication with the study director.

THE QUALITY AND INTEGRITY of laboratory and field studies submitted in support of pesticide registration are guided by Good Laboratory Practice (GLP) Standards and standard operating procedures (SOPs).

The U.S. Environmental Protection Agency (EPA) issued GLP Standards under the Federal Insecticide, Fungicide, and Rodenticide Act (FIFRA) to specify minimum practices and procedures that testing facilities, such as analytical laboratories, are required to follow in support of the registration of a pesticide product under FIFRA (*1*). However, these GLP Standards initially were designed to fit the needs and criteria of the analytical laboratory, which can be considered a controlled environment.

In August 1989, the EPA expanded the final FIFRA ruling on GLP Standards to require compliance with these regulations of all studies performed in support of pesticide product registration, including field studies (*1*). Complying with the new FIFRA GLP regulations, however, has presented field investigators with a unique challenge because the environment in which field work is performed can be highly variable, and field conditions can change rapidly.

Although compliance with the regulations is now required for all field monitoring studies, the regulations have not been altered or revised in any way to address specific GLP issues as related to field studies. Consequently, field investigators have been forced to adapt to the field environment GLP Standards that were designed for the laboratory environment.

Because no specific GLP requirements identify or address field monitoring issues, the following list offers a few examples of adaptations of GLP Standards needed for field-scale studies:

- collecting on-site precipitation data

- monitoring on-site test substance storage conditions

- calibrating chemical application equipment

- monitoring and documenting supplemental irrigation

- documenting in a field notebook date, time, personnel, and the procedures followed to collect soil, soil–water, and groundwater samples for residue analysis

- documenting weather and field conditions for every sampling event

- monitoring the direction of groundwater flow and seasonal changes in water-level fluctuations

- collecting shallow and deep soil cores for characterization

- labeling all monitoring installations (e.g., wells or suction samplers) and subsequent samples collected from these installations with unique identifiers

- documenting chain-of-custody forms to accompany all samples shipped from the field site to the analytical laboratory
- documenting cultural agronomic practices
- monitoring wind speed, air temperature, and relative humidity at the time of application

Types of Groundwater Monitoring Studies

Field-scale groundwater monitoring studies in support of product registration are performed for two purposes: either to determine the potential for a pesticide to leach to groundwater from a specific use or to determine whether a currently registered pesticide with a history of use already has leached to groundwater (2). To address these two objectives, two types of field-scale groundwater monitoring studies have been designed: small-scale prospective studies and small-scale retrospective studies.

Small-scale prospective studies are designed to monitor the movement and to characterize the fate of pesticide residues from the time and point of application to the land surface to a time and point where almost 100% of the chemical has dissipated. Small-scale prospective studies are a requirement for registration for new pesticides, for pesticides with a short history of use, and for existing pesticides with proposed new uses. Monitoring and characterizing the leaching patterns of pesticide residues on a site-specific basis requires the collection of soil samples; soil-pore water samples from suction lysimeters; and groundwater samples from shallow, small-diameter monitoring wells. Additionally, site-specific field data are collected to characterize the hydrogeologic environment and groundwater conditions of the study site.

Small-scale retrospective studies are designed to determine if, after years of normal use and agricultural practices, residues of pesticides have leached to groundwater. Similar to prospective studies, retrospective studies also are conducted on a site-specific basis; however, the types of data collected to evaluate the impact of pesticide residues on the local groundwater system differ. Field data collected for small-scale retrospective studies include soil cores collected from the unsaturated zone down to the water table and groundwater samples collected from shallow, small-diameter monitoring wells. Soil cores are collected from the unsaturated zone at the beginning of a retrospective study to characterize the physical nature of the soil comprising the unsaturated zone, to evaluate the concentration of pesticide residues that may be present in the soil as a result of previous applications, and to determine the maximum depth of pesticide leaching. Similar to prospective studies, site-specific hydrogeologic data also are collected for retrospective studies to characterize not only existing groundwater conditions, but also the hydrogeologic vulnerability of the study site.

SOPs and Study Protocol

Two integral parts of a groundwater monitoring field study that are requisite to ensure the integrity and quality of study results are SOPs and study protocols. SOPs set forth in writing the study methods to be used in the field to ensure the collection of accurate, reliable data. The study protocol describes clearly and succinctly the objectives of the study and all methods to be used to conduct the study (*1*).

The study protocol, which is developed before beginning any field work, delineates a series of well-defined investigative tasks; the methods and procedures outlined in the protocol are based entirely upon the SOPs. Developing the study protocol before the onset of field work thus enables the registrant (manufacturer of the agricultural chemical), the team conducting the study (often a consultant), and the EPA to review and critically evaluate the approach and effectiveness of the proposed monitoring program (*3*).

Regardless of whether the study is prospective or retrospective in nature, the series of investigative tasks outlined in the study protocol address the following five phases of field work:

• selecting a study site

• characterizing the study site

• installing monitoring equipment

• sampling soil and water (includes applying chemicals for a prospective study)

• reclaiming the study site

The protocol serves as the definitive guide for conducting the study. Therefore, the protocol must not only follow the GLP format, but should also be written with sufficient flexibility to adapt procedures and methods to site-specific conditions without compromising the integrity of the study.

The best approach to developing a study protocol is to involve key field scientists in all stages of protocol and SOP writing and reviewing before the onset of a study. Participation of field scientists trained in GLPs in the early stages of protocol development takes advantage of their familiarity and expertise with field techniques and field conditions. All too often, researchers make the mistake of assuming that the protocol should be written in "cookbook" fashion with step-by-step instructions so that anyone can go into the field and conduct a groundwater monitoring study. Consequently, the final protocol and SOPs may be written without sufficient flexibility. For instance, in an SOP that describes soil sampling procedures, indicating that soil samples should be composited into a 13 × 22-inch polyethylene sample bag would be a mistake. Stating specific sizes is not only impractical in this instance, but eventually will lead to protocol deviations whenever these exact size bags are not used because of some unforeseen circumstance in the field. A more practical statement would be to indicate in the SOP that sample bags used in the field must be of sufficient size to contain the samples. This statement still addresses the issue of the sample bags, but it now provides the field scientist with more latitude in making modifications or adaptations in the field whenever necessary, without affecting the integrity of the study.

Training

Regardless of how much care is taken in developing the study protocol and SOPs, unforeseen problems still can occur in carrying out investigative tasks in the field. This is the nature of all field work because the field environment can be highly variable and conditions commonly are unpredictable.

From time to time, even the most seasoned field scientist will face unexpected events that result in problems that may require immediate, on-site assessment and adjustments of routine field procedures. As a result, all field personnel involved in GLP-related groundwater monitoring studies should receive extensive training in working with GLPs, in performing field procedures, and in operating field equipment and instruments. GLP training teaches the field scientist how to make and record properly any deviations from or amendments to protocols and SOPs when the situation presents itself in the field.

All training that field personnel receive, whether in-house or in the field, must be documented on training records and kept on file in the archives. Training records should include the following:

- seminars and courses
- certifications
- project work histories

- field equipment operation

- technical procedures proficiency

Training records serve not only to document training received by personnel working on GLP-related field studies, but also to indicate who needs additional training in specific areas.

In-house GLP training provides field scientists with a basic understanding of what GLP Standards are, why they are required, and how they are implemented in the field. No substitute can be made, however, for actual hands-on experience in the field. Field training in the use of specific types of field instruments and in the performance of routine and nonroutine field procedures should always be conducted by an experienced field scientist or field manager. Field training under the watchful eye of an experienced field scientist will teach the inexperienced field technician the proper techniques and methods to use in the field and how to record field notes and data in a field notebook in accordance with GLP Standards.

The field scientist can never receive too much training. For new or inexperienced field personnel, exhaustive field training serves to build confidence in the performance of their duties. It also teaches by example the types of problems that can occur without warning, how to evaluate these problems, when and with whom the field scientist should communicate in the event of a problem, always to anticipate the unexpected, and how to adapt or modify on-site routine procedures to resolve problems.

Thus, thoroughly educating field personnel in proper field methods and techniques and making everyone aware of the importance of implementing and adhering to GLP Standards are important elements in conducting and producing a high-quality field-scale study.

Examples of GLP Issues in the Field

The following are examples of problems that occurred in the field because the field SOP or study protocol was too inflexible or written without flexibility to accommodate unpredictable field conditions such as weather problems, lack of proper training and experience, and poor communication between groups involved in the study.

SOPs Written Without Sufficient Flexibility. Protocols for small-scale groundwater monitoring studies follow the EPA document "Draft Guidance for Ground-Water Monitoring Studies" (2) that, as of this writing, are still in draft form. One of the more important phases in these studies is site selection and characterization. Selection of a site that meets the EPA vulnerability criteria can be difficult, and obtaining the EPA's approval of a site can be time consuming. Because of the pressures of registering or reregistering a particular chemical and the minimum 2-year duration of small-scale groundwater studies, work commonly is begun at sites that the registrant believes will be acceptable to and approved by the EPA.

One such study was begun at a site that had been characterized by the registrant. The protocol for the study and the site-characterization SOPs were designed to meet the EPA guidelines. Soil borings were collected in the field at locations identified from a grid pattern designed for the site. Initial analyses of soil samples collected in these borings indicated that the site was underlain by sandy soils, a permeable unsaturated zone, and a water table within 20 ft of land surface (vulnerable hydrogeologic conditions). On the basis of these initial findings, the registrant then proceeded to the next phase of the study, which consisted of installing clusters of monitoring wells. Four well clusters were to be installed along the edge of the field site. While drilling the third well cluster, a substantial thickness of clay was found below the root zone and above the water table. The presence of clay in the unsaturated zone caused some concern because of its proximity to the edge of the field site. Concern increased when clay was found at about the same depth below the land surface where the fourth well cluster was installed. To comply with the study protocol and to meet the definition of a hydrogeologically vulnerable site as per the EPA guidelines, no restrictive layers (i.e., clay) can be present in the unsaturated zone. The field investigator at this point felt helpless because initial field results for characterization of the site indicated that the field was suitable as a test site and also because all SOPs had been strictly adhered to during the characterization phase of the study. Unfortunately, at this point in the study, the monitoring well clusters were in place, and more than $100,000 had already been spent by the registrant.

Upon independent review, an experienced soil scientist identified evidence at the land surface of the subsurface clay lens. The soil scientist questioned why soil characterization borings had not been collected in areas of the field where ponded water was in evidence. The registrant's field scientist explained that he had followed the grid design (as laid out in the study protocol) and did not feel he had authority or license to alter the program even if he had recognized a potential problem. The independent soil scientist recommended that additional soil borings be collected to determine the lateral extent of the clay lens. These additional borings confirmed that the lateral extent of the clay was limited. As a result, the site and the study were salvaged by merely adjusting the boundaries of the actual test site beyond the lateral extent of the clay lens, thus meeting the requirements of the study protocol and the EPA guidelines.

Education, Experience, and Training. Another important issue concerns who in the field is designated to supervise the work. SOPs are mistakenly written specifically to ensure that anyone with a college degree can follow the procedure. Real problems occur, however, when things do not go as planned or when the person in the field does not understand the intent of the SOPs or the study protocol. A professional with very little field experience should not supervise the installation of monitoring wells or the collection of groundwater samples, yet some registrants have openly complained that this has happened. Furthermore, a geologist who is qualified to supervise the installation of monitoring wells and to collect soil samples at Superfund sites is not automatically qualified to supervise or conduct groundwater studies for product registration. Taking this idea one step further, a chemist may know how to sample and analyze water samples but should not be put in the position of characterizing the hydrogeologic conditions at potential study sites. Likewise, a geologist should not be in the laboratory analyzing water samples.

Proper education and training are necessary to ensure that none of these things happens. The field portion of the protocol and the accompanying SOPs should be written by people with field experience. The EPA provides some guidance, but many gray areas in the guidance documents still need to be addressed by the registrant.

Communication Between Field and Laboratory. The analytical portion of the study protocol should be written to accommodate the field portion of the protocol. There should not be a lack of communication between the field and the laboratory. For example, suction lysimeters are required for the small-scale prospective groundwater study. Obtaining a substantial volume of water from these samplers is often difficult. The field scientists should express their concerns to their laboratory counterparts on this issue before the analytical portion of the study protocol is written. In this manner, the issue of analyzing small sample volumes can be addressed in the protocol, and may in fact result in method development that will be compatible with sample volumes. Likewise, sample collecting, packing, shipping, and specific chain-of-custody SOPs are usually specific to certain chemicals and should be agreed upon in advance by both parties so that inconsistent protocols and SOPs between the field and laboratory do not emerge.

Weather and Schedule. The EPA provides guidelines for the timing of soil and groundwater sampling in small-scale studies. These time frames are written into the study protocol by the registrant; however, because these time frames conform to the EPA guidelines, they rarely are written with any degree of flexibility. Consequently, field personnel may be required to obtain samples in bad weather and even at night. Under these conditions, the risk of human error and cross-contamination increases dramatically. Likewise, proper documentation under these conditions can be difficult. The sampling schedule in the study protocol needs to be flexible without compromising the integrity of the study. In most cases, postponing sampling by several days and adjusting subsequent sampling rounds accordingly will not affect the study.

Flexibility To Accommodate the Cooperator. Flexibility is the key to successful field work. For example, when the farmer or cooperator asks for the wells to be finished flush with land surface so that his equipment can drive by unobstructed, this should not be a major GLP issue. The well-installation SOP should be written to allow for the wells to be finished appropriately, based on site conditions, rather than so many feet above or below land surface.

Sensitivity Issues. Field personnel may at times feel resentment toward the study director or manager (who usually is trapped in the office), especially if the field personnel had no say in writing or revising the SOPs and the study protocol. Complaints by field investigators that the people who wrote the protocol and SOPs have no idea of what conditions are really like in the field are common. This complaint often is legitimate and may lead to friction and poor communication between the two groups. Giving experienced field personnel major responsibilities in designing and preparing SOPs and a role in the writing of study protocols can minimize these difficulties.

Conclusion

The key to ensuring that field-scale monitoring studies comply with GLP Standards is the amount of flexibility provided in the written study protocols and SOPs. Because field-scale studies are conducted in a highly variable and often unpredictable environment, on-site modifications to SOPs and deviations from protocols will occur. However, these corrective actions need not affect either the integrity or the quality of the study. Identifying and using experienced field scientists who are trained in and have a thorough understanding of GLP Standards will ensure that a field-scale study is conducted appropriately and that the results of the study are accurate and reliable.

References

1. "Federal Insecticide, Fungicide, and Rodenticide Act (FIFRA); Good Laboratory Practice Standards; Final Rule", *Code of Federal Regulations* Title 40, Pt. 160, *Federal Register* 54:158 (August 17, 1989).

2. Eiden, C.; Lorber, M.; Holden, P. W.; DeBuchananne, G. "Draft Guidance for Ground-Water Monitoring Studies", U.S. Environmental Protection Agency, Office of Pesticide Programs, Washington, DC, 1988.

3. DeMartinis, J. M. Ground Water Monitoring Review **1989**, *9*(4), 167–176.

RECEIVED for review September 17, 1990. ACCEPTED revised manuscript August 17, 1991.

Runoff Studies

Peter N. Coody

PTRL East, Inc., 3945 Simpson Lane, Richmond, KY 40475

Field studies designed to quantitate the pesticide runoff from agricultural fields are sometimes required in support of pesticide registration. These studies may involve monitoring pesticide residues in a receiving body of water such as a farm pond after a runoff event or the direct measurement of pesticide runoff at the edge of a field. Because runoff is a sporadic event that occurs only under conditions of heavy rainfall, these studies offer unique challenges in terms of complying with the U.S. Environmental Protection Agency's Good Laboratory Practice (GLP) Standards as is required for these studies to be used in the pesticide registration process. This chapter introduces the reader to the basics of runoff studies and the special considerations that must be observed when conducting such a study in compliance with the GLP Standards.

MOST OF THE 450,000 METRIC TONS OF PESTICIDES used annually in the United States (1) is applied either directly to soil or to an overlying crop. A portion of the pesticide active ingredient (AI) deposited on the crop is subject to transfer to the soil surface by direct deposition or wash off. Upon reaching the soil, pesticides are subject to a number of transport and transformational processes, including chemical and biochemical decomposition, volatilization, sorption, leaching, and runoff. The transport of pesticides from treated fields via surface runoff can be a significant pathway for pesticides. When runoff occurs, the pesticide in the runoff water can reach aquatic ecosystems where the chemical concentration may pose a hazard to many nontarget organisms. This chapter describes the requirements imposed by the current Good Laboratory Practice (GLP) Standards (2) on the conduct of field studies designed to quantitate "edge of field" losses of pesticide via surface runoff. Given the nature of this broad-based book, it will undoubtedly be helpful to many readers to describe the nature of pesticide runoff as well as the current regulatory implications surrounding runoff study results.

Factors Affecting Pesticide Runoff

Pesticide runoff requires the movement of surface water from a treated field. This surface flow is generally derived from discrete storm events, making runoff studies perhaps the most weather sensitive of all environmental studies conducted to support pesticide registration. Because most pesticides are labile in soil, the timing between pesticide application and a subsequent runoff-producing rainfall (or snow melt) is thought to be the most significant factor controlling the amount of pesticide transported via runoff (3–5, 15). Soil factors such as slope, slope length, texture, infiltration capacity, antecedent moisture, and plant and residue cover also affect the amount of sediment, water, and pesticide that leave a field as runoff (6).

Pesticide runoff occurs as aqueous and sediment-bound fractions. The water solubility and organic partition characteristics of a particular pesticide and soil characteristics, such as percent clay and organic carbon content, will determine which fraction contains the greater mass of AI leaving the field. When the sediment-bound fraction is significant (which is generally the case with hydrophobic pesticides), a sampling system that is capable of sampling runoff water with a significant sediment load must be used. Equally important in this case is the need to measure and sample the flowing runoff water without significantly changing its velocity in the field as this change could result in artificial scour or deposition of sediment containing bound pesticide.

The pesticide formulation and method of application are also factors affecting pesticide runoff (7). In the absence of significant gully erosion, the pesticide in only the upper 2 or 3 cm of the soil profile is generally subject to loss via surface runoff (4). Therefore, application methods that result in a significant amount of applied pesticide being placed below the soil surface may protect the AI from loss via runoff. The pesticide formulation affects the runoff potential by modifying the chemical and physical nature of the AI. When a runoff event occurs before the AI is released from the formulation, significant differences in pesticide transport may be observed between different formulations (*8*). Modification of these two agronomic practices is currently viewed as an important factor that can be studied in an attempt to reduce pesticide runoff and subsequent contamination of surface waters.

Runoff studies can generally be divided into two broad categories: those that require natural rainfall to generate surface runoff and those that use simulated rainfall to force runoff. Runoff studies relying on natural rainfall are generally conducted on large (ca. 2–40 ha) fields. A rainstorm with sufficient intensity and duration to meet regulatory criteria (ca. 1 in 5-year minimum return frequency; that is, a storm that, on average, occurs at a given location one time in five or more years) is unlikely to occur within a limited window of time after the pesticide application is made. In fact, often a field will be instrumented and treated only to have no runoff produced when pesticide residues in the field are significant.

Simulated rainfall has often been used to provide "designed" rainstorms after pesticide application to ensure that specific severe rainstorm events are evaluated in runoff experiments (*7–9*). This approach has the distinct advantage of ensuring that runoff data will be generated during the experimental period and that high-intensity storms will be evaluated with respect to runoff potential. However, given the difficulty of simulating rainfall over large areas, the field size is generally limited to approximately 0.1 ha for these experiments. Coody et al. (*10*) presented a "small plot" methodology for quantitating pesticide runoff using simulated rainfall on 0.1 ha test plots.

Runoff Studies in the Regulatory Arena

Runoff studies are frequently performed as part of an exposure assessment, in which the potential for contamination of an aquatic environment with a given AI applied at the maximum labeled rates is examined. The treated land area required for these studies can range from a small field to a large watershed. Likewise, the aquatic environment of concern can range from small farm ponds or primary streams to large lakes and marine ecosystems that might receive agricultural runoff.

In 1985, the EPA published the *Field Agricultural Runoff Monitoring (FARM) Manual* (*11*), which serves as an excellent introduction to the soil, hydrologic, and climatic factors controlling agricultural runoff. This manual also details many of the important aspects of a field runoff study.

To date, however, there are no study guidelines or Standard Evaluation Procedures (SEPs) to aid the study director in planning an acceptable runoff study. The EPA is currently working to provide general direction in this regulatory area. In December 1990, the EPA published in the *Federal Register* (*12*) a request for comments on specific aspects of runoff studies and on the need to include exposure assessments in the next edition of their "Pesticide Assessment Guidelines". Currently, only a few established triggers identify a specific requirement for runoff and exposure assessment work to supplement the pesticide registration data submission. However, these studies are needed to provide exposure information to address the potential hazard to nontarget species and particularly to aquatic and marine ecosystems. As a result, the need for and design of runoff studies to support product registration is generally addressed by the EPA on a case-by-case basis.

Given the fact that runoff studies are often conducted to define the exposure of nontarget aquatic organisms to a test compound, the EPA typically requires that these field studies be conducted on sites where the runoff flows directly into a body of water, such as a farm pond (i.e., the runoff does not cross much land before entering the body of water). In this way, it is possible to study the resulting pesticide concentrations and aquatic biology in the pond to determine directly if there are specific aquatic hazards associated with the use of the test compound on that field. However, given the great variability that exists in runoff potential among agricultural fields and the natural diversity in aquatic ecosystems, the EPA has found it difficult to obtain usable data about exposure and effects from these unreplicated test systems (field and adjacent pond).

The current approach taken by the EPA for addressing aquatic hazards resulting from the use of a given pesticide is to request replicated pond studies, such as mesocosms (Chapters 25 and 27), to determine the biological effects resulting from specific levels of chemical exposure. In this case, an exposure assessment is then required to determine what pesticide residue levels may result from product use in the real world in order to dose the test ponds and to interpret properly the mesocosm results. Thus, the need to incorporate a receiving body of water into runoff study design is now a subject of considerable debate in the scientific and regulatory arenas. Because the need to incorporate a receiving body of water in a runoff study is not a standard requirement and because the GLP aspects of aquatics testing are described elsewhere in this text, the discussion here will be limited only to measuring runoff as it exits the field. That is to say, the test system will be limited to agricultural fields and not fields with adjacent ponds.

The results of an exposure assessment are related to the aquatic toxicology data for a given AI in an aquatic risk assessment. The final risk assessment is used to determine the overall hazard that a given AI has on the environment. The EPA's interpretation of the risk assessment can lead to support for registration, request for product suspension or cancellation, or the need for further testing, which can include the generation of additional exposure (runoff) data and aquatic effects data.

GLP Requirements in the Conduct of Runoff Studies

The GLP requirements for field and laboratory studies demand that the study be well planned and properly defined before being conducted. The effort required to initiate a runoff study under GLP is significant, but the quality of the results and the structure provided for maintaining complete experimental records justifies the effort. Following is a description of key elements involved in conducting a runoff study under GLP Standards with emphasis given to certifying the performance of the equipment commonly used and to recording critical field data.

The Study Protocol. With the exception of studies conducted using simulated rainfall, runoff studies are unpredictable with respect to the timing of sample generation (when it rains) and the number of samples that may be collected for a given runoff event. In addition, the critical sample collection period can occur at any time of day and usually when weather conditions are at their worst. Under these conditions maintaining the needed control over sample collection, storage, and chain of custody is often difficult. A well-developed study protocol is crucial to maintain study integrity during these intense periods, and its value as a training tool cannot be overstated.

The development of protocols meeting GLP requirements for field studies is addressed in detail in Chapters 5 and 6, and details such as general information to be included, pesticide application documentation, geographic description of the test site, and requirement for field records will not be repeated here. However, specific details are unique to the conduct of runoff studies and deserve consideration when developing a protocol. For example, the protocol should specify that the field slope, erosion slope length, and other hydrologic parameters be determined during the course of the study. Of particular concern is the need to describe how the test system (field or runoff plot) will be modified to permit the containment and measurement of runoff water. In the case of large fields, this may involve creating earthen dikes, reshaping certain problematic areas, or restricting planting and treatment to define the drainage area better. Runoff plots for simulated rainfall studies generally require the installation of a collection system along the lower edge to divert runoff into a measuring device (*10*). General details regarding land shaping, runoff containment, and plot layout can be described in the protocol, as a means of ensuring that proper consideration will be given to defining the test system during the study design period.

The study protocol must also address how the runoff flow will be measured and how samples of runoff and sediment water will be collected, stored, and shipped to an analytical laboratory. When preliminary sample processing is required at the site of sample collection (i.e., solvent extracting, filtering, modifying pH, or adding a preservative) these activities must be detailed as well. The means of maintaining sample chain of custody should also be clearly defined in the study protocol because this aspect of the study can be particularly problematic when samples are collected at unpredictable times. An example chain-of-custody form, such as that presented in Figure 28.1, is helpful to include in the protocol so that it can be used during the training process to ensure that sample handling and tracking are performed properly with a continuous chain of custody as the result.

When applicable, the collection and handling of control runoff water for use in preparing field spikes also must be clearly described. Control runoff can be difficult to obtain (particularly when not using simulated rainfall), so it is sometimes prepared by adding a known amount of surface soil to irrigation or surface water known to be free of the test compound. The preparation of the actual field spikes is a critical part of the quality control process surrounding runoff studies, and the procedure must be described either in the protocol or in the standard operating procedure (SOP).

When simulated rainfall is used to provide designed rainstorms, the intensity and duration of the rainstorm being simulated must be defined. The water source to be used and a means of establishing its quality (by taking samples for analysis) must also be defined. Details on monitoring the performance of the rainfall simulator are addressed in the following sections.

Specification, Calibration, and Maintenance of Equipment.

The basic equipment required to contain, measure, and sample agricultural runoff may include a primary device used to quantitate open channel flow (flume or weir), a flow meter, and a sampling device. Commonly, the natural drainage in a field will not force surface runoff directly into the primary device, so some means of containment in the field must be provided as well. Because runoff generally occurs during periods of severe weather, the equipment chosen to perform the necessary tasks must be highly reliable. Because the types of storms that will be encountered when

PTRL-East Chain of Custody Form

PTRL Study Number _____

<-------------- To Be Completed in the Field -------------->

--- Sampling ---

Sample Number	Sample Description	Date	Time	Time to Field Cooler	Ice in Cooler (Y/N)	*Signature*	Date Shipped	Condition Shipped	Date Received	Condition Received

TO BE COMPLETED BY ANALYTICAL LAB

Observations when Transferring Contents to Freezer

Date _____ Time _____ of Transfer _____

Is Ice Still Present in Cooler? (Y/N): _____

Signature: _____ Date: _____

Signature for Shipping Release:

Signature: _____ Date: _____

Signature for Analytical Receipt:

Signature: _____ Date: _____

PTRL Form # 10
Created 6/90
Revision None

Figure 28.1. Example chain-of-custody form completed to verify shipment of a single field sample.

quantitating surface runoff are not known in advance, equipment with a capacity to measure runoff from storms having a minimum return frequency of perhaps 10–20 years might be specified.

It is not within the scope of this chapter to discuss the selection process and design constraints for the basic equipment used in runoff studies. Rather, discussion will be limited to the ability to calibrate, install, and maintain typical equipment in accordance with GLP requirements.

Primary Device for Flow Measurement. A primary device is an engineered structure installed in the flow channel whereby the depth of flowing water can be related to its volume by a known (nonlinear) relationship. Primary devices of many standard designs are commercially available and commonly used to measure open channel flow. Calibration of a primary device (which may have a capacity of <1–100 cubic feet per second) against a standardized flow is a difficult task. Given this limitation and the fact that the engineered device is designed and built specifically to measure flow, my view is that the primary device, if accompanied by specific design specifications, generally need not be calibrated under the GLP Standards. The analogy is often made that the primary device is similar to a graduated cylinder in that it is designed and built to measure a volume; neither measuring device requires calibration (*13*). Of course, it is good practice to ensure that the primary device is sufficiently rigid to prevent any distortion that might take place during the installation process and result in a system that does not perform according to specifications.

Flow Meter. Unlike the primary device, a flow meter is a dynamic measuring device that requires periodic calibration. Flow meters generally measure the depth of water flowing through a primary device; some flow meters can calculate flow from the measured depth using the relationship defined by the primary device and programmed by the user.

In calibrating a flow meter, one must first demonstrate that the instrument properly measures water depth and then accurately relates that measurement to flow. Calibration measurements are made by measuring known water depths over the dynamic range of the primary device. This calibration can be accomplished by measuring the depth of a water column using the flow meter and a standard rule. However, when sonic devices are used (as in the ISCO Model 3210 meter), it is important to demonstrate that the calibration vessel used does not interfere with the transmission of the sound waves used in the measurement. Once it is confirmed that the flow meter properly measures depth, the corresponding flow must be calculated manually to demonstrate that the flow meter has been properly programmed to relate the measured depth to the calculated flow.

Flow meter calibration and programming are recurring procedures that are best described by using an SOP. Recalibration of the flow meter should be performed according to a schedule defined either by an SOP or by the study protocol. Because a runoff study can last an entire growing season and the timing of a runoff event is often unpredictable, the calibration must be performed regularly to ensure accurate measurement at any time during the study. As is the case with any repetitive measurement, a flow meter calibration form (Figure 28.2) helps to ensure that this critical calibration is performed properly in the field.

Runoff Water Sampler. The sampler is generally a pumping device that removes an aliquot of runoff water passing through the primary device and delivers it to a sample container. Because eroded sediment is of particular importance in quantitating pesticide runoff, the scientist must verify that the pumping velocity is sufficiently high to ensure that the sediment entrained in the runoff water will be delivered to the sample container. Some samplers can take samples on fixed time intervals or as a function of flow (flow-proportional sampling). The sample may also be collected into discrete containers or into a large compositing vessel. The study protocol must address the method and frequency of sampling. Most samplers require some degree of user programming, so an SOP to address the programming requirements for the desired sampling regime is needed.

Pump samplers require calibration to ensure that the desired sample volume is being delivered. These samplers are particularly sensitive to the height above the water being sampled (hydraulic head), so the unit must be calibrated once it is positioned in the field and the height from the sampling point is established. The user's manual for the pump sampler will contain instructions describing the calibration process. An SOP describing this procedure and actions taken when calibration cannot be achieved is necessary for GLP compliance.

Rainfall Simulator. Using a rainfall simulator to provide the desired rainfall intensity and duration introduces additional calibration and maintenance requirements. Although the designs of rainfall simulators used in pesticide studies vary greatly, one common requirement is to document that the proper amount of "rainfall" defined in the protocol is evenly applied to the test area. Uniformity tests can be conducted by placing a sufficient number of collection cups randomly around the test area during the simulation event and then recording the volume collected in each container during irrigation. The number of collection cups required will depend on the area being subjected to the simulated rainfall. The size spectrum and impact energy of the water droplets produced by the simula-

PTRL–East Data Reporting Form

Flow Meter Calibration

Study Number: _____

Model: _____   Serial Number: _____

Actual Depth (mm)	Reported Depth (mm)	Reported Flow (LPM)	Calculated Flow (LPM)
_____	_____	_____	_____
_____	_____	_____	_____
_____	_____	_____	_____
_____	_____	_____	_____
_____	_____	_____	_____
_____	_____	_____	_____
_____	_____	_____	_____
_____	_____	_____	_____
_____	_____	_____	_____
_____	_____	_____	_____
_____	_____	_____	_____

Notes:

Performed By: Reviewed By:

Date: _____   Date: _____

PTRL Form #1025   Created 5/90 Revision: None

Figure 28.2. Example Flow Meter Calibration form for use in runoff studies.

tor may also need to be documented, but characterization of these design criteria are beyond the scope of this chapter.

Soil Sampling. When conducting a runoff study the soil characteristics of the test system must be documented. This documentation will typically involve taking samples for chemical and physical analysis, with samples collected as either intact soil cores or surface grab samples. Fortunately, this aspect of the runoff study can be planned in advance, and the sample collection process can be detailed in the study protocol. Samples should be taken from random locations within the test area. This randomness can be accomplished by developing a grid system to designate all possible sampling locations (Figure 28.3), then using a random number table to select the actual grids to be sampled. Specific details concerning random sampling in the field are addressed in Chapter 21.

The number of unique samples taken and composited into analytical samples will depend largely on the spatial variability of the site. Without the benefit of the EPA guidelines, ensuring that the sampling regime used will be favorably reviewed by the EPA is sometimes difficult. However, the SEP published by the EPA for terrestrial field dissipation (*14*) suggests that 15–20 samples be collected with up to 5 samples composited to provide each analytical sample. The soil samples are generally used to quantitate pesticide concentrations before or after the runoff event or to characterize the soil under study (the test system). The field dissipation SEP suggests that the soil characterization of the test site include pH, moisture capacity, texture analysis, percent organic matter (organic carbon), bulk density, and cation exchange capacity.

An additional soil parameter that is useful in evaluating runoff data is the antecedent soil moisture present immediately before the runoff event. All soil analyses should be conducted in compliance with the GLP Standards, and sufficient documentation and raw data must be provided to ensure that the analyses were performed properly.

Data Collection and Verification. Runoff studies require a considerable amount of field data to be collected to document the tillage practices, pesticide application, weather, and crop development. These data can be collected on data sheets with specific areas for data entry as well as locations for signature by the data collector. In addition, the flow meter may produce some form of data record, such as a hydrograph on a strip chart or a stored electronic data file relating flow to time. These data must be provided as hard copy that is dated and signed to be included in the raw data file. The chain-of-custody form (Figure 28.1) must accompany the samples to the analytical laboratory. However, an exact copy of

Top Of Plot

| 1 | 1 | 2 | 3 | 4 | 5 | 6 | flag # |

2	*1*	*2*	*3*	*4*	*5*	*6*
3	*7*	*8*	*9*	*10*	*11*	*12*
4	*13*	*14*	*15*	*16*	*17*	*18*
5	*19*	*20*	*21*	*22*	*23*	*24*
6	*25*	*26*	*27*	*28*	*29*	*30*
7	*31*	*32*	*33*	*34*	*35*	*36*
8	*37*	*38*	*39*	*40*	*41*	*42*
9	*43*	*44*	*45*	*46*	*47*	*48*
10	*49*	*50*	*51*	*52*	*53*	*54*
11	*55*	*56*	*57*	*58*	*59*	*60*
12	*61*	*62*	*63*	*64*	*65*	*66*

PTRL Study Number _____

Random Sampling

<u>Location</u>   <u>Completed</u>

41 ____
18 ____
26 ____
54 ____
38 ____
50 ____
3 ____
11 ____

flag #

Bottom Of Plot

Sampling Event Performed/Notes

Performed by: Confirmed by:

_____   _____

Signature   Date   Signature   Date

PTRL Form # 1026
Created 5/90
Revision: None

Figure 28.3. Example sampling grid and random number sequence used to locate random soil sampling locations.

the form should be maintained by the study director to aid in the sample tracking process.

Study audits and inspections by the quality assurance unit must be conducted as a routine part of the field study and subsequent residue analyses. These audits should include inspection of the runoff and sample generation phase as well as the analytical aspects. A final data audit must be conducted as part of GLP compliance.

Sample Shipment and Analysis. The shipping, processing, and analysis of the samples obtained during the in-life portion of the study must be detailed. For example, how the samples will be stabilized (i.e., frozen or preserved) in the field for shipment and how they will be shipped to an analytical laboratory should be stated in the protocol. Of particular importance in runoff studies is the need to detail how the sediment and water phases of the runoff will be separated and analyzed (if separation and analysis are part of the overall study design). The analytical methodology must also be part of the protocol and may be included in the form of an amendment.

Study Termination. Because the conduct of a runoff study is extremely sensitive to natural rainfall, the study may have to be terminated once it is under way. When natural rainfall is required, an extended dry period may result in study termination without useful data being obtained.

Conversely, the study may also be terminated shortly after the pesticide application is made, should a runoff-producing rainfall occur. The opposite situation exists when simulated rainfall is used, whereby dry weather is desirable to prevent unintentional runoff before the simulated rainfall is applied. In any case, the protocol must address when and how the study is to be terminated due to uncontrollable, weather-related problems.

Conclusion

Conducting field runoff studies under the GLP Standards requires considerable planning before the study is initiated. Because much of the activity involved in instrumenting a runoff site is case specific, the protocol must deal with many of the details that might be addressed by SOPs in some field studies. Because sample generation depends largely on uncontrollable environmental conditions (rainfall intensity and duration and the antecedent soil moisture), the number of samples that will be generated and require analysis cannot be predicted. For these reasons, a well-defined study protocol is the best means of ensuring that the necessary planning takes place before study initiation.

The use of data-recording forms greatly simplifies the data collection and auditing processes. With proper planning, forms can be created to document nearly all aspects of the study, including instrument calibration, treatment, sample collection, and sample storage. These forms help ensure that the proper data are collected and are a great aid when preparing a final report. An added benefit is that these forms also help ensure that the study integrity is maintained according to GLPs.

Experience has demonstrated that the GLP-related overhead of a runoff study accounts for approximately 30% of the total experimental cost. Because the added cost cannot be avoided, it is helpful to recognize that the extra effort required to perform a study under GLP will almost always result in a study that is better planned, better documented, and easier to interpret than one that does not have such stringent control measures imposed.

References

1. *Statistical Abstract of the United States, 1990: The National Data Book*, Census Bureau, U.S. Department of Commerce, Washington, DC, 1990.

2. "Federal Insecticide, Fungicide, and Rodenticide Act (FIFRA); Good Laboratory Practice Standards", *Code of Federal Regulations* Title 40, Pt. 160, *Federal Register 54*:158 (August 17, 1989).

3. Smith, S.; Reagan, T. E.; Flynn, J. L.; Willis, G. H. *J. Environ. Qual.* **1983**, *12*, 534–537.

4. Wauchope, R. D. *J. Environ. Qual.* **1978**, *7*, 459–472.

5. Willis, G. H.; Hamilton, R. A. *J. Environ. Qual.* **1973**, *2*, 463–466.

6. "Predicting Rainfall Erosion Losses: A Guide to Conservation Planning", *Agriculture Handbook 537*, U.S. Department of Agriculture, Washington, DC, 1972, p 58.

7. Kenimer, A. L.; Mitchell, J. K; Felsot, A. S. *Proc. International Summer Meeting of the Canadian Society of Agricultural Engineers*, paper number 89–2118, 1989.

8. Wauchope, R. D. *J. Environ. Qual.* **1987**, *16*, 212–216.

9. Hubbard, R. K.; Williams, R. G.; Erdman, M. D.; Marti, L. R. *Trans. ASAE*, **1989**, *32*, 1250–1257.

10. Coody, P. N.; White, J. W.; Graney, R. L. Abstracts of Papers, 11th National Meeting of the Society of Environmental Toxicology and Chemistry, Abstract Number P342.

11. Smith, C. S.; Brown, D. S.; Dean, J. D.; Parrish, R. S.; Carsel, R. F.; Donigian, A. S., Jr. *Field Agricultural Runoff Monitoring (FARM) Manual*; U.S. Environmental Protection Agency, Washington, DC, EPA/600/3–85–043, 1985; p 230.

12. *Federal Register* **1990,** *55,* 10394–50395.

13. Garner, W. Y., Garndal Associates, personal communication 1990.

14. "Standard Evaluation Procedure: Terrestrial Field Dissipation", U.S. Environmental Protection Agency, Washington, DC, EPA–540–09–90–073, 1989; p 29.

15. Willis, G. H.; Rogers, R. L.; Southwick, E. M. *J. Environ. Qual. 1975, 4,* 399–402.

RECEIVED for review September 17, 1990. ACCEPTED revised manuscript September 13, 1991.

Regulatory Impact

Auditing Field Studies

Government Perspective

Dean F. Hill

National Enforcement Investigations Center, U.S. Environmental
Protection Agency, Denver Federal Center, Building 53, Box 25227,
Denver, CO 80225

Study audits are conducted on complete and submitted regu-
latory studies through review of raw data, records, and other
documents generated during the planning and conduct of
such prescribed testing as part of official inspections by the
U.S. Environmental Protection Agency (EPA). An audit
ensures that reported findings are supported by the raw data,
that the study is fully reconstructible, that Good Laboratory
Practice Standards requirements were followed, and that
sound scientific judgment was employed. Typically a field
study, such as the determination of residue levels in support
of a tolerance petition, will be performed at more than one
test site, and multiple test sites impose burdens on the audi-
tors with respect to conducting firsthand interviews and
review of all equipment, facilities, and standard operating
procedures. Guidelines spelled out in this chapter address
some of the areas that the EPA inspection personnel will
review during audits of field and chemical-fate studies and
some of those areas for which compliance by the regulated
community has been troublesome.

THE REVISED AND EXPANDED Good Laboratory Practice (GLP) Standards regulations were published on August 17, 1989, in the *Federal Register* for pesticide studies regulated by the Federal Insecticide, Fungicide, and Rodenticide Act (FIFRA) and for toxic-substance studies required by the Toxic Substances Control Act (TSCA). These regulations became effective on September 16, 1989, and October 16, 1989, for TSCA and FIFRA, respectively. This chapter focuses on those requirements pertaining to FIFRA; however, much of the discussion and many of the issues pertain to TSCA studies as well.

One major change of the new regulations on FIFRA testing requirements was the extension of GLP jurisdiction to non-health-effects studies (i.e., to those primarily conducted in field, greenhouse, and chemical laboratory environments). Some FIFRA efficacy testing has also been included under the new GLP requirements, particularly for products whose use directly affects public health. In addition, the GLP Standards were extended to product chemistry studies and testing conducted on human subjects. Examples of the additional field and laboratory studies now covered under FIFRA include

- nature and magnitude of residue

- plant and animal metabolism

- photodegradation

- soil dissipation

- rotational crop uptake

- volatility

- biomonitoring

- effect on nontarget organisms

The Environmental Protection Agency (EPA) requires that these and other studies (including those to determine health effects) be conducted by pesticide registrants (or their agents) to provide a reasonably accurate assessment of the chemical's effect on human health and the environment. Thus, the EPA audits and reviews these studies to ensure their overall reliability and usefulness. Specifically, the EPA audits data, records, and reports associated with a study to ensure

- consistency of raw data and other records with study report findings and conclusions

- ability to reconstruct the study

- accountability of all raw data
- compliance of the study with FIFRA GLP Standards requirements

  — quality assurance unit (QAU)

  — protocols, standard operating procedures (SOPs), and final reports

  — personnel, facilities, and equipment

  — adequacy of test systems and test substances

  — data recording and archiving

- consistency with test guidelines
- scientific soundness

An EPA study audit is usually performed as part of an overall GLP inspection on one or more studies that have been submitted to the EPA. If the study to be audited was conducted at more than one testing site, as is often the case with field studies, the test system locations or the analytical laboratory will likely be the sites of the audit. However, there may be occasions when a sponsor facility, a test-substance-characterization laboratory, or a commodity-processing facility will be the site of an audit. Usually, an EPA audit will be conducted through

- review of raw data, records, reports, and correspondence
- interviews with the study director, technical staff, QAU personnel, and "management"[1]
- inspection of facilities and equipment

For purposes of this discussion and to provide some general guidance on the EPA's inspection and study-audit procedures, a typical field study may be simplified into five distinct phases:

- protocol development
- provision of test substance
- field phase (test substance application and specimen sampling)
- laboratory phase (specimen analyses)
- report preparation and archiving

---

[1]Management is defined here as the person or persons responsible for setting policy at a facility. For a description of the duties of testing facility management, see § 160.31 of the GLP Standards.

There may occasionally be additional phases or more than one component for each of the listed phases. The field portion of a study may be conducted at as many as 20 distinct geographical locations for a residue study conducted in support of a tolerance for a pesticide used on a widely grown commodity.

The next five sections identify the typical data points and elements of the GLP Standards that the EPA inspectors and auditors will evaluate. Deficiencies for these elements will be documented during the reviews performed by the inspectors and auditors.

Protocol Development

The protocol is essential to any required pesticide study as it prescribes the specific experimental design as well as defines other requirements related to the test, reference, and control substances to be used; testing facilities; and critical experimental dates. The protocol should supplement standard operating procedures (SOPs) with study-specific requirements. The protocol should also address each phase of a study, and all study objectives and methodologies are to be included or adequately referenced. The experimental design given in the protocol should reflect the minimal requirements specified in the appropriate EPA Pesticide Assessment Guidelines, as cited in 40 CFR Part 158.

The EPA auditor or inspector will review the protocol to ensure that it contains each of the 15 required elements, as well as the objectives of the study and all methodologies. In addition to verifying that the protocol was actually followed during the conduct of the study, the auditor or inspector will verify that necessary deviations or oversights were improved in writing by the study director at the time the deviation occurred or are listed as GLP deviations in the study report GLP compliance statement, as required by 40 CFR § 160.12(b).

Provision of Test Substance

The requirements of 40 *Code of Federal Regulations* (CFR) 160.105 are important for field studies, just as they are for toxicological studies. Whether the test (or control) substance is a commercial formulation, a radiolabeled compound, a technical-grade pesticide, or another material, its source, characterization, and stability must be established and documented. The assay, characterization, and stability determinations must be generated according to GLP requirements. Each test (and control) substance should be fully traceable back to its origin. The EPA recognizes that the test-substance information and supporting raw data usually will be located at the sponsor's facility and may contain proprietary information. Thus, if the necessary information cannot be provided at the inspection site, the inspector may request that it be sent to the EPA after the on-site inspection. If questions arise regarding the validity or completeness of the provided test-substance data and records, a follow-up inspection may be scheduled at the sponsor's laboratory or office.

The following questions are typically asked by the EPA auditors or inspectors regarding the test or control substance during review of a study that was conducted according to the GLP regulations.

- What was the described form of the test substance and was it consistent with the appropriate pesticide assessment guidelines?

- What information regarding test or control substance identity, source, method of synthesis, purity or strength, stability, solubility, and other identifying characteristics is given in the protocol and final report? Are the protocol and final report consistent with each other with regard to test or control substance information?

- Is the protocol and final report information consistent with the raw data, records, and reports in the study file or as provided by the sponsor? Is there adequate documentation regarding synthesis, characterization, solubility, and stability, if not given in the protocol or final report? Were these data and information developed under GLP purview?

- Are the characterization data and other information adequate with respect to determination of strength or purity for all batches of test or control substances used in the study? Were pesticide formulations tested to ensure that other product specifications were met?

- Are there adequate synthesis and analytical data to verify the radiolabeled positions and specific activity of radioactive test substances?

- Has the stability of the test or control substance been demonstrated over the period of its use in a study and under conditions of its storage and use at the test site?

- Was each batch of test or control substance fully traceable in the study from its point of origin, through analysis, and through its use in the study?

- Was a reserve sample retained for each test or control substance used in studies of more than 4 weeks experimental duration?

Field Phase

In addition to verifying compliance with the GLP Standards with respect to personnel, facilities, equipment, and SOPs, the EPA auditor or inspector reviews the completeness of data and records to ensure the ability to reconstruct a field phase of a study. The review will focus on the three key aspects (in addition to test substance provision) that are critical to most field studies:

- site location, management, and operations

- test substance applications

- specimen sampling

The EPA auditor or inspector will likely address most of the topics in Lists 29.1 and 29.2 regarding the field phases of a study under review as part of an official compliance inspection. Test-substance identification was addressed in the previous section. These GLP issues apply at all inspected facilities, including analytical and agricultural commodity processing laboratories, as well as field sites.

Upon completion of the field phase of a field study (i.e., the application of the test substance and sample collection), the specimens are usually transferred to an off-site laboratory for analysis; only in a few instances will there be an on-site analytical capability. Thus, to complete a particular study audit, the EPA auditor or inspector must conduct an inspection at the chemical laboratory where the analytical work was performed, have the data (or exact copies) brought or transmitted to the appropriate field site for review, or have the data (or exact copies thereof) transferred to the auditor's or inspector's office after the field-site inspection.

In practice, the first option is preferable because the volume of raw data and records is likely to be large. Also an audit is always more effective if associated technical and QAU personnel are available for interviews and the facilities and equipment can be examined. If any problems are discovered during the review of data and records at the field site or at the EPA auditor's office, an official inspection will probably be conducted at the analytical laboratory to document fully and properly the apparent discrepancies and data gaps.

Laboratory Phase

The inspection of the analytical phase of a field study will, in principle, be somewhat similar to the field-phase review. The completed or ongoing

List 29.1. GLP Compliance Issues

The following GLP issues should be addressed when reviewing any ongoing or completed regulatory study:

- Were management, the study director, technical staff, and the QAU adequately identified with respect to their GLP-defined functions?
- Did management fulfill its responsibilities with respect to appointing the study director and providing personnel, equipment, facilities, and properly characterized test and control or reference substances?
- Did the study director fulfill his or her required function with respect to protocol development and amendment, overall study control, and final report approval?
- Did a QAU officer inspect study phases and the final report and report any findings satisfactorily to management and the study director? Did the QAU have adequate operational SOPs, and did the QAU follow the SOPs? Were GLP master schedule requirements fulfilled with respect to required elements and upkeep?
- Were adequate numbers of qualified personnel available for the effective and timely conduct of the study?
- Were facilities adequate with respect to test system supply and care; test, control, and reference substance handling and storage; laboratory operations; and data and specimen storage?
- Was equipment (including that used for environmental control) of adequate design and capacity to meet protocol requirements, and were calibration, inspection, and maintenance properly carried out and documented?
- Were SOPs that addressed all GLP-required elements in place, and were all deviations and changes properly authorized? Was a historical record of SOPs maintained?
- Were test system care requirements met, and was GLP-required reagent labeling in place?
- Was test, control, and reference substance information adequate to meet the data and information requirements as cited previously under Provisions of Test Substance?
- Were test, control, and reference substance mixtures adequately tested and documented with respect to stability, homogeneity, and strength?
- Did an approved protocol address all GLP-required elements, and were changes or revisions properly approved by the study director? Was the protocol followed?
- Were all manually entered data recorded promptly in ink and properly signed (or initialed) and dated, and were all corrections made according to the GLP requirements?
- Were data that were directly entered into automated collection systems properly identified, validated, and protected?
- Was the final report approved by the study director, and did it include all required elements, including the GLP certification statement? Were report amendments made by the study director?
- Were data, records, and specimens properly stored in the archives and retained for the required periods of time?

List 29.2. Data Requirements for the Field Phase of a Study

Raw data and other records must be complete with respect to reconstructibility of the following aspects of each field study audited, as required by the protocol, SOPs, or testing guidelines:

1. identification of each batch of test substance used and complete documentation regarding the amount received, its storage location and conditions, its distribution, and the final disposition of any excess
2. specific geographic location of the overall test facility, as well as detailed descriptions and plot maps showing specific and relative locations and sizes of treated plots, control plots, and buffer zones
3. identification and roles of all technical personnel involved in the study and documentation of their qualifications
4. availability of approved protocols, amendments, and SOPs
5. maintenance and calibration for all equipment, including balances, scales, freezers, refrigerators, and application rigs
6. general environmental conditions over the course of the study, including daily air and ground temperature (maximum and minimum), rainfall, and extreme variations from normal conditions
7. layout of plot (i.e., length of rows and spacing) as well as all cultural practices followed during the study such as weeding, thinning, irrigation, pesticide treatments, and fertilization
8. soil characteristics (e.g., type, pH, chemical and physical characteristics, and percent organic matter) as necessary, sampling sites, and identification of the soils laboratory
9. methods and other details of each test substance application, including
   - type and specific application equipment
   - calibration procedure and data
   - type of application (e.g., band, broadcast, injection, or ultra low-volume)
   - specific dates and times of mixing, applications and storage conditions of dilutions (tank mix), if not applied immediately after mixing
   - dose rate and mixing data to achieve target concentration, including identification and amounts of additives, and pH, hardness, and temperature of dilution water, as appropriate
   - analytical and stability data and records to verify concentration of test-substance dilutions at the time of application
   - spray-volume and application-speed measurements to achieve target application rate
   - specific weather conditions during application, including air and soil temperature, wind speed and direction, and relative humidity
   - growth stage at time of application
10. sampling data for each specimen, including the following:
    - identification of the personnel involved, and dates and times of sampling
    - specific procedures used for sampling, including precautions taken to preclude cross contamination
    - sample identification (including labeling and encoding procedures) for each specimen so as to preclude mixup at the field site, during transport, and at the laboratory
    - growth stage of each raw agricultural commodity at the time of sampling and any trimming or other field preparation
    - conditions and duration of storage prior to shipment to the laboratory, including interval between sampling and preservation
    - method and dates of sample packaging and preservation and date and means of sample shipment to the laboratory
    - all other circumstances and observations that may have had an effect on the reported findings or recorded data
11. agricultural commodity processing information (if applicable), including transfer of samples, processing procedures, storage conditions, and subsequent transfer of processed samples to the laboratory for analyses

work will be reviewed both for GLP Standards compliance and for adequacy of the raw data. The laboratory work will be reviewed for consistency of the raw data and records with the protocol and final report, completeness of the raw data, consistency with SOPs and pesticide assessment guidelines, and overall scientific soundness.

The GLP review of the analytical facilities, whether related to ongoing work or to a completed regulatory study, should include those elements given in List 29.1, as appropriate. Particular emphasis will be given to the following:

- *Facilities*: Were areas where high- and low-level work was performed adequately separated to minimize cross contamination in the low-level areas?

- *Equipment*: Were all instruments properly calibrated? In particular, were spectrophotometers and chromatographs verified for response over the entire range of sample responses, and were balances appropriately verified for accuracy?

- *Reference standards*: Did all analytical reference standards used in GLP-related work meet the requirements for source, characterization, stability, labeling, and control, as given in 40 CFR 160.105?

- *Protocol*: Was the protocol complete with respect to methodology, methods used to control bias (quality control), frequency of analyses, and records to be maintained?

- *Personnel*: Was there an adequate number of qualified personnel to conduct the analyses properly and in a timely fashion (i.e., were there no undue delays such that degradation or changes in distribution pattern of analytes might have occurred between sample collection or receipt and analysis)?

- *Conduct of study*: Were all specimens labeled by test system, study, nature of the specimen, and date of collection?

- *Archives*: Were all raw data (e.g., notebooks, work sheets, chromatograms, and spectra), other pertinent records, correspondence, and reference substances and specimens properly stored in the archives at the completion of the study (i.e., date that study report was signed by the study director)?

In addition to the preceding, analytical data for audited studies should be reviewed for the data elements given in List 29.3 to ensure the overall scientific soundness and completeness of the analytical phase of a study. Of particular concern is the possible degradation, loss, or changes in metabolic distribution of samples during the retention period between

List 29.3. Major Concerns for the Analytical Phase
of Field Studies

Raw data and other records must be adequate with respect to completeness and reconstructibility of the following aspects related to the analytical phase of audited field studies:

• specimen receipt, condition, and storage

• identification of all technical personnel, particularly technicians, associated with the conduct of the analytical portion of the study

• specific identification of all instrumentation and specialized equipment used in the specimen analyses, including balances, chromatographs, spectrophotometers, spectrometers, temperature-recording equipment, and pH meters

• clear identification of storage locations and conditions for specimens, extracts, reference standards and solutions, and other related materials before, during, and after analysis

• stability data (under conditions of storage) of specimens extracted and other related preparations to verify integrity from time of collection or preparation until time of analysis

• evidence that reported sample results reflect instrumental responses within the calibration-response range established for reference standards

• validation (method recovery) data to show applicability of the protocol method to specific specimen matrices and the laboratory's ability to achieve reliable results with the specified method

• clear identification of the analytical reference standards used (and their preparations), as well as data and records to meet fully GLP requirements for reference substances (40 CFR 160.105)

• quality control data for each specimen analysis or batch of analyses to ensure evaluation of precision, accuracy, freedom from interferences, and confirmation of analyte identity

• documentation showing peer and supervisory review of data and results

collection and analysis. Thus, the EPA auditor will usually ascertain whether appropriate stability data are available, particularly if significant delays appear to have occurred between sampling and analysis. The data elements given in List 29.3 may also be considered relevant to the conduct and audit of laboratory chemistry studies, such as the determination of chemical and physical properties, photolysis, hydrolysis, and soil metabolism, although additional data elements generally are required, according to the study design and other protocol requirements.

Report Preparation and Archiving

The fourth step of a data audit conducted as part of an EPA inspection is the GLP review of the final report that was submitted to the EPA. This review may be conducted at any phase of the audit and may precede the actual inspection if the inspector or auditor has received an advance copy of the final report from the EPA Office of Pesticide Programs. In addition to verifying that the final report is consistent with raw data and records, each report will ordinarily also be reviewed by the auditor for the following:

- inclusion of the 14 elements required by 40 CFR 160.185(a), including the QAU inspection statement and the location of all specimens and data

- the GLP certification statement, signed by the study director, sponsor, and applicant

- the signature of the study director

- assurance that report amendments (corrections and additions) have been properly submitted by the study director

Observed GLP Problems with Field Studies

Based on the limited experience that the EPA has gained in its inspection of field sites and associated specimen-analysis laboratories, the regulated community appears to have made notable strides toward achieving compliance with the GLP regulatory requirements. Most of the large pesticide manufacturers and formulators, in their capacities as sponsors, have played a major role in this effort, largely through education of contractors and in-house staff before implementation of the FIFRA regulations in October 1989.

Some of the GLP-related requirements found by the EPA auditors and inspectors to be the most difficult for industry to comply with are:

- General Provisions (Subpart A)

    — lack of a proper GLP compliance statement as required by 40 CFR 160.12(b) in that all GLP deviations were not listed

- Organization and Personnel (Subpart B)

    — lack of adequate or up-to-date training and experience records for personnel as required by 40 CFR 160.29(b)

— lack of an adequate number of qualified personnel in analytical laboratories to achieve timely and complete specimen analysis as required by 40 CFR 160.29(c)

— lack of one study director for pesticide residue and other studies as required by 40 CFR 160.31(a) and 160.33

— lack of a completely independent QAU as required by 40 CFR 160.35(a)

• Facilities (Subpart C)

— lack of adequate separation between relatively high-concentration test-substance mixing and storage areas and low-level specimen handling areas at field sites and laboratories as required by 40 CFR 160.41

— lack of adequate archives as required by 40 CFR 160.51 (e.g., uncontrolled access, storage of paper data with flammable chemicals, or lack of test substance archives)

• Equipment (Subpart D)

— lack of adequate or up-to-date records of inspection, maintenance, testing, calibration, and standardization as required by 40 CFR 160.63(c)

• Testing Facilities Operation (Subpart E)

— deviations from SOPs not authorized by the study director or not completely documented in the raw data as required by 40 CFR 160.81

— SOPs not immediately available to necessary personnel as required by 40 CFR 160.81(c)

• Test, Control, and Reference Substances (Subpart F)

— data not adequately generated or retained for test, control, and reference substances as required by 40 CFR 160.105

— identity, strength, purity, composition, and other characterization data for test, control, and reference substances not generated under full GLP purview

— stability data not available or generated for test substances (or test-substance mixtures) under storage conditions encountered at field sites as required by 40 CFR 160.105(e) and 160.113(a)(3)

— lack of adequate testing or other supporting data to satisfy the periodic analysis and uniformity requirements for test-substance mixtures (usually tank mixes) to fulfill the requirements of 40 CFR 160.113(a)(1)

- Protocol for and Conduct of a Study (Subpart G)

  — lack of complete methodology, particularly analytical methodology as required by 40 CFR 160.120(a)

  — lack of identification of specific records to be maintained for a study as required by 40 CFR 160.120(a)(13)

  — lack of approval (or timely approval) by the study director for deviations from the protocol as required by 40 CFR 160.120(b)

  — study not conducted according to the protocol as required by 40 CFR 160.130(a)

  — all specimens not properly identified by test system, study, nature, and date of collection as required by 40 CFR 160.130(c)

  — data not recorded promptly (on day of entry) or in ink as required by 40 CFR 160.130(e)

  — corrections to raw data improperly made contrary to the requirements of 40 CFR 160.130(e)

- Records and Reports (Subpart J)

  — final report without signed and dated reports of all scientists involved as required by 40 CFR 160.185(a)(12), or the locations of all raw data and specimens as required by 40 CFR 160.160(a)(13)

  — all raw data, documentation, records, specimens, and pertinent correspondence not retained as required by 40 CFR 160.190(a)

  — archives that do not meet the requirements of 40 CFR 160.190(c) or 160.190(d) because a single individual is not responsible, or unauthorized personnel have access

  — records of data and other information not maintained as true copies (e.g., use of opaquing fluid and penciled data are not reproduced or indicated as such) as required by 40 CFR 160.195(i)

NOTE: FIFRA Books and Records regulations [40 CFR 169.2(k)] also require the retention of all original raw data for underlying research data supporting a pesticide registration for the life of the registration.

Conclusion

An EPA audit of a field study can be an arduous task, involving expertise in product chemistry, agronomy, sampling procedures, and specimen analyses, including a knowledge of radiochemistry. The audit may also require inspection of a number of associated, but geographically remote, facilities where different phases of the study were conducted to resolve study design, final report, QAU, or other related study questions. These facilities include (1) research facilities where the test substance was synthesized, formulated, and characterized; (2) one or more field sites where test substances were applied and specimens collected; (3) one or more processing facilities where agricultural commodities were processed; (4) one or more laboratories where the specimens were analyzed; and (5) the sponsor's or study director's office or facility. Thus, the single study audit may require the close coordination of a number of the EPA inspectors and auditors and may cover facilities located over a wide geographical area.

Acknowledgments

Special thanks are due to Diane Bradway and John Gillis of the EPA's National Enforcement Investigations Center (NEIC) and G. Thomas Gale of Ciba–Geigy for their review and constructive comments, and to Illa Schipporeit of NEIC for preparing the text.

RECEIVED for review September 17, 1990. ACCEPTED revised manuscript September 13, 1991.

Chapter
30

Is the Federal Insecticide, Fungicide, and Rodenticide Act Good Laboratory Practices Program at a Crossroads?

David L. Dull and Francisca E. Liem

U.S. Environmental Protection Agency, EN–342, 401 M. Street, S.W., Washington, DC 20460

Since October 1989, virtually all testing of pesticide chemicals for health and environmental effects and environmental fate characteristics (including studies conducted in the field) required by the Federal Insecticide, Fungicide, and Rodenticide Act has had to be performed according to Good Laboratory Practice (GLP) regulations. Experience with these regulations (dating back to 1983) for a limited group of covered laboratory studies (human health effects) suggests that they are very difficult to enforce. Inclusion of field studies under the GLP banner is likely to complicate the enforceability problem. Is it time to admit that the Environmental Protection Agency's (EPA) GLP program is fundamentally a laboratory accreditation program without a formal accreditation, or can (or should) enforcement cases with monetary penalties and other sanctions be prosecuted? This issue and related issues are discussed in an EPA manager's review of the current program.

IN THE 1970S, A SERIES OF AUDITS by the Food and Drug Administration (FDA) and the Environmental Protection Agency (EPA) revealed serious deficiencies in toxicological tests conducted by Industrial Bio-Test Laboratories to support the registrations, in the United States and other countries, of numerous pesticides and drugs. The Industrial Bio-Test Laboratories case raised concerns about the quality and integrity of health and safety data submissions to regulatory agencies. These concerns led the FDA and the EPA to initiate and adopt principles of Good Laboratory Practices (GLPs) for the conduct of these tests. These GLP Standards have been recognized and adapted by the agricultural chemical and pharmaceutical regulatory authorities in most of the world's developed countries.

To implement these principles, the EPA promulgated GLP regulations under both the Federal Insecticide, Fungicide, and Rodenticide Act (FIFRA) and the Toxic Substances Control Act (TSCA) in November 1983. These regulations were amended and expanded in August 1989. The 1983 GLP regulations under FIFRA were applicable to only animal laboratory studies used to obtain data on the expected effect of human exposure to pesticides. The TSCA 1983 GLP regulations, in contrast, were applicable to people conducting human health effects, ecological effects, and environmental fate studies required by rules promulgated under Section 4(a) of TSCA.

The 1989 amendments to the FIFRA GLP regulations broadened their scope in a dramatic fashion. Since the August 1989 amendments became effective on October 16, 1989, virtually all testing of pesticide chemicals for health and ecological effects, environmental fate characteristics (including studies conducted in the field), and chemical characterization data, required by FIFRA, has had to be performed according to these GLP regulations. The FIFRA GLP regulations are intended to ensure the quality and integrity of laboratory and field test data required to evaluate health and ecological effects and the fate of chemical substances subject to the provisions of FIFRA.

To implement the goals of the GLP principles and the regulations on GLPs, the FDA and the EPA have conducted a vigorous inspection and data audit program for many years. A memorandum of understanding (MOU) between the two agencies ensures a cooperative approach that recognizes that many laboratories conduct testing to evaluate both pesticides and drugs and that attempts to avoid inconsistent and duplicative requirements.

In December 1983, the EPA reorganized its laboratory inspection and data audit program to support the agency's concern for data quality and to implement the newly promulgated GLP regulations. The Office of Pesticides and Toxic Substances established an inspection unit in the Office of Compliance Monitoring (OCM) for the data audit and laboratory inspection programs under both FIFRA and TSCA.

Scope of the Inspection and Audit Program

This combined inspection and audit program currently has 477 domestic testing laboratories in its inventory that have submitted studies to the EPA in support of a marketing or experimental-use permit petition under FIFRA or test rules (Section 4) and orders (Section 5) under TSCA. Of these laboratories, 188 domestic laboratories are viewed as currently active in conducting tests of FIFRA and TSCA chemicals and are thus under the EPA's routine GLP surveillance. Under this surveillance, the OCM attempts to inspect the domestic testing facilities at least once every 2 years. The program currently conducts approximately 70 GLP inspections each year and audits more then 400 studies during these inspections.

The EPA also conducts an active international GLP program through bilateral MOUs with five countries (Japan, the United Kingdom, Germany, the Netherlands, and Switzerland) and participates in multilateral efforts in organizations such as the Organization for Economic Cooperation and Development and the World Health Organization. Under the bilateral agreements, the EPA recognizes the GLP monitoring systems in the other countries and accepts data from these countries for the registration of pesticides. The same considerations are applied by the bilateral partners to data generated in the United States. For data generated abroad, the EPA relies largely on the monitoring and compliance systems of other countries, but for domestic data the agency has in principle a number of administrative and regulatory sanctions that can be brought to bear on a testing facility that is in violation of the FIFRA GLP regulations. In summary these sanctions are, in order of increasing severity

1. issuance of a notice of warning

2. civil administrative penalties (monetary) under Section 12 of FIFRA for failure to maintain raw data as required by Section 8 of FIFRA and for signing a false GLP certification statement

3. rejection of study results for regulatory purposes (40 *Code of Federal Regulations* (CFR) 160.15 and 160.17)

4. criminal sanctions for willful or knowing violations (Sections 12(a)(2)(M), 12(a)(2)(Q), and 12(a)(2)(R) under FIFRA)

Since consolidation of the GLP program in December 1983, approximately 500 inspections have been conducted. Of these, at least 44 have resulted in some type of enforcement action under FIFRA:

- Twenty-six laboratories were sent notices of warning.

- The EPA has filed civil complaints against 14 registrants for lack of raw data to support their pesticide registrations. These cases have been settled out of court with monetary penalties ranging from $1500 to $30,000.

- A criminal suit has been prosecuted against a laboratory for willfully and knowingly falsifying data. The company was fined $100,000; the director was fined $1000 and was sentenced to 30 days imprisonment and 1 year probation.

- Regulatory action has been taken on at least three studies in which the studies have been rejected or additional data have been requested on the basis of GLP deficiencies.

Problems with the EPA GLP Program

In spite of the successes of the program, and the common perception that GLPs have been widely adopted by the testing industry and have led to enhanced data quality and integrity, the FIFRA GLP program currently faces a number of new challenges. Since the amendment of the regulations in 1989, the program has been faced with applying the concept of GLPs to studies conducted in the field, far from the control and easily manipulated conditions of the laboratory. Field studies as currently conducted by the agricultural chemicals testing industry are managed in a complex fashion and frequently involve a sponsor (usually the pesticide manufacturer or a consulting firm hired by the manufacturer), an agricultural chemicals testing contractor, a separate analytical chemistry laboratory, and one or more field site cooperators (e.g., the farmer who owns the field or the independent contractor who applies the pesticide). Applying the GLP regulatory requirements (e.g., for one study director and a centralized quality assurance unit) to such complex institutional arrangements for these types of studies will be problematic for some time to come.

However, with time and experience in applying the GLP regulations to field sites, these problems will be overcome. Currently, other problems involve the manner in which the program is operated and, more importantly, raise the question of whether a regulatory enforcement program is the best way of ensuring data integrity and quality. These challenges suggest that the basic structure of the GLP program may require reevaluation. The following specific issues will be discussed in turn in the remainder of this chapter:

- persistent complaints from the testing industry on the lack of timeliness of the case referral and closure process

- a relatively low output of formal enforcement actions in the FIFRA program since 1983, given the number of inspections that have been performed

- the intrinsic weakness of any systematic effort to enforce the FIFRA GLP regulations, in that it is not a violation of FIFRA to fail to follow the GLPs, as long as the compliance statement required by the GLP regulations states that the regulations were not followed

- an inability to state easily and in a timely fashion whether a given laboratory is in compliance with GLPs as required by our international commitments under our existing bilateral agreements with Japan, the United Kingdom, Germany, Switzerland, and the Netherlands

Lack of Timeliness. According to this system, it takes a minimum of 210 days from the start of an inspection until the closure of the file, and this time frame assumes that no enforcement action will result from the inspection. These delays are in part a result of the fact that the current program is decentralized and involves a number of the EPA's regional offices. In addition, each case is treated the same way irrespective of the presence or absence of GLP violations, so a full inspection report is prepared for each case. This inability to state easily and in a timely fashion the outcome of a specific inspection makes it more difficult to fulfill provisions of our international bilateral agreements, which call for this information to be made available to our partners.

Low Output of Enforcement. In spite of a high annual level of inspection and data audit activity, the number of enforcement cases involving violations of the GLPs is currently minimal. In fact until October 1991, a civil penalty had never been assessed for violation of the FIFRA GLP regulations. Civil penalty cases have only resulted from cases of missing raw test data required by Section 8 of FIFRA. These cases are relatively easy to document and prosecute.

The reasons for this absence of enforcement activity are multifold, involving the inherent complexity of the GLP regulations, the fact that the regulations are in many ways general standards rather than specific "dos and don'ts", the absence until September 1991 of policies for assessing penalties under the regulations, and the fact that the testing industry is largely in compliance with the standards.

One of the most significant reasons for this low output of formal actions enforcing the GLP regulations may involve FIFRA itself. People often say that unless specifically enumerated in Section 12 of FIFRA ("Unlawful Acts"), failure to obey a regulation under FIFRA is not itself a violation of FIFRA. In short there is no general authority to assess monetary penalties for violation of FIFRA regulations. For example, Section 12(a)(2)(B)(i) makes it an unlawful act to refuse to maintain records required under Section 8 of FIFRA (i.e., data on testing of pesticides). No such explicit statement exists in Section 12 regarding the GLP regulations. However, a number of provisions in Section 12 of FIFRA deal with acts of falsification, including falsification of any part of an application for registration of a pesticide [§ 12(a)(2)(M)] and falsification of any information related to the testing of a pesticide [§ 12(a)(2)(Q)].

Enforcement Problems. Because of these statutory peculiarities, enforcement of GLPs must focus on establishing that the certification statement required by the GLP regulations is false. This provision requires a submitted study to contain a statement that one of three things is true: that the study (or parts of it) was done according to GLPs, that the study (or parts of it) was not done according to GLPs, or that the person signing the certification statement does not know whether the study was done according to GLPs. Successful enforcement actions will result from an affirmative certification statement plus inspection results indicating that provisions of the GLP regulations were not followed. This certification statement must be signed by the submitter of the study, the sponsor of the study, and the study director. However, as long as the certification statement required by the GLP regulations reflects the fact that certain requirements of the regulations were not followed, there is no enforceable violation.

This inability to cite testing facilities for failure to observe specific requirements of the GLP regulations is a major weakness of the current GLP program, especially in light of the fact that the EPA does not have an explicit policy of rejecting studies not done according to GLPs. Some pieces of draft legislation amending FIFRA have addressed this problem, but none has yet been enacted into law by Congress. Comparison of this situation with TSCA, for which a violation of any provision of the TSCA GLPs is automatically a violation of Sections 4 and 15 of TSCA, is useful.

Inability To Verify Compliance. Numerous testing facilities have complained about the inability to obtain closed inspection files in a timely manner through Freedom of Information Act requests and in general to obtain a clear statement from the EPA of whether the facility is in compliance with GLPs after an inspection or data audit. The agency's current procedures for processing inspection and data audit files do not afford an expedient closure of files.

Addressing Problems

These problem areas have been placed in sharp focus by two recent developments affecting the EPA's GLP program. The first is a General Accounting Office (GAO) investigation of the efficacy testing of disinfectants that criticizes the GLP program and suggests that the routine inspection process accompanied by data audits does not ensure the quality of registration data for disinfectants and does not produce completed laboratory evaluations in a timely fashion.

The second event is the formation of an EPA task force to explore the possibility of a laboratory accreditation program for all laboratories submitting test data to the EPA. These two developments will be discussed separately.

GAO Investigation. Many of the aforementioned deficiencies of the current program have surfaced during a recent investigation by the GAO. The GAO, at the request of Congress, has conducted an investigation of the EPA's regulation of antimicrobial pesticides; the focus of the investigation is on issues related to the testing of disinfectant products to establish their efficacy. This investigation has also focused on the OCM's past GLP inspection and audit of laboratories conducting product performance (efficacy) studies of antimicrobial pesticides.

The GAO criticisms involve procedural issues such as timely issuance of inspection reports. But more significantly, GAO staff apparently question whether a neutral-scheme inspection program that relies heavily on auditing completed studies that have been submitted to the EPA in support of pesticide registration is able to ensure the quality of data for disinfectants. A neutral-scheme inspection is one in which the selection of laboratories to be audited is made according to nonbiased, objective criteria.

Laboratory Accreditation. Since the inception of the concept of GLPs, the EPA and the FDA have been asked by different groups to accredit laboratories that are generating environmental and human health (toxicology) data. As an example of such activity, in 1980, the Toxicology Laboratory Accreditation Board was created through initiatives taken by

382   GOOD LABORATORY PRACTICE STANDARDS

the Society of Toxicology. The board operated as an independent nonprofit organization. Accreditation was voluntary and was accomplished by a peer review confirming satisfactory achievement. The peer review covered completed studies that demonstrated competency in meeting established standards and also scientific aspects of the studies.

In general, however, both the FDA and the EPA have in the past rejected the notion of a certification or accreditation program as a substitute for the current enforcement approach based on GLP regulations. In 1986, the EPA specifically rejected such an option as a means of addressing problems in testing the efficacy of antimicrobial pesticides.

The EPA's approach to laboratory accreditation, however, may be about to change. In 1989, the EPA was again urged to eliminate duplicative or inconsistent requirements for laboratories conducting tests for more than one of the agency's regulatory programs by accrediting all laboratories conducting tests to be submitted to the EPA. As a result, the EPA has established a task force (Environmental Management Monitoring Council [EMMC], Ad Hoc Panel on National Laboratory Accreditation) to explore the feasibility of a national laboratory accreditation program. Such an effort clearly has significant implications for the existing GLP program.

A number of fundamental issues need to be addressed in determining whether the EPA should adopt a laboratory accreditation program for laboratories submitting test data under FIFRA. Some of these questions are

1. Does the EPA have the legal authority under FIFRA to pursue a national program for accreditation of testing facilities? What are the implications of a national laboratory accreditation program for the EPA's statutory mandates and existing regulatory programs, such as the GLP program?

2. Would a national accreditation program replace the EPA's existing FIFRA GLP enforcement oriented program, or would the certification program exist independently of the EPA's enforcement programs? If the latter is the case, what are the potential problems of conflict between the GLP enforcement programs and the accreditation program?

3. In the event that an accreditation program is established independently of the GLP enforcement program, should the accrediting authority be delegated to independent third parties as has been suggested by the task force? Who would serve on this independent accrediting body and how would potential conflicts of interest be addressed?

4. What criteria would be used to accredit or certify the various types
    of testing facilities, and what would be the relationship between
    these criteria and the GLP regulations? Could the same accrediting
    body address both conventional toxicology laboratories and field
    sites? Could a negative accreditation decision be repealed or could
    a positive one be revoked?

5. What are the EPA and the FDA roles and responsibilities in a
    national laboratory accrediting program, as the EPA and the FDA
    currently run their GLP programs in cooperation?

The answers to these questions are not yet available, so it is possible
only to speculate on some of the benefits and disadvantages of a national
accreditation program for FIFRA laboratories.

Disadvantages of Accreditation. Two of the more significant problems of a laboratory accreditation program for FIFRA studies deserve
mention, however. An accreditation program for laboratories conducting
complex toxicological, ecotoxicological, and environmental fate studies,
including field applications at cooperator farms, would present difficult
issues of deciding what criteria to use in determining whether to accredit a
particular laboratory or field site. Proficiency testing does not seem practical for complex residue and field dissipation studies, not to mention
human toxicology testing. If the GLP Standards are used as standards for
accreditation, unnecessary duplication would be apparent between the
EPA's program and that of the accrediting body (in the event that the
accreditation program is operated by a third party independent of the
EPA's GLP program).

In addition, conflicts would be likely between the GLP enforcement
program and the accreditation program with regard to the status of
specific laboratories, and the result would be confusion and lowered credibility for both programs.

Advantages of Accreditation. After considering some of the benefits
and problems associated with a national laboratory accreditation program,
an accreditation program does not have any clear benefits over the existing
system of enforceable GLP regulations under FIFRA as far as ensuring
the quality and integrity of test data is concerned. Furthermore, an
accreditation program lacks the inherent flexibility of a neutral,
enforcement-oriented inspection program, where laboratories can be targeted on the basis of their past inspection history, the volume of work
they are doing, or the regulatory importance of the studies they have submitted. However, an accreditation program could go a long way toward

addressing some of the existing procedural and legal problems in the GLP program and will be given careful consideration and analysis in the months ahead by the EMMC ad hoc panel.

In the event that the accreditation program replaces the existing GLP program or is run in concert with it, a clear statement that a laboratory is in compliance with GLPs would be available in a timely fashion. The timeliness of the case referral or closure process would no longer be an issue.

A laboratory accreditation program would provide to the public a list of approved facilities. Marginal facilities would be eliminated. In the international arena, our major trading partners, with whom we have bilateral GLP MOUs, would also have access to an approved list of accredited facilities from whom they could with confidence accept test data.

GLP Review Committee. In the interim, efforts are being made to address some of the procedural problems of the GLP program discussed in this chapter. An ad hoc decision-making group (called GLP/Data Audit Review Committee or GRC) has been formed. The GRC's main objective is to streamline the case review process. The GRC will review cases as soon as possible after completion of the inspection or audit and recommend appropriate categorization for enforcement actions.

Cases will be presented to the GRC as follows. The inspector (in collaboration with the auditors) will write a brief summary report within 2 weeks after completion of the GLP inspection or data audit. A complete inspection report will be prepared only if the GRC determines that the case presents significant violations of the GLP regulations. Otherwise the brief summary report will serve as a report of the inspection, and the case file will be quickly closed.

The GRC will meet approximately once a month, or as often as necessary, to act on pending cases for which summary reports have been completed. Cases resulting from the inspections and data audits will be classified by the GRC into the following four categories:

I. significant violations

    a. missing data (FIFRA § 8)

    b. falsification of data or a GLP certification statement [FIFRA § 12(a)(2)(Q)]

    c. knowing falsification of data or a GLP certification statement [FIFRA § 12(a)(2)(R)]

    d. accumulation of one or more significant, enforceable GLP violations

II. minor GLP violations not justifying a civil penalty but justifying a notice of warning

III. scientific deficiencies

IV. no action warranted

The GRC will then determine in what category the case belongs and instruct appropriate staff to proceed as indicated:

Category I: The full inspection report will be completed on a fast track and submitted to the director of the GLP program, who will promptly refer to compliance personnel for further review and possible enforcement action (e.g., civil administrative complaint, criminal referral, or followup inspection).

Category II: A letter of warning will be promptly issued.

Category III: A summary of the scientific problems in the study that may have an effect on study validity will be referred for appropriate action to the office that registers pesticides. Such action would involve rejection of the study or a requirement that certain parts of it be repeated.

Category IV: The file will be closed promptly. A brief postcard notice will be provided to the laboratory and sponsor. A complete inspection report will not be prepared.

This activity of the GRC will not address problems such as whether FIFRA provides adequate authority to enforce the GLP regulations. However, it will go far toward improving the process by which GLP inspections and data audits lead to determinations that a facility conducting tests for FIFRA purposes is or is not in compliance with the FIFRA GLP regulations.

RECEIVED for review September 17, 1990. ACCEPTED revised manuscript August 17, 1991.

Chapter
31

Good Laboratory Practice Standards Policies and Interpretations

Phyllis E. Flaherty and Stephen J. Howie

Office of Pesticides and Toxic Substances, U.S. Environmental Protection Agency, EN–342W, 401 M Street, S.W., Washington, DC 20460

Numerous policy questions have been raised in regard to the Federal Insecticide, Fungicide, and Rodenticide Act Good Laboratory Practice (GLP) Standards, especially since the 1989 revisions added field tests, environmental effects, and environmental fate testing. How the Environmental Protection Agency (EPA) develops its policies, how major policy decisions have been made, how the EPA disseminates such information, and how to get policy questions answered are discussed, along with revisions being considered for GLP Standards violations.

THE U.S. ENVIRONMENTAL PROTECTION AGENCY (EPA) promulgated its first Good Laboratory Practice (GLP) Standards regulations pursuant to the Federal Insecticide, Fungicide, and Rodenticide Act (FIFRA) in 1983 (*1*). These regulations were initiated as a result of problems found with data that had been submitted to the agency in support of registration of pesticides. The EPA amended these regulations in 1989 (*2*) to expand the GLP requirements to include field testing, environmental effects testing, and environmental fate testing. Because of the broad nature of these regulations, numerous policy questions have been directed to the agency. The EPA has responded to many of these questions and continues to work on resolving a number of issues.

Managing the development of the revised FIFRA GLP Standards regulations to expand the coverage to the majority of studies submitted to the EPA under FIFRA and preparing responses to questions that have arisen have provided a unique challenge. Part of the challenge arose from the development of regulations for a diverse set of testing facilities, and the remaining challenge arose from setting consistent policy on how the EPA interprets and enforces the regulation. To understand how the EPA establishes policies, it is important to look at why GLP regulations were issued, what a policy does, how policy issues are identified, the process within the agency for developing a policy, the factors affecting policy decisions, and how the EPA makes its policies known. A separate but related issue is how the EPA responds to violations of the regulations.

Why GLP Regulations Were Necessary

The EPA relies largely on data submitted by registrants to make its regulatory decisions on pesticides. The importance of these studies to the EPA decision-making demands that all studies be conducted according to scientifically sound protocols with detailed attention to quality control. These studies must be performed as reported and the results must be correctly reported. Written with this goal in mind, the GLP Standards are general standards applicable to almost every type of study, and when correctly followed they ensure the integrity of data submitted.

To realize why GLP regulations are considered necessary, one must recognize that the impetus was the discovery of some extremely serious deficiencies in data that had been submitted to the Food and Drug Administration (FDA) and to the EPA. Prior to the mid-1970s, the EPA and the FDA generally accepted with little oversight the accuracy and validity of data submitted for regulatory decision making. However, in 1974 and 1975, problems surfaced in some laboratories generating data submitted to the FDA. As a result, the FDA initiated an inspection and audit program that provided for an in-depth examination of questionable studies. The FDA identified significant problems in the way some studies were being performed and reported to the administration.

The FDA findings caused the EPA concern over the validity of the data on which it was relying to make its decisions. Some data used by the EPA had originated from studies that were performed at the same laboratories where the FDA found problems. Moreover, the studies the EPA was relying on were similar to the studies for which the FDA had identified difficulties. Thus, the EPA began a laboratory data audit program similar to that of the FDA and closely coordinated with the FDA to determine if, indeed, data submitted to the EPA were also flawed. Although the EPA's experience indicates that most testing facilities were not involved in fraudulent submissions, one laboratory that had been responsible for fraudulent or misleading data submissions had supported studies for almost 600 chemicals. As a result of problems with this laboratory, the agency implemented a validation program to review approximately 1600 studies in which the laboratory had been involved. About 65% of those studies were deemed invalid. Thus, the impact of one laboratory with major problems was substantial, both in terms of the resources necessary to ensure that valid data supported registrations and in the public's confidence in the data and related regulatory decisions. Audits at other laboratories turned up a number of problems. One of the more common problems was a lack of supporting raw data that would allow the auditor or inspector to determine if the study had been performed as indicated.

To do a credible job of protecting the public and the environment, the EPA must be able to look at the supporting raw data and files and verify that studies were conducted as reported and that results are accurately reflected in the data that it uses to make its regulatory decisions. Thus, the agency felt it necessary to issue GLP regulations that specify certain minimal standards that make it possible for an inspector or auditor to go in during or after a study and know how that study was conducted with some certainty.

The initial FIFRA GLP Standards were proposed in 1980 (*3*), and final regulations became effective in May 1984, several years after the FDA promulgated GLP Standards under its statute (*4*). The 1984 regulations did not address all studies, although the EPA recognized the need to do this at some point. At the time the regulations were being prepared, the focus of the EPA's efforts was on health effects data because the FDA's investigative findings dealt solely with these data. The agency felt it needed more experience with some of the nonhealth effects studies in order to write GLP regulations applicable to them. The GLP regulations had a fairly narrow scope of types of studies, but the FIFRA § 8 record-keeping regulations [40 *Code of Federal Regulations* (CFR) 169.2(k), which require that the data supporting all studies be kept for the life of the pesticide's registration, did not. Thus, whereas FIFRA GLP requirements initially applied to a limited range of studies, that is, health effects studies, the agency's data audit program encompassed all types of studies submitted to the EPA to support registrations.

In August 1989, the EPA published its final revised FIFRA GLP Standards rule, which expanded the coverage of the GLP Standards to almost all studies submitted to the EPA under FIFRA. The revisions also made language changes to reflect changes that the FDA made to its GLP regulation in 1987 (5). The EPA's revisions thus ensure consistency between the two rules, as well as with the GLP rule issued by the EPA under the Toxic Substances Control Act (TSCA) (6). This revision was done to minimize the regulatory burden on laboratories that may conduct studies under all three statutes. The revised FIFRA GLP Standards rule became effective on October 16, 1989. The GLP Standards is a regulation, and its requirements are enforceable under FIFRA. The purpose of the GLP Standards as discussed in the preamble to the rules (1–3) is to ensure that studies are conducted as indicated with certain safeguards in place regarding the quality of the studies and that the required records and reports will allow reconstruction of the studies.

What a Policy Does

The GLP requirements are very general and apply to a wide array of studies. They address requirements regarding laboratory organization and personnel (e.g., requirements for a study director and quality assurance unit); requirements for the facility such as test, control, and reference substance handling and data storage; equipment requirements such as routine maintenance and calibration; facility operation requirements such as the need for standard operating procedures; requirements on test, control, and reference substances; protocol and conduct of study requirements; and record-keeping and reporting requirements. The GLP Standards provide flexibility to the laboratory by providing general requirements as opposed to the very detailed, specific requirements spelled out by the EPA.

Questions arise partly because of the general nature of the requirements and partly as a result of the uncertainty felt when requirements are implemented for the first time. Sometimes the question is a request for permission not to comply with the regulations. Policies issued by the agency or policy determinations made by the EPA such as those in response to letters do not change the regulation; they clarify what is intended and how the EPA reads the regulation. Answers to many questions that the regulated community has can be found by reading the preamble to the regulation because numerous questions were raised during the comment period after the rule was proposed. Some questions take a great deal of consideration to answer because all of the factors involved cannot be anticipated when a regulation is written. Although all issues may not have been specifically considered at the time the regulation was developed, a regulation cannot be changed by a policy; revisions to regulations require going through the full rule-making process, which can take years. Therefore, the policy is based on what the regulation states, even if it is in general terms.

The agency may respond to violations with policies that in effect allow deviations from the regulations without those deviations being penalized. For example, on a case-by-case basis the agency may allow a laboratory to discard test substance containers prior to the end of the study if thousands of containers are involved, provided other measures are taken, as specified by the agency, that ensure that the intent of the GLPs are met. Written authorization from the Policy and Grants Division of the Office of Compliance Monitoring is necessary.

How Policy Questions Arise

Policy questions often arise from incoming correspondence from the regulated community. Responses to such correspondence may be a matter of a straightforward reading of the regulations, or they may involve a great deal of policy analysis. In correspondence in which the EPA has responded to a question on the regulations, the answer given may be applicable only to a specific set of circumstances and not necessarily to all situations, which may be similar but not identical in all aspects.

Policy issues are also raised during conferences, training sessions, and meetings with the regulated community; by inspectors; or by case development officers, at headquarters or in the regions alike.

The Process Within the Agency

Although official policy is issued by the Director of the Policy and Grants Division, Office of Compliance Monitoring, Office of Pesticides and Toxic Substances of the EPA, policy is not made by one or two people. It involves many individuals and several offices, depending on the complexity of the question. Many questions are variations of questions that have already been answered and may be handled easily by relying on previous answers. Other policy questions require in-depth analysis and coordination among the Office of Pesticide Programs, the Office of Compliance Monitoring, the Office of General Counsel, and the Office of Enforcement. In addition, the EPA often consults with the FDA to ensure consistent interpretations between its GLP regulations and the EPA's, both of which contain almost identical language on many requirements. Answers may involve scientific, technical, and legal decisions.

Factors Affecting Policy Decisions

Several key factors shape the answer to a policy question. First of all, FIFRA and the regulations promulgated pursuant to FIFRA must be considered as the governing factors. These regulations, including the GLP regulations, can only be changed by revising them and going through a formal comment and review period. In answering questions on the FIFRA GLP regulation, the EPA first looks at what the regulation states. The rule or the preamble to the rule may provide answers directly, and an in-depth analysis may not be necessary.

If someone suggests that a different interpretation is possible or asks for a different response from the agency's previous statements based on circumstances that were not fully anticipated at the time the regulations were written, EPA looks carefully at the purpose of the regulations to see if there is more than one way to satisfy reasonably the requirements in the context of the regulatory language. In addition, the EPA is very concerned about the need for consistency between how the EPA regulations, both FIFRA and TSCA GLP regulations, and the FDA GLP regulation are interpreted where the language is identical among the various regulations.

How EPA Disseminates Its Policy Determinations

Given the number of inspectors and other people involved within the EPA and the FDA in implementing the FIFRA GLP regulations, it is important that these people be aware of any policy statements or interpretations that the EPA makes for these regulations. One way of ensuring consistency is to direct all new policy questions to the Policy and Grants Division in the Office of Compliance Monitoring and to make sure that the EPA personnel involved in the GLP inspection and audit program receive information on new decisions. The EPA does this by sending copies to appropriate offices when correspondence is issued on GLPs, by making sure other divisions see such correspondence before it goes out and by running training sessions and conferences.

With regard to disseminating information to the regulated community, the EPA has used a number of mechanisms in the past and some that are being implemented now. The EPA has held a number of meetings with the regulated community to answer questions on the regulations; attended various seminars, symposia, and training sessions at the request of the regulated community; sponsored training sessions that were attended by representatives of the public and the regulated community; provided copies of responses to correspondence to the public and the regulated community; and responded to incoming correspondence. Currently, the EPA is preparing a question-and-answer document specifically for the GLP regulations. This document will contain many of the questions to which the EPA has already responded, and new questions and answers will be added periodically until it represents a fairly complete set of FIFRA GLP policy issues to which the agency has responded. When this document is complete, the EPA plans to disseminate it to the professional organizations that have been involved in GLP issues and to the public.

The EPA is interested in how to make such information more readily available to the regulated community and continues to seek more effective ways of doing so. A better understanding of how the EPA views the regulations facilitates compliance.

Examples of Policy Issues

Archiving. The regulations require data to be archived during or at the close of the study. Some people have indicated that the complexity of studies performed at multiple locations causes delays in transferring material to be archived at the end of technical work. A grace period was suggested instead of requiring that all data be archived at the end of the study, that is, before or when the study director signs the final report.

The EPA reviewed the language in the rule, present EPA and FDA policy, and the technical aspects of studies newly subject to GLP Standards. The rule is clear in stating that data must be archived during or at the close of a study. Any change in this requirement would have to be made by amending the rule. Previous EPA and FDA policies are that the archiving must explicitly be completed before the study report is signed. Finally, there was not a difference in the technical needs of studies newly under GLP Standards that would suggest that they be treated differently than studies that the FDA monitors or the studies previously subject to the EPA GLPs. The benefit from archiving data at the end of the study is unchanged, and any practical difficulties arising from increased complexity of studies suggested that it would be, if anything, more beneficial to secure all raw data in archives before signing the final report.

Thus, on the basis of the most straightforward reading of the rule, the history of interpretation both at the EPA and at the FDA, and a practical evaluation of the GLP Standards in question, the EPA decided that this standard must be interpreted consistently for all studies, even in the case of more complex studies that are required to comply with GLP Standards because of the 1989 revisions.

In considering the concerns raised, the EPA examined the case of complex studies that occur at several locations and involve more than one discipline. It concluded that there is flexibility in the location of the archives of raw data and specimens. In 40 CFR 160.190(b), the GLP Standards state that retention of records at alternate locations is acceptable, provided that there is specific reference to those locations in the archives. Such off-location archives must still meet the full requirements of 40 CFR 160.190. Whether records are archived at a central location or at separate contractors' locations, the study director must ensure that all raw data and specimens have been archived before the study report is signed. In the case that the study director cannot ensure that records at a particular location are archived correctly, he or she should not sign a compliance statement that indicates that this standard has been met.

Quality Assurance Inspections. When the EPA amended the FIFRA GLP Standards, the EPA modified the requirement that each phase of each study be inspected to require simply that each study be inspected at adequate intervals. The wording that was used was adapted directly from the FDA's 1987 amendments. As stated, the wording is explicit in requiring that each study must be inspected at least once. Some people have suggested that this requirement is not consistent with the FDA's interpretation, which they state does not necessarily require each study to be inspected. The EPA reviewed this requirement and concluded that the wording explicitly required an inspection of each study.

Certain studies include work that is difficult to schedule according to a strict timetable because the work depends on conditions such as weather. The EPA was asked whether it would be acceptable to schedule quality assurance unit inspections of certain activities when there was no actual study in progress. These inspections were meant to help monitor that tasks were performed correctly without inspections being subject to last-minute rescheduling because of changes in weather.

The EPA reviewed the language of the rule to determine whether it was required that all inspections occur during study operations. The rule is in fact silent on this, which the EPA interprets as allowing flexibility. However, the requirement remains that each study be inspected at least once. Consequently, the EPA responded that it would be permissible for inspections of processes to be made when no actual work was in progress, although it would be necessary to demonstrate that the process subsequently used on a study was identical to the process that was inspected. Some part of the study must still be inspected when the study is in progress. This inspection could be arranged for a phase or location of the study that could be conveniently scheduled for inspection.

This interpretation was discussed with the FDA to ensure that it was not contrary to their policy regarding quality assurance inspections and was found to be acceptable.

Future Revisions to the Regulations

The current regulations are anticipated to remain in effect for some time. As concerns are raised, these concerns will be kept in mind if the agency initiates additional revisions.

Enforcement Policy

Violations of the GLP Standards may be discovered in a number of ways. Generally, the following represent the means by which most GLP violations are likely to be found: during laboratory inspections and data audits conducted by the EPA and the FDA; during the EPA's review of data submissions; through investigations of tips or complaints; or as a result of data called in. Under FIFRA (7), it is unlawful for any person

1. to knowingly falsify all or part of any application for registration, application for experimental use permit, . . . any records required to be maintained pursuant to the Act, any report filed under the Act, or any information marked as confidential and submitted to the Administrator under any provision of the Act [FIFRA Section 12(a)(2)(M)]

2. to falsify all or part of any information relating to the testing of any pesticide (or any ingredient, metabolite, or degradation product thereof), including the nature of any protocol, procedure, substance, organism, or equipment used, observation made, or conclusion or opinion formed, submitted to the Administrator, or that the person knows will be furnished to the Administrator or will become part of any records required to be maintained by the Act [FIFRA Section 12(a)(2)(Q)]

3. to submit to the Administrator data known to be false in support of a registration [FIFRA Section 12(a)(2)(R)]

FIFRA provides the authority to issue notices of warning for minor violations and to assess administrative civil penalties of up to $5000 per violation. Any person who knowingly violates FIFRA may also incur criminal penalties of up to $50,000 or 1 year imprisonment. The level of response, and the appropriate penalty within that level, is determined by the nature and severity of the violation. The violator's size of business and ability to pay must be considered when administrative civil penalties are issued. Falsification of submissions or records may be the basis for a criminal referral under the U.S. Code of Violations (18 U.S.C. 1001). Actions may be taken against the registrant, the laboratory, or individuals for falsifying the certification statements, records, or reports. Generally, the EPA will examine the specific case to determine who is most appropriately subject to an enforcement action.

In addition to the direct penalties provided under FIFRA, failure to submit a true and correct compliance statement may be used as the basis for rejection of studies. Where a study is not conducted in accordance with GLP Standards, the EPA may refuse to consider the data as meeting requirements. Submission of a false statement of compliance may also be the basis for the cancellation or suspension of a registration, modification of the research or marketing permit, or denial of an application for such a permit.

In its July 2, 1990, FIFRA Enforcement Response Policy (ERP) (8), the EPA presented its policy regarding penalty determinations for FIFRA violations. Copies of the FIFRA ERP may be obtained by writing FIFRA ERP, Pesticide Enforcement Policy Branch, Office of Compliance Monitoring (EN–342), The Environmental Protection Agency, 401 M Street S.W., Washington, DC 20460.

The ERP includes tables that list the gravity levels for each violation of FIFRA and matrices for determining appropriate penalty amounts based on the gravity level, size of business, history of noncompliance, culpability of the violator, and harm to human health or the environment. The ERP applies to any violations of FIFRA, including those concerning the GLP regulations. Specific FIFRA charges given in the ERP are in reference to § 12(a)(2)(M), for knowing falsification of reports submitted to the EPA; § 12(a)(2)(Q), for falsifying information related to testing; and § 12(a)(2)(R), for submission of data known to be false.

The July 2, 1990, FIFRA ERP addresses GLP violations in the following manner: A high-level GLP violation has a maximum civil penalty of $5000 per violation; a middle-level GLP violation has a maximum civil penalty of $4000 per violation; and a low-level GLP violation has a maximum civil penalty of $3000 per violation. On September 30, 1991, the EPA issued a GLP ERP supplement (9) to the July 2, 1990 FIFRA ERP that defines which violations of the GLPs constitute high-, middle-, or low-level violations.

The supplement describes in greater detail the appropriate enforcement response for specific GLP violations. (See Appendix C of this book for the text of the EPA's penalty policy.) For drafting the supplement, the EPA reviewed specific GLP Standards for relative gravity associated with the violation.

Conclusion

The EPA relies on FIFRA and the regulations promulgated pursuant to this statute to determine responses to policy questions. At the same time, it strives to allow flexibility under those requirements consistent with maintaining the goals of the GLP Standards. In responding to policy questions, the agency examines the concerns raised and tries to apply the law in a reasonable and fair manner. The efforts by the regulated community to identify issues and to obtain resolution as soon as possible are much appreciated.

The EPA is prepared to apply enforcement remedies to violators. An active compliance and enforcement program is important to prevent credibility problems for those in the regulated community who comply as well as for the EPA. However, the EPA prefers to work with the regulated community to obtain compliance up front instead of relying solely on enforcement actions. The agency recognizes that ensuring the validity and integrity of data developed and submitted is a goal shared by the regulated community and the federal government. Opportunities to further this goal through meetings, correspondence, and other outreach efforts are welcomed.

References

1. "Federal Insecticide, Fungicide, and Rodenticide Act (FIFRA) Good Laboratory Practice Standards; Final Rule"; *Federal Register* 48:53964 (November 29, 1983).

2. "Federal Insecticide, Fungicide, and Rodenticide Act (FIFRA) Good Laboratory Practice Standards; Final Rule"; *Federal Register* 54:34052 (August 17, 1989).

3. "Federal Insecticide, Fungicide, and Rodenticide Act (FIFRA

Good Laboratory Practice Standards; Proposed Rule", *Federal Register* 45:26373 (April 18, 1980).

4. "Nonclinical Laboratory Studies: Good Laboratory Practice Standards: Final Rule", *Federal Register* 43:59986 (December 22, 1978).

5. "Good Laboratory Practice Standards: Final Rule", *Federal Register* 52:33768 (September 4, 1987).

6. "Toxic Substances Control Act (TSCA) Good Laboratory Practice Standards; Final Rule", *Federal Register* 54:34034 (August 17, 1989).

7. "Federal Statute: Federal Insecticide, Fungicide, and Rodenticide Act (FIFRA)", Title 7 U.S. Code 136 et seq., as revised October 1988.

8. "Federal Insecticide, Fungicide, and Rodenticide Act Enforcement Response Policy", Office of Pesticides and Toxic Substances and Office of Compliance Monitoring, U.S. Environmental Protection Agency, Washington, DC, July 2, 1990.

9. "Enforcement Response Policy for the Federal Insecticide, Fungicide, and Rodenticide Act Good Laboratory Practice (GLP) Regulations", Office of Pesticides and Toxic Substances and Office of Compliance Monitoring, U.S. Environmental Protection Agency, Washington, DC, September 30, 1991.

RECEIVED for review September 17, 1990. ACCEPTED revised manuscript August 17, 1991.

Chapter
32

Economic Impact of Regulations on Field Contractors and an Agrochemical Company

James L. Platt, Jr.[1]

Audit and Inspection Division, International Chemical Consultants, Inc., 5520 Cherokee Avenue, Suite 200, Alexandria, VA 22312–2319

This chapter presents the results of a 1990 International Chemical Consultants, Inc., survey of the impact of the FIFRA Good Laboratory Practice (GLP) Standards on field contractors. Results are compared with a 1988 survey of the anticipated impacts of GLP Standards on field contractors. The impact of GLP on a sponsor agrochemical company is included. During May 1990, questionnaires were sent to 138 field contractors. The results provide a profile of field contractors as a business community, their scope of operation, and the initial impact of GLP on resources, costs of conducting studies, and the quality of research. Operating cost increases are primarily associated with staff time needed for documenting field work and

[1]Current address: Dynamac Corporation, Environmental Services, 2275 Research Boulevard, Rockville, MD 20850–3268

laboratory operations according to GLPs. Field contractors report the lack of standardization of the interpretation of GLP requirements among sponsors is a major problem. Field contractors overwhelmingly agree that standardization of requirements is critical to making GLP more cost effective. They recommend training courses, workshops, and manuals to harmonize sponsor and the EPA definitions of GLPs as they apply to field studies.

THE GOOD LABORATORY PRACTICE (GLP) STANDARDS (40 CFR 160) administered by the U.S. Environmental Protection Agency (EPA) under the Federal Insecticide, Fungicide, and Rodenticide Act (FIFRA) were published as proposed regulations in November 1987. They became law on October 16, 1989. The GLP Standards specify good "management"[2] practices for the planning, conduct, recording, reporting, and archiving of test data from laboratory and field studies. GLPs require qualified personnel, management involvement, adequate facilities and equipment, and an independent quality assurance (QA) program.

Between 1987 and 1989, many companies initiated steps to comply with GLP regulations. Estimates of the cost increases associated with pesticide registration ranged from 10 to 700% for field trials and from 20 to 25% for the analytical and report-writing phases (*1*).

Field Contractor Survey

In May 1990, 138 field contractors were surveyed to assess the impacts of GLP requirements after approximately one season and to obtain their comments on the cost and quality of field research. The survey (Appendix 32.I) also solicited field contractors' comments on how to improve or simplify operating procedures under GLPs. This chapter

- presents results of the contract field laboratory survey

- compares the results with a 1988 survey conducted by a sponsor agrochemical company

- summarizes the economic impact for the same period on a sponsor company providing study directors, analysis of field samples, and characterization of test substances

[2]Management is defined here as the person or persons responsible for setting policy at a facility. For a description of the duties of testing facility management, see § 160.31 of the GLP Standards.

Results and Discussion

During May 1990, questionnaires were sent to 138 companies either conducting or commissioning contract field research regulated by GLP Standards. By the June 20, 1990, deadline, 43% of the companies had responded. The survey was designed to

- profile facilities of the respondents
- profile the respondents' GLP-regulated programs
- assess the impact of GLPs on resources
- assess areas of field research that improved or deteriorated due to GLPs
- define needs in resources and training
- compile recommendations for improving operations under GLPs

Contractors participating in the survey were selected on the basis of having successfully performed contract research for an agrochemical company. Members of the National Alliance of Independent Crop Consultants (NAICC) who identified themselves as contract researchers were also included in the survey list.

The questions and discussion of the responses are presented below. Figure 32.1 shows the distribution of the survey respondents. All respondents were in the continental United States.

1. How many people does your facility employ (casual, full, and part time)?

Figure 32.2 shows that 58% of the respondents have fewer than 10 employees, 81% have fewer than 20 employees.

2. How many studies does your facility have primary responsibility for annually (conducted on-site or subcontracted)?

Figure 32.3 shows that 54% of the respondents conduct fewer than 50 studies, 79% conduct fewer than 100 studies. The number of trials per study was not included in the data.

3. What portion of the studies is regulated by FIFRA GLP (40 CFR 160)?

Figure 32.4 shows the portion of GLP-regulated studies is evenly divided among the respondents. Approximately 20% of the respondents are in each of the categories: 0–19, 20–39, 40–59, 60–79, and 80–100% GLP regulated.

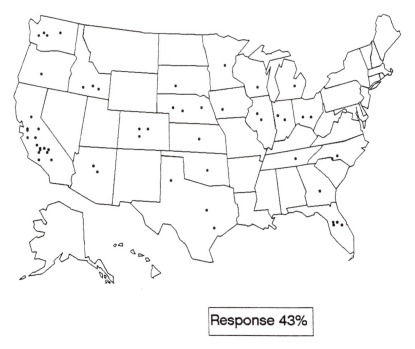

Response 43%

Figure 32.1. Distribution of respondents to the 1990 contract field laboratory survey.

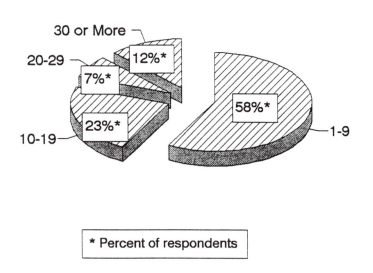

30 or More

20-29

7%*

12%*

58%*

23%*

10-19

1-9

* Percent of respondents

Figure 32.2. Number of employees at facilities.

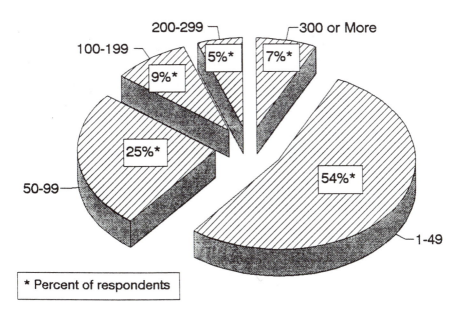

Figure 32.3. Number of studies conducted annually.

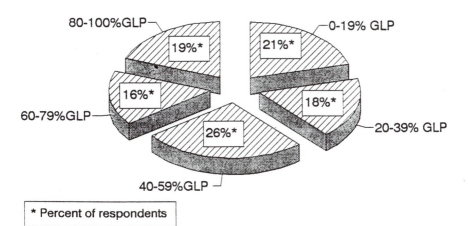

Figure 32.4. Portion of studies that are regulated by GLP Standards.

4. What is the approximate distribution of the studies in Question 3?

Table 32.I and Figure 32.5 show the categories listed in data requirements for registration (40 CFR 158) and summarize the responses. Residue studies make up 13–100% (mean 62%) and Environmental Fate Studies 0–75% (mean 17%). The mean for other categories was 2% or less.

Question 4 included product performance as defined in 40 CFR 158.640 (product performance submissions associated with human health

Table 32.I. Question 4 Responses:
Distribution of GLP-Regulated Program

Study Type	Minimum	Maximum	Mean
Residue	13	100	62
Environmental fate	0	75	17
Wildlife aquatic	0	20	1
Nontarget plants and insects	0	10	1
Reentry protection	0	75	2
Spray drift evaluation	0	5	<1

NOTE: All values are range percent; 53 people responded to this question.

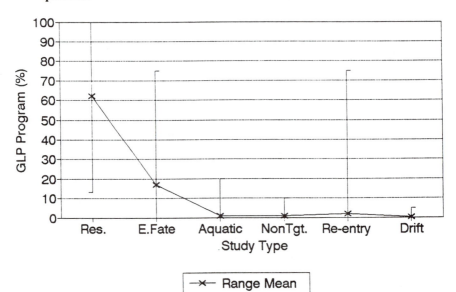

Figure 32.5. Distribution of studies regulated by GLP Standards.

effects). Some respondents interpreted this to refer to efficacy studies being used for label development and product evaluation. Because of this confusion, product performance was not included in the evaluation of response data.

5. What is the distribution of the non-GLP studies?

The non-GLP-regulated program is predominantly efficacy work. The survey showed efficacy at 90% and others at 10%. "Others" included varietal comparisons, application methodology, food-processing development, pesticide damage investigations, and interaction trials.

6. Using 1988 as a baseline (pre-GLP), what is your estimate of the cost increase due to GLP for your TOTAL PROGRAM outlined in Question 4?

The ranges listed were 0–9%, 10–19%, 20–39%, 40–59%, 60–79%, 80–99%, and 100% or more. Figure 32.6 shows the distribution of cost increases. The largest fraction (40%) is in the 20–39% increase range. The median for the responses is 40%. Note that a significant minority (13%) are experiencing a 100% increase or more. A correlation between GLP programs (i.e., type of study, number of trials, facility size) and those respondents indicating ≥100% cost increases was not evident. An alternate explanation is that unlike their colleagues this group were far from

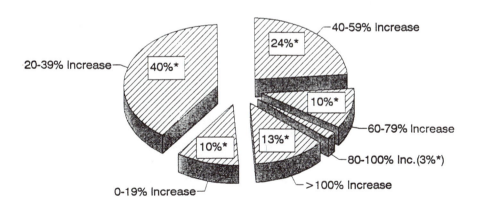

Figure 32.6. Distribution of cost increases due to GLP implementation.

GLP compliance, and the transition to full GLP compliance was a major change in facility operations.

7. Using 1988 as a baseline (pre-GLP), how much more of your staff time is spent on GLP-related documentation?

Figure 32.7 shows the demand on staff time is very similar to the overall cost increase distribution given in Figure 32.6, that is, 44% of the respondents are experiencing a 20–39% increase. A significant minority (17%) are seeing 80–100% increases. The similarity between Figures 32.6 and 32.7 suggests that the major impact of GLP compliance is an increased demand on staff time.

8. What areas have improved, or do you expect to improve, because of GLP?

The categories listed were reconstructibility, documentation, accuracy, precision, none, and other. Figure 32.8 shows the categories in which the respondents recognized improvements. Improvements listed in the "other" category were uniformity in conducting study trials, improved training, and use of standard operating procedures (SOPs).

9. What areas, if any, have deteriorated since GLP was implemented?

The categories listed were the same as for Question 8. The "none" category received 68% of responses, and the "comments" category received 32%. "Comments" implied GLP increased study costs, reduced

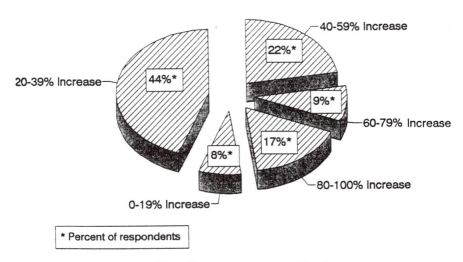

Figure 32.7. Staff time spent on GLP issues.

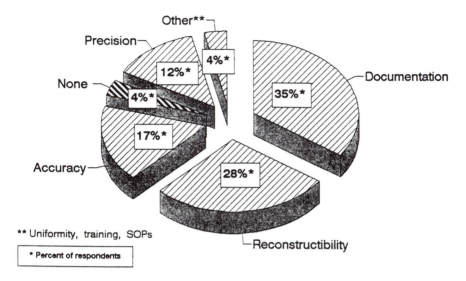

Figure 32.8. Improvements due to GLP implementation.

creativity, reduced local control, and lowered productivity. Impressively, the category "none" was marked by 68% of the respondents. "None" implies no perceived negative results due to GLP compliance. Some of the comments suggested concern that GLPs did or would reduce flexibility in dealing with field variables, and the effect would be lower quality research. Respondents also suggested lower productivity meant fewer projects could be accomplished and hence contractor income would be reduced.

10. Please rank the following activities according to the staff time required to ensure GLP compliance (1 = most, 6 = least).

The activities are field work documentation, personnel training, chain-of-custody procedures, quality assurance procedures, calibration documentation, and other. Field work documentation was the overall top ranked staff time consumer. It was twofold greater than any other. The differences between the other categories were less distinct. Figure 32.9 shows the distribution of the number 1 ranked items, that is, it summarizes the distribution of contractors' top concerns. It shows

- field work documentation at 56%
- quality assurance procedures at 20%
- calibration documentation at 12%

- chain-of-custody procedures at 6%
- personnel training at 6%
- others at <1%

Typical "others" were sponsor communication, archiving, test scheduling, applying the test substance, and documentation associated with shipping samples.

11. How are you handling GLP Quality Assurance requirements?

Internal QAU received 62%, and contract QA received 38% of the responses. Respondents indicated they were expecting to do more contracting of quality assurance activities as the demands in other areas increased.

12. Are you receiving sufficient guidance from your sponsors' QAU or Study Directors on how to comply with GLP regulations?

Seventy percent of survey respondents answered yes, and 30% answered no. The "yes" response (70%) indicates that contract researchers find their sponsor personnel accessible. GLP information is available for conducting studies in a manner the sponsor expects.

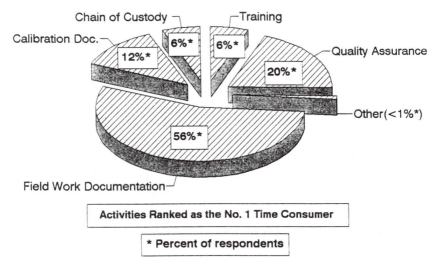

Figure 32.9. Activities ranked as the Number 1 staff time consumer.

13. Is the GLP compliance information you receive consistent among sponsors?

Question 13 elicits almost the reverse of the response to Question 12: 35% of the respondents answered yes, and 65% answered no. The results of Questions 12 and 13 indicate that sponsors are accessible, but there is wide disparity in their interpretation of GLP Standards as presented in 40 CFR 160. This disparity places a major burden on contractors. They are forced to tailor studies to the sponsor's perception of GLP Standards. This extra effort makes the cost of a study higher than if a clear set of standards for conducting and documenting field trials existed among sponsors.

14. Has GLP forced you to subcontract studies that you might have otherwise done yourself?

A majority of the respondents (91%) did not need to subcontract studies. An important comment that appeared several times was that qualified subcontractors frequently were not available, and the result was that contractors had to turn down projects that in previous years they could have accommodated.

15. What additional resources do you need to comply with GLP?

The responses to this question were almost equally divided among more staff (39%), more equipment (31%), and other (30%). Typical "other" categories were archiving space, storage space, time, and money.

16. What training do you feel your staff needs to ensure that studies are conducted in compliance with GLP?

Training needs identified include specific information and examples of what the EPA expects, specific information on what sponsors expect, annual updates on new EPA information, and a good GLP manual. This response reinforces the concern expressed in Question 13 about lack of consistent interpretation of GLPs. Contractors are looking for consistent information on the expectations of sponsors and the EPA. Respondents suggest having annual workshops to provide updates on the EPA policy.

17. Who should supply this training to contract facilities?

Figure 32.10 shows the distribution of responses was the EPA, 17%; sponsor, 30%; professionals, 25%; and internal, 28%. The "EPA" and "professionals" included the comment that annual workshops from the

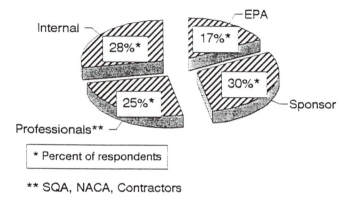

Figure 32.10. Suggested GLP training sources.

EPA, the National Agricultural Chemicals Association, or the Society of Quality Assurance would be helpful. Responses in the "sponsor" category usually included the comment that because the sponsor was ultimately responsible for the results of a study, it should provide the training. Responses to the "internal" category suggested that the on-site QA officers were expected to stay current on the regulations and to train staff.

18. In your opinion, what would make conducting GLP studies more cost effective?

This question did not suggest categories. The specific answers were different but the message was the same: We want standardization. Typical comments were

• Standardize the interpretation of the GLP Standards.

• Standardize sponsor forms.

• Standardize reporting requirements.

• Provide field notebooks.

19. Will the quality assurance procedures required for GLP studies be used for efficacy studies at your facility?

Almost two-thirds of the respondents (61%) replied that they would not apply GLP QA procedures to efficacy studies. Efficacy studies as referred to by contractors are those studies not currently regulated by GLP Standards. These studies are targeted for a sponsor's internal use. They include studies to establish label rates, to assess phytotoxicity, and to determine relative performance of test compounds. The "yes" respondents' (39%) two common themes were

- They preferred to have a single system for documenting field studies.

- Eventually efficacy trials would be regulated so contractors might as well start GLPs now.

The "no" respondents' common theme was that GLPs cost money, and when sponsors will support GLP efficacy trials, then contractors will provide them.

20. Would you recommend that a separate set of "Good Field Practice" (GFP) Regulations be written to better describe the uniqueness of field studies vs. laboratory studies?

The majority ("yes", 87%) want separate regulations; however, they added that GFPs must be written by people who are experienced in field work and who appreciate the variables associated with field work. The EPA's Office of Compliance Monitoring, Policy, and Grants Division, has indicated that they will not issue a separate set of regulations for field studies. The minority ("no", 13%) are generally concerned that separate, specific GFPs would be too restrictive. One response noted that the regulations as published were generic enough to be applied readily to field work.

Summary of Field Contractor Survey

The survey of economic impacts of GLP Standards shows that

- study costs are up about 40%

- field work documentation is the major staff time consumer

- benefits of GLP include improved documentation, accuracy, precision, and reconstructibility

- standardization of data collecting and reporting procedures is a critical need

Previous Survey

In 1988, Ciba–Geigy reported the results of a survey of a group of field contractors prior to regulations going into effect (2). In summary, this survey showed that

- about 45% of staff time was spent on GLP documentation

- field work documentation was the major time consumer

- a major improvement in field work would be documentation of what actually occurred in the field

- lack of consistency among sponsors was expected to be a problem
- a separate set of regulations for good field practices would be needed

Agrochemical Company Experience

One major agrochemical company was examined to determine the impact of GLP on the cost of operations after approximately one year fully under GLP Standards. In this company, the study director typically was a staff member who wrote the protocol, placed the study with a field contractor, analyzed samples from the field, and wrote the final report. Test substance characterization was done within the company's analytical department.

In summary, after one year the cost of operating under GLP Standards was estimated to be 20–25% greater than the pre-GLP period. The basis for this estimate was that the cost of laboratory studies increased 10–15% primarily due to increased staff time spent on documentation required, and the cost of adding a quality assurance unit increased operating costs by approximately 10%.

Company management projected that because both cost elements had significant start-up costs (e.g., training and creating sets of SOPs), costs would be expected to decrease over time. Management affirmed that the quality of registration packages generated under GLPs was higher than prior to GLPs.

Conclusions

The lessons as of mid-1990 seem to be

- Cost increases due to GLPs are real (this survey found a median increase of 40% for field contractors and one sponsor agrochemical company reported 20–25%). Sponsors and field contractors expect GLP costs to decrease over time as initial training programs and start-up projects are completed.

- Establishing a standard interpretation of GLP documentation and reporting requirements is a top priority of researchers. The organization to establish this standard has not been defined.

- Improvements in studies attributed to GLPs are in the areas of documentation, reconstructibility, accuracy, and precision. Overall benefits also include improved employee training and use of SOPs.

Acknowledgments

I thank the field contractors who participated in the survey, and M. L. Allyn and C. R. Morris of International Chemical Consultants, Inc., who contributed to and supported this project.

References

1. Platt, J. L. Presented at the National Meeting of the Society of Quality Assurance, San Diego, CA, October 1989.

2. Haigler, J. Presented at the National Meeting of the Society of Quality Assurance, Nashville, TN, October 1988.

RECEIVED for review September 17, 1990. ACCEPTED revised manuscript September 17, 1991.

Appendix 32.I: Industry Survey of Contact Field Laboratories

The purpose of this industry survey is to estimate the impact of FIFRA Good Laboratory Practice Regulations (GLP) on contract laboratories. The sources of responses will be confidential. Detailed results and analyses will be mailed to survey participants providing address information (Page 4). If you have questions call: (703) 658–8888.

1. How many people does your facility employ (casual, full, and part time)?

__ 1–9
__ 10–19
__ 20–29
__ 30 or more

2. How many studies does your facility have primary responsibility for annually (conducted on-site and/or subcontracted)?

__ 1–49
__ 50–99
__ 100–199
__ 200–299
__ 300 or more

3. What portion of the studies is regulated by FIFRA GLP (40 CFR 160)?

__ 0–19%
__ 20–39%
__ 40–59%
__ 60–79%
__ 80–100%

4. What is the approximate distribution of the studies in Question 3?

___%, Residue	(40 CFR 158.240)
___%, Environmental Fate	(40 CFR 158.290)
___%, Wildlife, Aquatic	(40 CFR 158.490)
___%, Non-Target Insects	(40 CFR 158.590)
___%, Reentry Protection	(40 CFR 158.390)
___%, Non-Target Plants	(40 CFR 158.540)
___%, Spray Drift Evaluation	(40 CFR 158.440)
___%, Product Performance	(40 CFR 158.640)

100% TOTAL PROGRAM

5. What is the distribution of the non-GLP studies?

___%, Efficacy
___%, Other. Please specify: _____

6. Using 1988 as a baseline (pre-GLP), what is your estimate of the cost increase due to GLP for your TOTAL PROGRAM outlined in Question 4?

__ 0–9%
__ 10–19%
__ 20–39%
__ 40–59%
__ 60–79%
__ 80–99%
__ 100% or more

7. Using 1988 as a baseline (pre-GLP), how much more of your staff time is spent on GLP-related documentation?

__ 0–19%
__ 20–39%
__ 40–59%
__ 60–79%
__ 80–100%

8. What areas have improved, or do you expect to improve, because of GLP?

__ Reconstructability
__ Documentation
__ Accuracy
__ Precision
__ None
__ Other. Please specify: _____

9. What areas, if any, have deteriorated since GLP was implemented?

__ Reconstructability
__ Documentation
__ Accuracy
__ Precision
__ None
__ Other. Please specify: _____

10. Please rank the following activities according to the staff time required to ensure GLP compliance (1 = most, 6 = least).

__ Field Work Documentation
__ Personnel Training
__ Chain-of-Custody Procedures
__ Quality Assurance Procedures
__ Calibration Documentation
__ Other. Please specify: _____

11. How are you handling GLP Quality Assurance requirements?

__ Internal QAU
__ Contract QAU

12. Are you receiving sufficient guidance from your sponsors' QAU or Study Directors on how to comply with GLP regulations?

__ Yes
__ No. Improvement needed in: _____
_____

13. Is the GLP compliance information you receive consistent among sponsors?

__ Yes
__ No. Most common discrepancy is: _____
_____

14. Has GLP forced you to subcontract studies that you might have otherwise done yourself?

__ No
__ Yes, due to: _____
_____

15. What additional resources do you need to comply with GLP?

__ More staff
__ More equipment
__ Other. Please specify: _____
_____

16. What training do you feel your staff needs to ensure that studies are conducted in compliance with GLP?

_____
_____

17. Who should supply this training to contract facilities?

_____
_____

18. In your opinion, what would make conducting GLP studies more cost effective?

_____

_____

19. Will the quality assurance procedures required for GLP studies be used for efficacy studies at your facility?

__ Yes
__ No

Why? _____

_____

20. Would you recommend that a separate set of "Good Field Practice" Regulations be written to better describe the uniqueness of field studies vs. laboratory studies?

__ Yes
__ No

Why? _____

_____

☐ **Yes, send me the detailed results and analyses of this survey.**

_____

Company Name

_____

Street/Route/P.O. Box

_____

City, State   ZIP Code

Attention: _____

May we contact your QAU Officer? Name: _____

Harmonization and Prospects for the Future

Frederick G. Snyder[1]

Hazleton Corporation, 13873 Park Center Road, Herndon, VA 22071

A great deal of activity has occurred in the past two to three years relative to good laboratory practice implementation worldwide. This activity is of particular importance in Europe as we approach 1992 with the changes spearheaded by the European Communities (EC). This chapter identifies major and minor efforts made by participating nations on a worldwide basis as well as the effect of international agreements among these nations. Such agreements or directives include, in addition to the EC, the Organization of Economic Cooperation and Development and various national, bilateral, and multilateral agreements. A comparison of these arrangements is made with a guess toward future developments.

As WE ENTER THE LAST DECADE OF THIS CENTURY and approach the year 2000, it is tempting to make far-reaching predictions as to the pathways national interests will pursue relative to good laboratory practice

[1]Current address: Glaxo, Inc., 5 Moore Drive, 5–5422, Research Triangle Park, NC 27709

2192–8/92/0419$06.00/0 © 1992 American Chemical Society

(GLP) compliance. This temptation is particularly evident in light of the activities occurring in Europe as European nations prepare to become a single market by 1992.

In that regard, I will attempt to concentrate on the activities of the Organization of Economic Cooperation and Development (OECD) and the European Communities (EC) and present a brief accounting of international GLP compliance programs and efforts of multilateral and bilateral agreements to comply with these regulations.

As countries have become more aware of the potential problems in instituting different testing requirements and different GLP standards, a major effort was needed to harmonize these various requirements to avoid potential nontariff barriers to trade. Since 1977, the United States and other members of the 24-nation OECD have been involved in extensive international consultations concerning the harmonization of chemical programs.

OECD Initiatives

The OECD is not a supranational organization but a center for discussion, where members have the opportunity to exchange information and clarify economic and social problems, analyze the effectiveness of economic and social policies, and search for common solutions and strategies. The primary purposes of the OECD are to promote sound economic growth, provide employment, improve the standard of living, and promote sound and harmonious development of the world economy with particular assistance to developing countries. As part of this philosophy, the OECD established several expert groups, including one to address the issue of writing, implementing, and monitoring international guidelines for GLPs.

Expert group meetings were held between 1978 and 1981, during which several international testing guidelines were presented to the participating countries for review, comment, and approval. Additionally, one of the expert groups that was established during this period had as its primary objective to develop a consistent and acceptable standard for GLPs as well as a mechanism for enforcement.

Through the international OECD expert group on GLPs, a major effort was directed toward the development of international guidelines for GLPs. The principal objective of these guidelines was to ensure, to the extent practicable under the laws of the OECD member countries, that data developed to meet one country's requirements would be acceptable to another country's requirements. The United States and other member countries placed a high priority on these activities because of the benefits for international pharmaceutical and chemical trade and for more effective health and environmental protection. The work performed by the GLP expert group was strongly endorsed at meetings of high-level national regulatory officials in May 1980 and in November 1982. In May 1981, the council of the OECD (*1*), which is the supreme authority of the OECD, adopted a decision concerning the mutual acceptance of data that, to the extent practicable under the laws of OECD, binds member countries to accept for assessment purposes data generated according to OECD test guidelines and the OECD principles of GLPs [C (81) 30 (final)] adopted on May 12, 1981.

In addition to the development of the OECD principles of GLPs, the OECD expert group on GLPs was given the responsibility of developing two additional guidance documents—one for the implementation of OECD principles of GLPs and one to serve as OECD guidelines for a national GLP inspection and study audits. This second initiative was realized as a first council recommendation concerning the mutual recognition of compliance with GLPs [C (83) 95 (final)] adopted on July 26, 1983. The implementation piece, which is the key document, encourages countries to adopt the OECD principles of GLPs into their legislative and administrative framework, calls for documentation of compliance programs by national authorities, and requires declaration or certification by each laboratory that studies were conducted in accordance with the OECD principles of GLPs or with national regulations or equivalents conforming to these principles. The guidelines recommend that national compliance programs use laboratory inspections and study audits as principal mechanisms for monitoring compliance with GLP principles. The guidelines further recommend that national authorities use properly trained personnel who are competent to assess laboratory compliance with the principles and to administer the GLP compliance programs. The implementation document also advocates the inclusion in each national GLP compliance program provisions to deal with noncompliance issues and to remedy deficiencies that may be found by the national authorities. In effect, this recommendation, with the decision concerning the mutual acceptance of data in chemical assessment, provides a foundation for ensuring data quality and confidence.

To assist OECD members in the implementation of these recommendations and decisions, a working group was formed in 1985 to deal primarily with the development of detailed guidance for administering applications for national GLP compliance monitoring procedures as well as a GLP compliance monitoring program, which is based on study audits and inspections. The result of the working group's efforts was published as an OECD Environmental Monograph No. 15 in March 1988 (*2*).

Three annexes that are of significance and worth mentioning are

1. Annex 4: Guides for the Compliance Monitoring Procedures for Good Laboratory Practices

2. Annex 5: International Acceptance of Compliance Monitoring Procedures

3. Annex 6: Guidance for the Conduct of Laboratory Inspections and Study Audits

The most recent council recommendation on compliance with GLP principles was published October 2, 1989 [C (89) 87 (final)]. This act supersedes and replaces the July 1983 recommendation [C (83) 95 (final)] and includes in the annex the guidelines developed by the ad hoc working group and presented via Environmental Monograph No. 15. This key document, dated October 2, 1989, is divided into three parts:

Part I. GLP principles and national compliance monitoring procedures

Part II. international aspects of mutual recognition of compliance among member states

Part III. future OECD activities and agenda

Of importance in this document is a directive in Annex 3 to the membership, encouraging the exchange of written information relative to members' procedures for and results of their monitoring of the compliance program with GLPs. It is envisioned that in the near future a data base for the exchange of this information will be readily available so that GLP monitoring authorities are aware of current situations within the member countries.

In January 1990, the OECD held their first training course on GLP inspections at the Ecole Nationale de la Santé Publique in Rennes, France. The next such course is being planned for early 1992.

The OECD also held its first consensus workshop October 16–18, 1990, in Bad Durkheim, Germany, and dealt with the following areas:

1. field studies: management from a GLP perspective

2. quality assurance: guidance for management's effective implementation of a quality assurance program

3. compliance of suppliers: whether GLPs should be applied to suppliers of products and services in support of regulated studies, as well as nonregulated studies, and if so, to what extent

A subsequent workshop is tentatively scheduled for October 1993 in Switzerland.

EC Initiative

In concert with these OECD initiatives but requiring more legal substance, and therefore additional legislation, for the EC purposes, the EC issued a series of directives. The EC, consisting of 12 member states (also members of the OECD), was established primarily to control tariffs and to enhance trade among its members. As this relationship involved laboratory testing by manufacturers and importers, the EC became actively involved in harmonizing the GLP guidelines and implementation programs within the EC. The commission of the EC actively participates in the OECD council meetings as well as in the expert group on GLPs. The initial legislation promulgated by the EC was the adoption of the Council Directive of December 18, 1986 [87/18/EC]. This piece of legislation addresses the harmonization of laws, regulations, and administrative procedures relative to the principles of GLPs and the verification of studies and tests performed on chemical substances conducted under these principles. A direct reference is made in this directive to the Principles of GLPs Annex 2 of the OECD Council Decision of May 12, 1981, as well as those recommendations adopted by the OECD council on July 26, 1983. Member states were to comply with the directive by June 30, 1988.

A second council directive of June 9, 1988, was issued on the inspection and verification of GLPs [88/320/EC]. This document addresses primarily the establishment and organization of national authorities, with emphasis on monitoring unit responsibilities for the verification of compliance with GLP principles. Similarly, this directive refers to the OECD GLP legislative piece as a guideline with additional references to the OECD Environment Monograph No. 15. The two sections that are highlighted from this monograph and presented in this directive are "Guidelines for the Compliance Monitoring Procedures for GLPs" and "Guidelines for the Conduct of Laboratory Inspections and Study Audits". Member countries were to comply with this directive by January 1, 1989.

As an aside, the EC council published a decision on July 28, 1989 (89/569/EC), on acceptance by the EC of a draft version of the OECD decision on compliance with principles of GLPs. This decision was initiated prior to the finalization of the OECD council decision of October 2, 1989.

The last significant piece of the EC legislation was the issuance of a commission directive on December 18, 1989 (90/18/EC), adapting to technical progress the Annex to Council Directive 88/320/EC on the inspection and verification of GLPs. In effect, this document incorporates the OECD principles and associated compliance language by replacing a reference with a full text version to facilitate uniform understanding throughout the community. This directive also extends the compliance date of Directive 88/320/EC from January 1989 to July 1, 1990. There may be further complications with adherence to this directive because several of the EC members, specifically Spain, Portugal, and Greece, do not have GLP programs in place at this time.

Summation of EC–OECD Initiatives

We may assume that all legislation relative to the OECD and the EC is complete, particularly when the OECD describes compliance to basic GLP principles, how to monitor compliance by national authorities, and also the mechanisms to ensure compliance monitoring procedures. The EC, which was directly involved with the OECD decisions, integrated these initiatives into European legislation either by approving the OECD texts, referring to the OECD guidelines, or adopting its own directives. Therefore, it appears that the role of the EC in the near future is to promote the implementation of existing legislation and to emphasize performance of the systems currently in place at the state, national, and international levels.

The effectiveness of both the OECD and the EC, separately and collectively, will be enhanced by their continual communication and exchange of information. Continual exchange of information will be achieved through the implementation of a sophisticated data-base system.

The role of the EC in future bilateral or multilateral negotiations is currently unclear, with the future directives to be dictated by the commission. Apparently no new memoranda of understanding (MOUs) will be forthcoming; however those termed as Phase II will remain intact for the present, and preliminary MOUs (Phase I) are ineffective.

Bilateral Agreements

Bilateral and multilateral agreements, or MOUs as they are more commonly called, continue to be a mechanism by which countries can mutually agree to accept data in support of product registration. The advantages of pursuing MOUs are numerous, including the effective conservation and use of resources and the sharing of information, experience, and expertise.

Bilateral agreements are generally conducted in two stages. Stage one is the introductory level, in which the agreement calls for bilateral discussions and active involvement regarding comparable GLPs, exchange of inspection information, and administration procedures recognized.

In the second stage, all parties become satisfied with the operation of their counterpart programs and mutually recognize and accept each other's data. Also, GLP standards are published, a compliance program is in place, and the implementation of procedures and mechanisms for continued cooperative efforts are established.

Although the 1992 EC initiative and the breakup of the Eastern European communist bloc countries have altered the fervor to pursue MOUs, presumably other nations, particularly Japan and the United States, will continue to pursue bilateral and multilateral arrangements for the mutual recognition of GLP compliance programs (Table 33.I).

International GLP Implementations

The principles of GLPs imposed by the government of the United States via the Food and Drug Administration (FDA) and the Environmental Protection Agency (EPA) have generated intense levels of activity and have put significant pressure on domestic and foreign firms that wish to submit safety data in support of product registration in the United States. Equally important in the effort, members of both the OECD and the EC have played, and continue to play, an integral part in the implementation and development of GLP principles. This participation was particularly evident in their assisting the various national authorities in establishing GLPs as universal standards. Some modifications to specific GLP implementations as well as those that have an impact on chemical testing will be highlighted in the following sections.

First and foremost is the EPA rewrite of the Toxic Substances Control Act (TSCA) GLP Standards effective September 18, 1989, and the Federal Insecticide, Fungicide, and Rodenticide Act (FIFRA), effective October 16, 1989.

Initially, it was envisioned that separate GLP Standards would be written under TSCA and FIFRA for ecological effects, residue chemistry, field studies, and the like, but that did not occur. Industry testing requirements, which must comply to the above revised regulations, must do so uniformly to a single generic EPA GLP Standards regulation (i.e., either under TSCA or FIFRA). No other revisions are expected in the near future. How these regulations are to be applied across the industry they affect will be the subject of many working sessions between the EPA and industry groups in the upcoming decade.

The majority of the EC countries have GLP standards published with a national compliance program in place. At this time, however, three countries, Portugal, Greece, and Spain, are in varying stages of GLP development. Spain, which has a GLP program, identified one of three agencies to have ultimate authority for GLP monitoring. Other European countries where GLPs are in the primary developmental stage are Belgium, Denmark, Ireland, Finland, and Norway.

Table 33.I. Summation of Agreements

Countries	Progress
U.S. FDA with	
Canada	Signed Phase I MOU[a] May 10, 1979
Japan (MOHW)[b]	Signed Phase I MOU April 15, 1983
	Phase II awaiting State Department approval
Sweden	Signed Phase I MOU May 25, 1979
Switzerland	Signed Phase I MOU March 5, 1980
	Signed Phase II MOU April 29, 1985
France	Signed Phase II MOU March 18, 1986
Italy	Signed Phase II MOU December 19, 1988
Netherlands	Signed Phase II MOU December 20, 1988
Spain	Discussion in progress
U.S. FDA and EPA with	
Germany	Signed Phase I MOU in conjunction with the U.S. EPA, December 23, 1988, that will be reassessed in December 1991
United Kingdom	A formal Phase II agreement was signed March 28, 1988, that included the U.S. EPA. Irrespective of this accomplishment, inspection and audit activities have been carried out between the two countries as though a formal agreement existed since 1979
U.S. EPA with agency counterparts	
Netherlands	Signed MOU October 18, 1988
Japan (MAFF)[c]	Signed Phase I MOU September 16, 1987
Switzerland	Signed MOU June 22, 1988
Other international bilateral agreements	
United Kingdom and Japan (MOHW)	Signed Phase I agreement March 1984
	Signed Phase II agreement November 16, 1988
France and Japan (MOHW)	Signed Phase II agreement September 24, 1986
Germany and Japan (MOHW)	Signed Phase I agreement September 4, 1986
Switzerland and Japan (MOHW)	Signed Phase I agreement late 1988
United Kingdom and Japan (MAFF)	Negotiations for agreement

[a]MOU stands for Memorandum of Understanding.
[b]MOHW stands for Ministry of Health and Welfare.
[c]MAFF stands for Ministry of Agriculture, Forestry, and Fisheries.

To date, no other national authority, including the United States, has promulgated more enforceable GLP regulations than Japan. Of their four sets of GLP regulations, three deal with agrochemicals and industrial chemicals. The agencies that are responsible for the administration of these regulations are the Ministry of Agriculture, Forestry, and Fisheries (MAFF), the Ministry of Trade and Industry (MITI), and the Ministry of Labor (MOL) (Table 33.II) (*3*). The remaining set of GLP regulations covers pharmaceuticals, and the agency responsible for the administration of these regulations is the Ministry of Health and Welfare (MOHW).

Of the three aforementioned agencies, MAFF is the most influential due to its unique administrative requirements and the interest expressed by foreign companies that want to register their products in Japan. Within the next 2–3 years the Japanese government is expected to amend the regulations to mirror more clearly their U.S. counterparts. Although revisions for MITI and MOL are possible, they do not appear to be a pressing issue. MAFF has several unique requirements that are worth mentioning. This ministry took a different approach from its Japanese counterpart agencies and from other international GLP governing authorities by implementing what is known as a petition for confirmation procedure. This procedure requires any person or organization who wishes to have confirmation by the director-general of MAFF to submit a petition via the director of the agricultural chemicals inspection station. The petition is a detailed document that must be in Japanese and in a format acceptable to MAFF. The contents of the petition should include

1. data specific to the testing facility (i.e., founder of the facility, detailed organizational charts, size of the testing facility, total floor area, building layout, environmental features, and equipment and instrument lists)

2. types and numbers of toxicology studies performed in the facility during the most recent 3-year period

3. the financial status of the testing facility

4. curricula vitae of all researchers, including management

5. any other data necessary to ensure testing facility compliance to the GLP Standards

Putting this document together is time consuming and costly. If the document deviates from the format required by MAFF, it will not be accepted. After the petition for confirmation has been formally submitted, reviewed, and accepted, the director-general of the Agricultural Protection Bureau will authorize an inspection of the testing facility. The director-general then will schedule a time for the inspection and submit the names

Table 33.II. Japanese Good Laboratory Practice Standards

Criteria	Agricultural Chemicals	Industrial Chemicals	
Administering authority	Ministry of Agriculture, Forestry, and Fisheries (MAFF)	Ministry of International Trade and Industy (MITI) supported by the Environmental Agency and Ministry of Health and Welfare (MOHW)	Ministry of Labor (MOL)
Applicable legislation	Agricultural Chemicals Regulation Law 82 (1948) (and as subsequently amended) and Notification No. 3850 August 10, 1984, issued by the Agricultural Production Bureau	Chemical Substance Control 117 (1973) (and subsequent amendments) and Notification No. 85 March 31, 1984, issued by the Basic Industries Bureau	Industrial Safety and Health Law 57 (1972) (and subsequent amendments) and Notification No. 261 May 18, 1985, issued by Labor Standards Bureau
Date GLP published	August 10, 1984	March 31, 1984	May 18, 1985
Date GLP effective	October 1, 1984	October 1, 1985	October 1, 1985
Definition and scope	All toxicological studies on agricultural chemicals	Tests related to new industrial chemicals imported or manufactured in quantities greater than 0.1 ton per company per year and not on the 1974 list of Existing Chemical Substances	Ref. Law 117, but in addition includes intermediates used in the production of drugs and agricultural chemicals

of the inspectors. After the inspection, the director-general will notify the petitioner of the results of the inspection and whether or not the petition has been confirmed. This process is repeated every 3 years.

Additionally, toxicological study data submitted to MAFF in support of agricultural chemical product registration also must be accompanied by an information package similar to that required for petitions of confirmation. Additional documents required include

1. documents identifying QAU personnel and their titles

2. a statement signed by testing facility management attesting that the study in question has been conducted according to the GLP standards

3. in the case of multifacility participation in a single study, presentation of evidence explaining the relationship of each facility to the study

MAFF is the Japanese counterpart to the U.S. EPA Office of Pesticide Programs, which is accountable under FIFRA. Those areas of the GLP principles that are significantly different from those of the EPA should be highlighted. MAFF's GLPs include

- a facility disqualification and reinstatement provision

- a statement of GLP compliance to be signed by the facility manager (there is no option of a noncompliance statement as required by the EPA GLP regulations)

- protocol signature by study director is not specified

- protocols and revisions to be prepared and approved by management

- retention of records by the testing facility

- permission by MAFF inspectors to review and copy QA records

MITI and MOL, with separate and somewhat different legislative requirements from MAFF, interact and use MAFF resources as well as those from the Environmental Agency to perform their functions. MITI is responsible for overseeing the safety of industrial chemicals. The Chemical Substance Control Law No. 117 (fl3), which has been in effect since 1973, is the statute that guides MITI. The purpose of this law is to confirm the reliability of test results regarding the toxicity or biodegradability of industrial chemicals.

Determination of biodegradability is the first stage of testing. If the chemical is biodegradable, no other testing is required; if it is not, toxicity and accumulation studies must be performed. If these results are positive, then mammalian toxicity tests must be performed. The Ministry of Labor is primarily interested in the manufacture and importation of industrial chemicals and concentrates primarily on mutagenicity testing as required by the Labor Standards Bureau under MOL.

Conclusion

The OECD has played, and continues to play, an integral role in worldwide development and implementation of GLP principles and compliance programs. However, each individual national or supranational authority must formulate these standards and compliance monitoring programs into usable and enforceable legislative documents. The United States continues to direct the GLP initiative. However, with the EC's thrust toward a single market for its members by 1992, the commission of the EC may be sitting in the driver's seat for the next decade because of its emphasis and uniform interest in developing GLP Standards under the EC 1992 banner.

References

1. Turnheim, F. "The Role of OECD in Promoting Communications Among GLP Authorities", paper presented at the 7th International Congress on GLP, Cannes, France, May 1–4, 1990.

2. Arany, J. "EC–OECD Interface", paper presented at the 7th International Congress on GLP, Cannes, France, May 1–4, 1990.

3. Morris, C.; Snyder, F. "Good Laboratory Practice Regulations, Chapter EPA and Foreign Country GLP Regulations", in *Drugs and the Pharmaceutical Sciences, Volume 38, Good Laboratory Practice Regulations*; Marcel Dekker Inc.: New York, 1989.

RECEIVED for review September 17, 1990. ACCEPTED revised manuscript August 17, 1991.

Appendixes

Appendix A: Federal Insecticide, Fungicide, and Rodenticide Act; GLP Standards

PART 160—GOOD LABORATORY PRACTICE STANDARDS

Subpart A — General Provisions

Sec.
160.1 Scope.
160.3 Definitions.
160.10 Applicability to studies performed under grants and contracts.
160.12 Statement of compliance or non-compliance.
160.15 Inspection of a testing facility.
160.17 Effects of non-compliance.

Subpart B—Organization and Personnel

160.29 Personnel.
160.31 Testing facility management.
160.33 Study director.
160.35 Quality assurance unit.

Subpart C—Facilities

160.41 General.
160.43 Test system care facilities.
160.45 Test system supply facilities.
160.47 Facilities for handling test, control, and reference substances.
160.49 Laboratory operation areas.
160.51 Specimen and data storage facilities.

Subpart D—Equipment

160.61 Equipment design.
160.63 Maintenance and calibration of equipment.

Subpart E—Testing Facilities Operation

160.81 Standard operating procedures.
160.83 Reagents and solutions.
160.90 Animal and other test system care.

Subpart F—Test, Control, and Reference Substances

160.105 Test, control, and reference substance characterization.
160.107 Test, control, and reference substance handling.
160.113 Mixtures of substances with carriers.

Subpart G—Protocol for and Conduct of a Study

160.120 Protocol.
160.130 Conduct of a study.
160.135 Physical and chemical characterization studies.

Subparts H and I—[Reserved]

Subpart J—Records and Reports

160.185 Reporting of study results.
160.190 Storage and retrieval of records and data.
160.195 Retention of records.

Authority: 7 U.S.C. 136a, 136c, 136d, 136f, 136j, 136t, 136v, 136w; 21 U.S.C. 346a, 348, 371, Reorganization Plan No. 3 of 1970.

Subpart A—General Provisions

§ 160.1 Scope.

(a) This part prescribes good laboratory practices for conducting studies that support or are intended to support applications for research or marketing permits for pesticide products regulated by the EPA. This part is intended to assure the quality and integrity of data submitted pursuant to sections 3, 4, 5, 8, 18 and 24(c) of the Federal Insecticide, Fungicide, and Rodenticide Act (FIFRA), as amended (7 U.S.C. 136a, 136c, 136f, 136q and 136v(c)) and sections 408 and 409 of the Federal Food, Drug and Cosmetic Act (FFDCA) (21 U.S.C. 346a, 348).

(b) This part applies to any study described by paragraph (a) of this section which any person conducts, initiates, or supports on or after October 16, 1989.

§ 160.3 Definitions.

As used in this part the following terms shall have the meanings specified:

"Application for research or marketing permit" includes:

(1) An application for registration, amended registration, or reregistration of a pesticide product under FIFRA sections 3, 4 or 24(c).

Reprinted from *Code of Federal Regulations* Title 40, Pt. 160, [OPP–300165A; FRL–3518–2] RIN 2070–AB68, *Federal Register*, Vol. 54, No. 158, Thur., Aug. 17, 1989, Pt. IV. This appendix is not subject to U.S. copyright. Published 1992 American Chemical Society.

(2) An application for an experimental use permit under FIFRA section 5.

(3) An application for an exemption under FIFRA section 18.

(4) A petition or other request for establishment or modification of a tolerance, for an exemption for the need for a tolerance, or for other clearance under FFDCA section 408.

(5) A petition or other request for establishment or modification of a food additive regulation or other clearance by EPA under FFDCA section 409.

(6) A submission of data in response to a notice issued by EPA under FIFRA section 3(c)(2)(B).

(7) Any other application, petition, or submission sent to EPA intended to persuade EPA to grant, modify, or leave unmodified a registration or other approval required as a condition of sale or distribution of a pesticide.

"Batch" means a specific quantity or lot of a test, control, or reference substance that has been characterized according to § 160.105(a).

"Carrier" means any material, including but not limited to feed, water, soil, nutrient media, with which the test substance is combined for administration to a test system.

"Control substance" means any chemical substance or mixture, or any other material other than a test substance, feed, or water, that is administered to the test system in the course of a study for the purpose of establishing a basis for comparison with the test substance for known chemical or biological measurements.

"EPA" means the U.S. Environmental Protection Agency.

"Experimental start date" means the first date the test substance is applied to the test system.

"Experimental termination date" means the last date on which data are collected directly from the study.

"FDA" means the U.S. Food and Drug Administration.

"FFDCA" means the Federal Food, Drug and Cosmetic Act, as amended (21 U.S.C. 321 et seq).

"FIFRA" means the Federal Insecticide. Fungicide and Rodenticide Act as amended (7 U.S.C. 136 et seq).

"Person" includes an individual, partnership, corporation, association, scientific or academic establishment, government agency, or organizational unit thereof, and any other legal entity.

"Quality assurance unit" means any person or organizational element, except the study director, designated by testing facility management to perform the duties relating to quality assurance of the studies.

"Raw data" means any laboratory worksheets, records, memoranda, notes, or exact copies thereof, that are the result of original observations and activities of a study and are necessary for the reconstruction and evaluation of the report of that study. In the event that exact transcripts of raw data have been prepared (e.g., tapes which have been transcribed verbatim, dated, and verified accurately by signature), the exact copy or exact transcript may be substituted for the original source as raw data. "Raw data" may include photographs, microfilm or microfiche copies. computer printouts, magnetic media, including dictated observations, and recorded data from automated instruments.

"Reference substance" means any chemical substance or mixture, or analytical standard, or material other than a test substance, feed, or water, that is administered to or used in analyzing the test system in the course of a study for the purposes of establishing a basis for comparison with the test substance for known chemical or biological measurements.

"Specimens" means any material derived from a test system for examination or analysis.

"Sponsor" means:

(1) A person who initiates and supports, by provision of financial or other resources, a study;

(2) A person who submits a study to the EPA in support of an application for a research or marketing permit; or

(3) A testing facility, if it both initiates and actually conducts the study.

"Study" means any experiment at one or more test sites, in which a test substance is studied in a test system under laboratory conditions or in the environment to determine or help predict its effects, metabolism, product performance (efficacy studies only as required by 40 CFR 158.640), environmental and chemical fate, persistence and residue, or other characteristics in humans, other living organisms, or media. The term "study" does not include basic exploratory studies carried out to determine whether a test substance or a test method has any potential utility.

"Study completion date" means the date the final report is signed by the study director.

"Study director" means the individual responsible for the overall conduct of a study.

"Study initiation date" means the date the protocol is signed by the study director.

"Test substance" means a substance or mixture administered or added to a test system in a study, which substance or mixture:

(1) Is the subject of an application for a research or marketing permit supported by the study, or is the contemplated subject of such an application; or

(2) Is an ingredient, impurity, degradation product, metabolite, or radioactive isotope of a substance described by paragraph (1) of this definition, or some other substance related to a substance described by that paragraph, which is used in the study to assist in characterizing the toxicity, metabolism, or other characteristics of a substance described by that paragraph.

"Test system" means any animal, plant, microorganism, chemical or physical matrix, including but not limited to soil or water, or subparts thereof, to which the test, control, or reference substance is administered or added for study. "Test system" also includes appropriate groups or components of the system not treated with the test, control, or reference substance.

"Testing facility" means a person who actually conducts a study, i.e., actually uses the test substance in a test system. "Testing facility" encompasses only those operational units that are being or have been used to conduct studies.

"Vehicle" means any agent which facilitates the mixture, dispersion, or solubilization of a test substance with a carrier.

§ 160.10 Applicability to studies performed under grants and contracts.

When a sponsor or other person utilizes the services of a consulting laboratory, contractor, or grantee to perform all or a part of a study to which this part applies, it shall notify the consulting laboratory, contractor, or grantee that the service is, or is part of, a study that must be conducted in compliance with the provisions of this part.

§ 160.12 Statement of compliance or noncompliance.

Any person who submits to EPA an application for a research or marketing permit and who, in connection with the application, submits data from a study to which this part applies shall include in the application a true and correct statement, signed by the applicant, the sponsor, and the study director, of one of the following types:

(a) A statement that the study was conducted in accordance with this part; or

(b) A statement describing in detail all differences between the practices used in the study and those required by this part; or

(c) A statement that the person was not a sponsor of the study, did not conduct the study, and does not know whether the study was conducted in accordance with this part.

§ 160.15 Inspection of a testing facility.

(a) A testing facility shall permit an authorized employee or duly designated representative of EPA or FDA, at reasonable times and in a reasonable manner, to inspect the facility and to inspect (and in the case of records also to copy) all records and specimens required to be maintained regarding studies to which this part applies. The records inspection and copying requirements should not apply to quality assurance unit records of findings and problems, or to actions recommended and taken, except that EPA may seek production of these records in litigation or formal adjudicatory hearings.

(b) EPA will not consider reliable for purposes of supporting an application for a research or marketing permit any data developed by a testing facility or sponsor that refuses to permit inspection in accordance with this part. The determination that a study will not be considered in support of an application for a research or marketing permit does not, however, relieve the applicant for such a permit of any obligation under any applicable statute or regulation to submit the results of the study to EPA.

§ 160.17 Effects of non-compliance.

(a) EPA may refuse to consider reliable for purposes of supporting an application for a research or marketing permit any data from a study which was not conducted in accordance with this part.

(b) Submission of a statement required by § 160.12 which is false may form the basis for cancellation, suspension, or modification of the research or marketing permit, or denial or disapproval of an application for such a permit, under FIFRA section 3, 5, 6, 18, or 24 or FFDCA section 406 or 409, or for criminal prosecution under 18 U.S.C. 2 or 1001 or FIFRA section 14, or for imposition of civil penalties under FIFRA section 14.

Subpart B—Organization and Personnel

§ 160.29 Personnel.

(a) Each individual engaged in the conduct of or responsible for the supervision of a study shall have education, training, and experience, or combination thereof, to enable that individual to perform the assigned functions.

(b) Each testing facility shall maintain a current summary of training and experience and job description for each individual engaged in or supervising the conduct of a study.

(c) There shall be a sufficient number of personnel for the timely and proper conduct of the study according to the protocol.

(d) Personnel shall take necessary personal sanitation and health precautions designed to avoid contamination of test, control, and reference substances and test systems.

(e) Personnel engaged in a study shall wear clothing appropriate for the duties they perform. Such clothing shall be changed as often as necessary to prevent microbiological, radiological, or chemical contamination of test systems and test, control, and reference substances.

(f) Any individual found at any time to have an illness that may adversely affect the quality and integrity of the study shall be excluded from direct contact with test systems, and test, control, and reference substances, and any other operation or function that may adversely affect the study until the condition is corrected. All personnel shall be instructed to report to their immediate supervisors any health or medical conditions that may reasonably be considered to have an adverse effect on a study.

§ 160.31 Testing facility management.

For each study, testing facility management shall:

(a) Designate a study director as described in § 160.33 before the study is initiated.

(b) Replace the study director promptly if it becomes necessary to do so during the conduct of a study.

(c) Assure that there is a quality assurance unit as described in § 160.35.

(d) Assure that test, control, and reference substances or mixtures have been appropriately tested for identity, strength, purity, stability, and uniformity, as applicable.

(e) Assure that personnel, resources, facilities, equipment, materials and methodologies are available as scheduled.

(f) Assure that personnel clearly understand the functions they are to perform.

(g) Assure that any deviations from these regulations reported by the quality assurance unit are communicated to the study director and corrective actions are taken and documented.

§ 160.33 Study director.

For each study, a scientist or other professional of appropriate education, training, and experience, or combination thereof, shall be identified as the study director. The study director has overall responsibility for the technical conduct of the study, as well as for the interpretation, analysis, documentation, and reporting of results, and represents the single point of study control. The study director shall assure that:

(a) The protocol, including any change, is approved as provided by § 160.120 and is followed.

(b) All experimental data, including observations of unanticipated responses of the test system are accurately recorded and verified.

(c) Unforeseen circumstances that may affect the quality and integrity of the study are noted when they occur, and corrective action is taken and documented.

(d) Test systems are as specified in the protocol.

(e) All applicable good laboratory practice regulations are followed.

(f) All raw data, documentation, protocols, specimens, and final reports are transferred to the archives during or at the close of the study.

§ 160.35 Quality assurance unit.

(a) A testing facility shall have a quality assurance unit which shall be responsible for monitoring each study to assure management that the facilities, equipment, personnel, methods, practices, records, and controls are in conformance with the regulations in this part. For any given study, the quality assurance unit shall be entirely separate from and independent of the personnel engaged in the direction and conduct of that study. The quality assurance unit shall conduct inspections and maintain records appropriate to the study.

(b) The quality assurance unit shall:

(1) Maintain a copy of a master schedule sheet of all studies conducted at the testing facility indexed by test substance, and containing the test system, nature of study, date study was initiated, current status of each study, identity of the sponsor, and name of the study director.

(2) Maintain copies of all protocols pertaining to all studies for which the unit is responsible.

(3) Inspect each study at intervals adequate to ensure the integrity of the study and maintain written and properly signed records of each periodic inspection showing the date of the inspection, the study inspected, the phase or segment of the study inspected, person performing the inspection, findings and problems, action recommended and taken to resolve existing problems, and any scheduled date for reinspection. Any problems which are likely to affect study integrity found during the course of an inspection shall be brought to the attention of the study director and management immediately.

(4) Periodically submit to management and the study director written status reports on each study, noting any problems and the corrective actions taken.

(5) Determine that no deviations from approved protocols or standard operating procedures were made without proper authorization and documentation.

(6) Review the final study report to assure that such report accurately describes the methods and standard operating procedures, and that the reported results accurately reflect the raw data of the study.

(7) Prepare and sign a statement to be included with the final study report which shall specify the dates inspections were made and findings reported to management and to the study director.

(c) The responsibilities and procedures applicable to the quality assurance unit, the records maintained by the quality assurance unit, and the method of indexing such records shall be in writing and shall be maintained. These items including inspection dates, the study inspected, the phase or segment of the study inspected, and the name of the individual performing the inspection shall be made available for inspection to authorized employees or duly designated representatives of EPA or FDA.

(d) An authorized employee or a duly designated representative of EPA or FDA shall have access to the written procedures established for the inspection and may request testing facility management to certify that inspections are being implemented, performed, documented, and followed up in accordance with this paragraph.

Subpart C—Facilities

§ 160.41 General.

Each testing facility shall be of suitable size and construction to facilitate the proper conduct of studies. Testing facilities which are not located within an indoor controlled environment shall be of suitable location to facilitate the proper conduct of studies. Testing facilities shall be designed so that there is a degree of separation that will prevent any function or activity from having an adverse effect on the study.

§ 160.43 Test system care facilities.

(a) A testing facility shall have a sufficient number of animal rooms or other test system areas, as needed, to ensure: proper separation of species or test systems, isolation of individual projects, quarantine or isolation of animals or other test systems, and routine or specialized housing of animals or other test systems.

(1) In tests with plants or aquatic animals, proper separation of species can be accomplished within a room or area by housing them separately in different chambers or aquaria. Separation of species is unnecessary where the protocol specifies the simultaneous exposure of two or more species in the same chamber, aquarium, or housing unit.

(2) Aquatic toxicity tests for individual projects shall be isolated to the extent necessary to prevent cross-contamination of different chemicals used in different tests.

(b) A testing facility shall have a number of animal rooms or other test system areas separate from those described in paragraph (a) of this section to ensure isolation of studies being done with test systems or test, control, and reference substances known to be biohazardous, including volatile substances. aerosols, radioactive materials, and infectious agents.

(c) Separate areas shall be provided, as appropriate, for the diagnosis, treatment, and control of laboratory test system diseases. These areas shall provide effective isolation for the housing of test systems either known or suspected of being diseased, or of being carriers of disease, from other test systems.

(d) Facilities shall have proper provisions for collection and disposal of contaminated water, soil, or other spent materials. When animals are housed, facilities shall exist for the collection and disposal of all animal waste and refuse or for safe sanitary storage of waste before removal from the testing facility. Disposal facilities shall be so provided and operated as to minimize vermin infestation, odors, disease hazards, and environmental contamination.

(e) Facilities shall have provisions to regulate environmental conditions (e.g., temperature, humidity, photoperiod) as specified in the protocol.

(f) For marine test organisms, an adequate supply of clean sea water or artificial sea water (prepared from deionized or distilled water and sea salt mixture) shall be available. The ranges of composition shall be as specified in the protocol.

(g) For freshwater organisms, an adequate supply of clean water of the appropriate hardness, pH, and temperature, and which is free

of contaminants capable of interfering with the study, shall be available as specified in the protocol.

(h) For plants, an adequate supply of soil of the appropriate composition, as specified in the protocol, shall be available as needed.

§ 160.45 Test system supply facilities.

(a) There shall be storage areas, as needed, for feed, nutrients, soils, bedding, supplies, and equipment. Storage areas for feed nutrients, soils, and bedding shall be separated from areas where the test systems are located and shall be protected against infestation or contamination. Perishable supplies shall be preserved by appropriate means.

(b) When appropriate, plant supply facilities shall be provided. As specified in the protocol, these include:

(1) Facilities for holding, culturing, and maintaining algae and aquatic plants.

(2) Facilities for plant growth, including, but not limited to greenhouses, growth chambers, light banks, and fields.

(c) When appropriate, facilities for aquatic animal tests shall be provided. These include, but are not limited to, aquaria, holding tanks, ponds, and ancillary equipment as specified in the protocol.

§ 160.47 Facilities for handling test, control, and reference substances.

(a) As necessary to prevent contamination or mixups, there shall be separate areas for:

(1) Receipt and storage of the test, control, and reference substances.

(2) Mixing of the test, control, and reference substances with a carrier, e.g., feed.

(3) Storage of the test, control, and reference substance mixtures.

(b) Storage areas for test, control, and/or reference substance and for test, control, and/or reference mixtures shall be separate from areas housing the test systems and shall be adequate to preserve the identity, strength, purity, and stability of the substances and mixtures.

§ 160.49 Laboratory operation areas.

Separate laboratory space and other space shall be provided, as needed, for the performance of the routine and specialized procedures required by studies.

§ 160.51 Specimen and data storage facilities.

Space shall be provided for archives, limited to access by authorized personnel only, for the storage and retrieval of all raw data and specimens from completed studies.

Subpart D—Equipment

§ 160.61 Equipment design.

Equipment used in the generation, measurement, or assessment of data and equipment used for facility environmental control shall be of appropriate design and adequate capacity to function according to the protocol and shall be suitably located for operation, inspection, cleaning, and maintenance.

§ 160.63 Maintenance and calibration of equipment.

(a) Equipment shall be adequately inspected, cleaned, and maintained. Equipment used for the generation, measurement, or assessment of data shall be adequately tested, calibrated, and/or standardized.

(b) The written standard operating procedures required under § 160.81(b)(11) shall set forth in sufficient detail the methods, materials, and schedules to be used in the routine inspection, cleaning, maintenance, testing, calibration, and/or standardization of equipment, and shall specify, when appropriate, remedial action to be taken in the event of failure or malfunction of equipment. The written standard operating procedures shall designate the person responsible for the performance of each operation.

(c) Written records shall be maintained of all inspection, maintenance, testing, calibrating, and/or standardizing operations. These records, containing the dates of the operations, shall describe whether the maintenance operations were routine and followed the written standard operating procedures. Written records shall be kept of nonroutine repairs performed on equipment as a result of failure and

malfunction. Such records shall document the nature of the defect, how and when the defect was discovered, and any remedial action taken in response to the defect.

Subpart E—Testing Facilities Operation

§ 160.81 Standard operating procedures.

(a) A testing facility shall have standard operating procedures in writing setting forth study methods that management is satisfied are adequate to insure the quality and integrity of the data generated in the course of a study. All deviations in a study from standard operating procedures shall be authorized by the study director and shall be documented in the raw data. Significant changes in established standard operating procedures shall be properly authorized in writing by management.

(b) Standard operating procedures shall be established for, but not limited to, the following:

(1) Test system area preparation.
(2) Test system care.
(3) Receipt, identification, storage, handling, mixing, and method of sampling of the test, control, and reference substances.
(4) Test system observations.
(5) Laboratory or other tests.
(6) Handling of test systems found moribund or dead during study.
(7) Necropsy of test systems or postmortem examination of test systems.
(8) Collection and identification of specimens.
(9) Histopathology.
(10) Data handling, storage and retrieval.
(11) Maintenance and calibration of equipment.
(12) Transfer, proper placement, and identification of test systems.

(c) Each laboratory or other study area shall have immediately available manuals and standard operating procedures relative to the laboratory or field procedures being performed. Published literature may be used as a supplement to standard operating procedures.

(d) A historical file of standard operating procedures, and all revisions thereof, including the dates of such revisions, shall be maintained.

§ 160.83 Reagents and solutions.

All reagents and solutions in the laboratory area shall be labeled to indicate identity, titer or concentration, storage requirements, and expiration date. Deteriorated or outdated reagents and solutions shall not be used.

§ 160.90 Animal and other test system care.

(a) There shall be standard operating procedures for the housing, feeding, handling, and care of animals and other test systems.

(b) All newly received test systems from outside sources shall be isolated and their health status or appropriateness for the study shall be evaluated. This evaluation shall be in accordance with acceptable veterinary medical practice or scientific methods.

(c) At the initiation of a study, test systems shall be free of any disease or condition that might interfere with the purpose or conduct of the study. If during the course of the study, the test systems contract such a disease or condition, the diseased test systems should be isolated, if necessary. These test systems may be treated for disease or signs of disease provided that such treatment does not interfere with the study. The diagnosis, authorization of treatment, description of treatment, and each date of treatment shall be documented and shall be retained.

(d) Warm-blooded animals, adult reptiles, and adult terrestrial amphibians used in laboratory procedures that require manipulations and observations over an extended period of time or in studies that require these test systems to be removed from and returned to their test system-housing units for any reason (e.g., cage cleaning, treatment, etc.) shall receive appropriate identification (e.g., tattoo, color code, ear tag, ear punch, etc.). All information needed to specifically identify each test system within the test system-housing unit shall appear on the outside of that unit. Suckling mammals and juvenile birds are excluded from the requirement of individual identification unless otherwise specified in the protocol.

(e) Except as specified in paragraph (e)(1) of this section, test systems of different spe-

cies shall be housed in separate rooms when necessary. Test systems of the same species, but used in different studies, should not ordinarily be housed in the same room when inadvertent exposure to test, control, or reference substances or test system mixup could affect the outcome of either study. If such mixed housing is necessary, adequate differentiation by space and identification shall be made.

(1) Plants, invertebrate animals, aquatic vertebrate animals, and organisms that may be used in multispecies tests need not be housed in separate rooms, provided that they are adequately segregated to avoid mixup and cross contamination.

(2) [Reserved]

(f) Cages, racks, pens, enclosures, aquaria, holding tanks, ponds, growth chambers, and other holding, rearing and breeding areas, and accessory equipment, shall be cleaned and sanitized at appropriate intervals.

(g) Feed, soil, and water used for the test systems shall be analyzed periodically to ensure that contaminants known to be capable of interfering with the study and reasonably expected to be present in such feed, soil, or water are not present at levels above those specified in the protocol. Documentation of such analyses shall be maintained as raw data.

(h) Bedding used in animal cages or pens shall not interfere with the purpose or conduct of the study and shall be changed as often as necessary to keep the animals dry and clean.

(i) If any pest control materials are used, the use shall be documented. Cleaning and pest control materials that interfere with the study shall not be used.

(j) All plant and animal test systems shall be acclimatized to the environmental conditions of the test, prior to their use in a study.

Subpart F—Test, Control, and Reference Substances

§ 160.105 Test, control, and reference substance characterization.

(a) The identity, strength, purity, and composition, or other characteristics which will appropriately define the test, control, or reference substance shall be determined for each batch and shall be documented before its use in a study. Methods of synthesis, fabrication, or derivation of the test, control, or reference substance shall be documented by the sponsor or the testing facility, and the location of such documentation shall be specified.

(b) When relevant to the conduct of the study the solubility of each test, control, or reference substance shall be determined by the testing facility or the sponsor before the experimental start date. The stability of the test, control, or reference substance shall be determined before the experimental start date or concomitantly according to written standard operating procedures, which provide for periodic analysis of each batch.

(c) Each storage container for a test, control, or reference substance shall be labeled by name, chemical abstracts service number (CAS) or code number, batch number, expiration date, if any, and, where appropriate, storage conditions necessary to maintain the identity, strength, purity, and composition of the test, control, or reference substance. Storage containers shall be assigned to a particular test substance for the duration of the study.

(d) For studies of more than 4 weeks experimental duration, reserve samples from each batch of test, control, and reference substances shall be retained for the period of time provided by § 160.195.

(e) The stability of test, control, and reference substances under storage conditions at the test site shall be known for all studies.

§ 160.107 Test, control, and reference substance handling.

Procedures shall be established for a system for the handling of the test, control, and reference substances to ensure that:

(a) There is proper storage.

(b) Distribution is made in a manner designed to preclude the possibility of contamination, deterioration, or damage.

(c) Proper identification is maintained throughout the distribution process.

(d) The receipt and distribution of each batch is documented. Such documentation shall include the date and quantity of each batch distributed or returned.

§ 160.113 Mixtures of substances with carriers.

(a) For each test, control, or reference substance that is mixed with a carrier, tests by appropriate analytical methods shall be conducted:

(1) To determine the uniformity of the mixture and to determine, periodically, the concentration of the test, control, or reference substance in the mixture.

(2) When relevant to the conduct of the study, to determine the solubility of each test, control, or reference substance in the mixture by the testing facility or the sponsor before the experimental start date.

(3) To determine the stability of the test, control, or reference substance in the mixture before the experimental start date or concomitantly according to written standard operating procedures, which provide for periodic analysis of each batch.

(b) Where any of the components of the test, control, or reference substance carrier mixture has an expiration date, that date shall be clearly shown on the container. If more than one component has an expiration date, the earliest date shall be shown.

(c) If a vehicle is used to facilitate the mixing of a test substance with a carrier, assurance shall be provided that the vehicle does not interfere with the integrity of the test.

Subpart G—Protocol for and Conduct of a Study

§ 160.120 Protocol.

(a) Each study shall have an approved written protocol that clearly indicates the objectives and all methods for the conduct of the study. The protocol shall contain but shall not necessarily be limited to the following information:

(1) A descriptive title and statement of the purpose of the study.

(2) Identification of the test, control, and reference substance by name, chemical abstracts service (CAS) number or code number.

(3) The name and address of the sponsor and the name and address of the testing facility at which the study is being conducted.

(4) The proposed experimental start and termination dates.

(5) Justification for selection of the test system.

(6) Where applicable, the number, body weight range, sex, source of supply, species, strain, substrain, and age of the test system.

(7) The procedure for identification of the test system.

(8) A description of the experimental design, including methods for the control of bias.

(9) Where applicable, a description and/or identification of the diet used in the study as well as solvents, emulsifiers and/or other materials used to solubilize or suspend the test, control, or reference substances before mixing with the carrier. The description shall include specifications for acceptable levels of contaminants that are reasonably expected to be present in the dietary materials and are known to be capable of interfering with the purpose or conduct of the study if present at levels greater than established by the specifications.

(10) The route of administration and the reason for its choice.

(11) Each dosage level, expressed in milligrams per kilogram of body or test system weight or other appropriate units, of the test, control, or reference substance to be administered and the method and frequency of administration.

(12) The type and frequency of tests, analyses, and measurements to be made.

(13) The records to be maintained.

(14) The date of approval of the protocol by the sponsor and the dated signature of the study director.

(15) A statement of the proposed statistical method to be used.

(b) All changes in or revisions of an approved protocol and the reasons therefore shall be documented, signed by the study director, dated, and maintained with the protocol.

§ 160.130 Conduct of a study.

(a) The study shall be conducted in accordance with the protocol.

(b) The test systems shall be monitored in conformity with the protocol.

(c) Specimens shall be identified by test system, study, nature, and date of collection. This information shall be located on the specimen container or shall accompany the specimen in a manner that precludes error in the recording and storage of data.

(d) In animal studies where histopathology is required, records of gross findings for a specimen from postmortem observations shall be available to a pathologist when examining that specimen histopathologically.

(e) All data generated during the conduct of a study, except those that are generated by automated data collection systems, shall be recorded directly, promptly, and legibly in ink. All data entries shall be dated on the day of entry and signed or initialed by the person entering the data. Any change in entries shall be made so as not to obscure the original entry, shall indicate the reason for such change, and shall be dated and signed or identified at the time of the change. In automated data collection systems, the individual responsible for direct data input shall be identified at the time of data input. Any change in automated data entries shall be made so as not to obscure the original entry, shall indicate the reason for change, shall be dated, and the responsible individual shall be identified.

§ 160.135 Physical and chemical characterization studies.

(a) All provisions of the GLP standards shall apply to physical and chemical characterization studies designed to determine stability, solubility, octanol water partition coefficient, volatility, and persistence (such as biodegradation, photodegradation, and chemical degradation studies) of test, control, or reference substances.

(b) The following GLP standards shall not apply to studies, other than those designated in paragraph (a) of this section, designed to determine physical and chemical characteristics of a test, control, or reference substance:

§ 160.31 (c), (d), and (g)
§ 160.35 (b) and (c)
§ 160.43
§ 160.45
§ 160.47
§ 160.49
§ 160.81(b) (1), (2), (6) through (9), and (12)

§ 160.90
§ 160.105 (a) through (d)
§ 160.113
§ 160.120(a) (5) through (12), and (15)
§ 160.185(a) (5) through (8), (10), (12), and (14)
§ 160.195 (c) and (d)

Subparts H and I—[Reserved]

Subpart J—Records and Reports

§ 160.185 Reporting of study results.

(a) A final report shall be prepared for each study and shall include, but not necessarily be limited to, the following:

(1) Name and address of the facility performing the study and the dates on which the study was initiated and was completed, terminated, or discontinued.

(2) Objectives and procedures stated in the approved protocol, including any changes in the original protocol.

(3) Statistical methods employed for analyzing the data.

(4) The test, control, and reference substances identified by name, chemical abstracts service (CAS) number or code number, strength, purity, and composition, or other appropriate characteristics.

(5) Stability and, when relevant to the conduct of the study the solubility of the test, control, and reference substances under the conditions of administration.

(6) A description of the methods used.

(7) A description of the test system used. Where applicable, the final report shall include the number of animals used, sex, body weight range, source of supply, species, strain and substrain, age, and procedure used for identification.

(8) A description of the dosage, dosage regimen, route of administration, and duration.

(9) A description of all circumstances that may have affected the quality or integrity of the data.

(10) The name of the study director, the names of other scientists or professionals and the names of all supervisory personnel, involved in the study.

(11) A description of the transformations, calculations, or operations performed on the data, a summary and analysis of the data, and a statement of the conclusions drawn from the analysis.

(12) The signed and dated reports of each of the individual scientists or other professionals involved in the study, including each person who, at the request or direction of the testing facility or sponsor, conducted an analysis or evaluation of data or specimens from the study after data generation was completed.

(13) The locations where all specimens, raw data, and the final report are to be stored.

(14) The statement prepared and signed by the quality assurance unit as described in § 160.35(b)(7).

(b) The final report shall be signed and dated by the study director.

(c) Corrections or additions to a final report shall be in the form of an amendment by the study director. The amendment shall clearly identify that part of the final report that is being added to or corrected and the reasons for the correction or addition, and shall be signed and dated by the person responsible. Modification of a final report to comply with the submission requirements of EPA does not constitute a correction, addition, or amendment to a final report.

(d) A copy of the final report and of any amendment to it shall be maintained by the sponsor and the test facility.

§ 160.190 Storage and retrieval of records and data.

(a) All raw data, documentation, records, protocols, specimens, and final reports generated as a result of a study shall be retained. Specimens obtained from mutagenicity tests, specimens of soil, water, and plants, and wet specimens of blood, urine, feces, and biological fluids, do not need to be retained after quality assurance verification. Correspondence and other documents relating to interpretation and evaluation of data, other than those documents contained in the final report, also shall be retained.

(b) There shall be archives for orderly storage and expedient retrieval of all raw data, documentation, protocols, specimens, and interim and final reports. Conditions of storage shall minimize deterioration of the documents or specimens in accordance with the require-

GOOD LABORATORY PRACTICE STANDARDS

ments for the time period of their retention and the nature of the documents of specimens. A testing facility may contract with commerical archives to provided a repository for all material to be retained. Raw data and specimens may be retained elsewhere provided that the archives have specific reference to those other locations.

(c) An individual shall be identified as responsible for the archives.

(d) Only authorized personnel shall enter the archives.

(e) Material retained or referred to in the archives shall be indexed to permit expedient retrieval.

§ 160.195 Retention of records.

(a) Record retention requirements set forth in this section do not supersede the record retention requirements of any other regulations in this subchapter.

(b) Except as provided in paragraph (c) of this section, documentation records, raw data, and specimens pertaining to a study and required to be retained by this part shall be retained in the archive(s) for whichever of the following periods is longest:

(1) In the case of any study used to support an application for a research or marketing permit approved by EPA, the period during which the sponsor holds any research or marketing permit to which the study is pertinent.

(2) A period of at least 5 years following the date on which the results of the study are submitted to the EPA in support of an application for a research or marketing permit.

(3) In other situations (e.g., where the study does not result in the submission of the study in support of an application for a research or marketing permit), a period of at least 2 years following the date on which the study is completed, terminated, or discontinued.

(c) Wet specimens, samples of test, control, or reference substances, and specially prepared material which are relatively fragile and differ markedly in stability and quality during storage, shall be retained only as long as the quality of the preparation affords evaluation. Specimens obtained from mutagenicity test,

specimens of soil, water, and plants, and wet specimens of blood, urine, feces, and biological fluids, do not need to be retained after quality assurance verification. In no case shall retention be required for longer periods than those set forth in paragraph (b) of this section.

(d) The master schedule sheet, copies of protocols, and records of quality assurance inspections, as required by § 160.35(c) shall be maintained by the quality assurance unit as an easily accessible system of records for the period of time specified in paragraph (b) of this section.

(e) Summaries of training and experience and job descriptions required to be maintained by § 160.29(b) may be retained along with all other testing facility employment records for the length of time specified in paragraph (b) of this section.

(f) Records and reports of the maintenance and calibration and inspection of equipment, as required by § 160.63 (b) and (c), shall be retained for the length of time specified in paragraph (b) of this section.

(g) If a facility conducting testing or an archive contracting facility goes out of business, all raw data, documentation, and other material specified in this section shall be transferred to the archives of the sponsor of the study. The EPA shall be notified in writing of such a transfer.

(h) Specimens, samples, or other non-documentary materials need not be retained after EPA has notified in writing the sponsor or testing facility holding the materials that retention is no longer required by EPA. Such notification normally will be furnished upon request after EPA or FDA has completed an audit of the particular study to which the materials relate and EPA has concluded that the study was conducted in accordance with this part.

(i) Records required by this part may be retained either as original records or as true copies such as photocopies, microfilm, microfiche, or other accurate reproductions of the original records.

[FR Doc. 89-19087 Filed 8-16-89; 8:45 am]

BILLING CODE 6560-50-M

Appendix B: U.S. Environmental Protection Agency FIFRA Advisories

Willa Y. Garner

Garndal Associates, 17485 Sierra Way, Monument, CO 80132

THE QUESTIONS AND ANSWERS in this appendix have been edited from the U.S. Environmental Protection Agency (EPA) Interpretation of the Good Laboratory Practice (GLP) Standards Regulation: GLP Regulation Advisories and are listed alphabetically by subject. To date (August 1, 1991), 36 advisories have issued by the EPA. When several of these documents are on the same subject, the questions and answers have been consolidated to eliminate redundancy and to provide the most complete coverage of the issue. In some cases, a specific question was not asked, but rather a statement was made by the correspondent, and the EPA addressed that statement.

Over the last two years, individuals have written to the EPA Office of Compliance Monitoring (OCM) requesting interpretations of the Federal Insecticide, Fungicide, and Rodenticide Act (FIFRA) and Toxic Substances Control Act (TSCA) GLP regulations. Originally, when the EPA replied to these individuals, the rest of the regulated community, including EPA regional inspectors, never heard the reply. To alleviate this problem, the Laboratory Data Integrity Assurance Division of OCM began to issue the questions along with the EPA's reply in documents called *Advisories*.

Anyone may obtain copies of these Advisories by writing LDIAD (EN–342), U.S. EPA, 401 M St., SW, Washington, DC 20460 or by calling (703) 308–8265. The EPA does not maintain a formal mailing list, and one must keep writing or calling to obtain the most recent issuances.

Archiving and Record or Sample Retention

(1) Advisory No. 12

Question: As a contract laboratory, many of our clients prefer to archive all the original raw data generated by us. Do the regulations

2192–8/92/0445$08.50/0 © 1992 American Chemical Society

require us to keep authenticated copies of the raw data in our own archives? If the answer to the preceding question is yes, it becomes harder for us to comply with the immediate archiving rule, especially where a large volume of data is generated. Would we be in violation if we do our archiving within 30 days of report completion?

Answer: Under the Federal Insecticide, Fungicide, and Rodenticide Act (FIFRA) Good Laboratory Practice (GLP) Standards, there must be archives for orderly storage and expedient retrieval of all raw data, documentation, protocols, specimens, and interim and final reports [40 CFR 160.190(b)]. Further, the study director is required to assure that all such records are transferred to the archives during or at the close of the study [40 CFR 160.33(f)]. This ensures that data are fully accounted for at the completion of the study. This must be completed prior to the signing of the compliance statement by the study director.

There is flexibility in the location of the archive for raw data and specimens. At 40 CFR 160.190(b), the GLPs state that retention of records at alternate locations is acceptable, provided that there is specific reference to those locations in the archives. Such off-location archives must still meet the full requirements of 40 CFR 160.190. Whether records are archived at the registrant's facility, at a contractor's central location, or at separate contractors' locations, the study director must assure that all raw data and specimens have been archived before the study report is signed. If the study director cannot assure that records at a particular location are archived correctly, he cannot sign a compliance statement that indicates that this standard has been met.

The sponsor is ultimately responsible for the raw data supporting the study [40 CFR 169.2(k)] and may require archived records to be shipped to the sponsor's archive at some date following a study. Alternatively, a testing facility may wish to consolidate archives at one location during or after a study. Such transfers of archives do not present a problem if the records are properly accounted for and if the testing facility maintains records of the archive location(s) at all times.

(2) Advisory Nos. 16 and 31

Question: According to 40 CFR 160.105(d), "For studies of more than 4 weeks' experimental duration, reserve samples from each batch of test, control, and reference substances shall be retained for the period of time provided by 160.195." I need a definition of "study duration", assuming the nature of the study is chemical analysis.

There are several factors to be considered when defining study duration for a chemical study. For example, is the duration of a chemical analysis study from initial sample preparation to final sample analysis? If so, what are the consequences on study duration if, during the course of the study, problems such as inconsistent results or equipment breakdowns develop? Those things could, of course, prolong study duration.

It is preferable to take a reserve sample at the onset of a study. In the previously mentioned situations, if the intended study duration was 2 weeks, a reserve sample would not be required, but if problems developed, then the study could have gone on for 6 weeks, in which case, a reserve sample would be required. If an adequate amount of test material was available at the end of that 6 weeks, a reserve sample could be taken; however, the amount of test material supplied by the sponsor is usually only enough for analysis. Would lack of a reserve sample in this instance mean noncompliance?

Answer: The term "four week experimental duration" is tied to the experimental start and termination dates. This is clarified in the preambles to the August 17, 1989 rules amending the FIFRA GLPs (54 FR 34052–34061) and the TSCA GLPs (54 FR 34034–34040). These dates are clearly defined and are not limited to the period of administration of the test substance.

Neither the 1983 FIFRA GLP Standards nor the 1987 proposed revisions suggest that study duration is anything but the actual time that the study takes. If a study is of more than 4 weeks' actual duration, reserve samples would be required to be taken under 40 CFR 160.105(d) provided the type of test is subject to that section. Unforeseen circumstances that increase the length of time that a study takes may affect whether a reserve sample is needed.

The requirement is for a reserve sample to be retained from each batch of test, control, and reference substances. A retention sample need not be taken from each subquantity supplied to the testing facility for each study. It may be taken from a different subquantity of the same batch, and taken after the beginning of a study, as long as it is retained for the period of time provided in 40 CFR 160.195. This also applies to Toxic Substances Control Act (TSCA) § 792.105(d).

(3) Advisory No. 2

Question: Retention of records and samples is largely dictated by the study protocol. In metabolism studies, only the major metabolites need be retained as reference sample. Samples analyzed during stability studies need not be retained.

Answer: Retention of records and samples is required by the regulations, which cannot be superseded by the protocol. The protocol will largely determine what records and samples are generated during the course of the study, and all such raw data must be retained. The GLPs do allow that certain samples need not be retained longer than they afford evaluation, and that certain specimens need not be retained after quality assurance verification [§§ 160.195(c) and 792.195(c)]. These provisions are made to allow the discarding of certain samples or specimens whose retention would not provide useful data integrity assurance.

(4) Advisory No. 7

Question: How long must soil samples, water samples, and plant samples generated from field-dissipation studies and subjected to analysis for their pesticide content be retained according to GLP § 160.195?

Answer: The GLPs state at 40 CFR 160.190(a) that specimens of soil, water, and plants do not need to be retained after quality assurance verification. The preamble to the rule published on August 17, 1989 (54 FR 34052) states that this means that such samples shall be retained until the quality assurance unit assures that their discarding does not negatively impact the integrity of the study. Unless such assurance is provided, such samples must be retained for the period of time specified under 40 CFR 160.195.

(5) Advisory No. 9

Question: What is the definition of *quality assurance verification* as it relates to sample retention? Can this be met through SOPs that are periodically inspected by the quality assurance unit (QAU)?

Answer: Quality assurance verification means that the material needs to be retained until the quality assurance unit (QAU) assures that discarding the material does not negatively affect the quality of the study. This clearly implies that the verification is a duty of the QAU, not the study personnel. It is not appropriate for the QAU to delegate its responsibilities to the personnel performing the study through SOPs or any other mechanism. Thus, the suggestion that verification be accomplished through SOPs periodically checked by the QAU is unacceptable.

(6) Advisory No. 31

Question: Would a reasonable time frame be allowed for archiving raw data and specimens after the study director signs the final report?

Answer: Under the Federal Insecticide, Fungicide, and Rodenticide Act (FIFRA) Good Laboratory Practice (GLP) Standards, there must be archives for orderly storage and expedient retrieval of all raw data, documentation, protocols, specimens, and interim and final reports [40 CFR 160.190(b)]. Further, the study director is required to assure that all such records are transferred to the archives during or at the close of the study [40 CFR 160.33(f)]. This ensures that data are fully accounted for at the completion of the study. This must be completed prior to signing the compliance statement.

There is flexibility in the location of the archives of raw data and specimens. At 40 CFR 160.190(b), the GLPs state that retention of records at alternate locations is acceptable, provided that there is specific reference to those locations in the archives. Such off-location archives must still meet the full requirements of 40 CFR 160.190. Whether records are archived at a central location or at separate contractors' locations, the study director must assure that all raw data and specimens have been archived before the study report is signed. In the case that the study director cannot assure that records at a particular location are archived correctly, he cannot sign a compliance statement that indicates that this standard has been met.

The sponsor is ultimately responsible for the raw data supporting the study [e.g., 40 CFR 169.2(k)], and may request archived records to be shipped to the sponsor's archives at some date following a study. Alternatively, a testing facility may wish to consolidate archives at one location during or after a study. Such transfers of archives do not present a problem if the records are properly accounted for and the testing facility maintains records of the archive location(s) at all times.

Please note that this position regarding archiving also applies to the Toxic Substances Control Act (TSCA) GLP standards at 40 CFR Part 792.

(7) Advisory No. 36

Question: Do the current GLP Standards permit disposal of specimens obtained from mutagenicity tests after quality assurance verification?

Answer: Please refer to the preamble of the final rule amending the GLPs (54 FR 34066, August 17, 1989). A commentor had asked what the term "quality assurance verification" meant. EPA responded that specimens must be retained until the quality assurance unit (QAU) assures that discarding the specimens does not negatively impact the integrity of the study. Therefore, once the testing facility QAU determines that discarding the specimens will not negatively impact the integrity of the study, the materials in question may be discarded. Under current EPA GLP requirements, there is no need to await the final EPA approval of the report or EPA inspection.

(8) Advisory No. 35

Question: I would like clarification of EPA's policy regarding retention of raw data as required under Section 8 of the Federal Insecticide Fungicide Rodenticide Act (FIFRA), specifically, the regulatory requirements for the retention of all underlying raw data at 40 CFR 169.2(k), the Good Laboratory Practice (GLP) Standards definition of "raw data" at 40 CFR 160.3, and the records retention requirement at 40 CFR 160.195 (i).

Answer: It is our understanding that it is your contention that these requirements do not conflict since: (1) FIFRA Section 8(a) provides that the administrator may prescribe regulations requiring retention of records; (2) although registrants are required at 40 CFR 169.2(k) to retain all underlying raw data, the term "raw data" is not therein defined; (3) "raw data" is defined in the GLPs at 40 CFR 160.3 which provides that "exact copies" of transcripts of raw data may be substituted for original raw data; and (4) it is stated further under GLPs at 40 CFR 160.195(i), that records "required by this part may be...true copies..."; that it is permissible for persons to substitute exact copies for original records to meet all regulatory needs for raw data retention. You further state that, since the term "exact copy" is not in itself defined, companies must set their own policy regarding what constitutes an "exact copy" and be prepared to defend it.

Please note that GLPs include specific archiving requirements for raw data (e.g., indexing for retrieval, minimization of deterioration, and protected access). Certain procedures may not be practicable with original records. For example, heat sensitive papers may deteriorate with time, while oversized charts and tape-recorded voice records may be difficult to index for expedient retrieval. After copies are made to assure compliance with GLP archiving requirements, the original records must still be retained to assure compliance with 40 CFR 169.2(k).

We agree that there is no conflict between GLPs and FIFRA Section 8 regulations. However, GLPs must not be viewed as superseding the records retention requirements stated at 40 CFR 169.2(k) or of providing regulatory clarification of terminology used in 40 CFR 169.2(k). Consequently, compliance with GLPs may be accomplished through retention of copies of raw data, but the destruction of original records would still be a violation of the provisions stated at 40 CFR 169.2(k).

Chemical Characterization

(1) Advisory No. 1

Question: Do organic solubility studies need to be conducted under the GLP Standards required by § 160.135(a), as do the water solubility studies, or may they be conducted under modified § 160.135(b)?

Answer: Section 160.135 does not distinguish between organic solvent and water solubility studies. Since there is no distinction in the regulations, you must assume that any solubility study, including organic solvent solubility, is subject to the full ιcgulations.

(2) Advisory No. 1

Question: In characterizing the test substance prior to the initiation of the study, can only those impurities present in amounts greater than 5% be identified?

Answer: Section 160.105(a) of the GLPs requires adequate test, reference, and control substance characterization to be performed and documented before study initiation. "Adequate" characterization (e.g., specific level of impurities) is study-specific and hence not defined further in the GLPs. It is appropriate to consult with the Office of Pesticide Programs (OPP) to determine if there are characterization requirements beyond those that your laboratory believes are sufficient. A decision to reject a study may be made by OPP independently of GLP compliance. Documentation as to when the levels of impurities were identified should be included in your submission to OPP.

(3) Advisory No. 10

Question: We understand that the studies to be conducted at contract laboratories must be conducted in complete compliance with GLPs (40 CFR 160), and the preparation and transmission of test articles to be used in those studies need not be done in compliance with GLPs.

Answer: The FIFRA GLP standards are intended to provide assurance that laboratory conduct during the performance of studies to be submitted to EPA is such that it does not jeopardize the integrity of the data generated. At 40 CFR 160.105, the GLPs specify that there must be determinations of the identity, purity, strength, composition, or other appropriate characteristics of test, control, or reference substances before their use in a study. Since these determinations must be made in the direct context of the use of the material in the study, they are part of study performance and must comply with GLPs. Further, there are provisions at 40 CFR 160.107 that procedures assuring proper storage, distribution, identification, and receipt of the test, control, and reference substances be followed. This includes handling during and after the time that the substance is characterized.

However, the methods of synthesis, fabrication, or derivation of the test, control, or reference substance need only be documented before the start of the study, with the location of documentation specified. The actual synthesis, fabrication, or derivation need not be performed under GLPs.

(4) Advisory No. 23

Question: What does the requirement under 160.105(e) of the GLPs, which states that the stability of test, control, and reference substances under storage conditions at the test site must be known for all studies, entail? Some of our sponsors want us to analyze the test chemicals and analytical reference standards for purity (as a check for stability) after their use in a study. Others feel that no analysis is necessary since they have conducted stability analyses on their test chemicals and analytical reference standards at their laboratories and have shown stability over a period of time (sometimes up to two years). These clients feel that if we store our chemicals under the same conditions as they did we will have met the intent of this section of the GLPs.

In short, is a stability analysis required to be conducted by the contract laboratory or the sponsor on each test chemical or analytical reference standard used in a study at the end of its use in that study or group of concurrent studies (which covers stability under storage conditions at the test site), or will the storage stability analysis conducted by the sponsor at a prior time suffice to meet the § 160.105(e) requirement?

Answer: The GLPs state at 40 CFR 160.105(e) that the stability of test, control, and reference substances under storage condition at the test site must be known for all studies. In a situation where stability testing has been previously performed under GLPs and such testing demonstrates stability under the storage conditions at the test site (i.e., comparable duration and temperature), then it is not necessary to retest for stability to comply with GLPs.

Where the stability is determined prior to the study under certain conditions, it is necessary to know that those condition apply to the study in question. If such conditions are not known, or if stability data are otherwise insufficient, (e.g., not done under GLPs), it is necessary to reaffirm by either completing stability testing under GLPs before the start of the study, or concomitant testing during the study.

(5) Advisory No. 30

Question: Please clarify the paragraph 792.113(a) "Mixtures of substances with carriers" in EPA's FIFRA GLP Standards as it pertains to efficacy testing for disinfectants. This section implies that analytical chemistry analysis is required each time the disinfectant is diluted for testing. I feel this is unnecessary and impractical. I believe analysis of the undiluted disinfectant is sufficient.

Answer: Clarification concerning mixtures of substances with carriers, for which the FIFRA GLP requirements are stipulated, can be found at 40 CFR 160.113. You stated that this standard appears to require that an analytical chemistry analysis be performed each time that a disinfectant is diluted for efficacy testing under the GLPs. You suggested that this is impractical due to the capabilities of laboratories that normally perform such testing and the fact that such dilutions must be prepared fresh for each test.

The GLPs state at 40 CFR 160.113(a) that for each test, control, or reference substance that is mixed with a carrier, tests by appropriate methods shall be conducted to determine, "periodically, the concentration of the test, control, or reference substance in the mixture." This requirement clearly applies to dilutions of disinfectants used in efficacy testing, just as it would for dilutions of agricultural pesticides used in studies involving field application.

However, this requirement does not apply to each batch of mixture prepared. The term "mixture" means a particular combination of ingredients, but is not restricted to mean only a particular batch or preparation involving those ingredients. Therefore, requirements regarding a mixture used in a particular test may be met by testing any preparation or batch of the mixture involving the same combination of ingredients. This testing must be sufficient to assure that the parameters stated in 40 CFR 160.113 are adequately supported, but need not be repeated for each batch or preparation unless there is a special reason to do so. Further, there is flexibility concerning where, when, and by whom such testing is performed, except that solubility and stability testing have time constraints as specified at 40 CFR 160.113(a)(2) and (3).

Compliance Statement

(1) Advisory No. 9

Question: Must the compliance statement be one sheet as defined by PR Notice 86–5, or is it acceptable to have the sponsor and applicant sign one statement and to have the study director sign the overall report which contains a section on compliance? Is it acceptable to have the sponsor and applicant sign one statement and the study director sign a separate statement? Finally, if the sponsor is the applicant, must there still be three signatures?

The agency has told our organization that the study completion date (by GLP definition, the date the final report is signed by the study director) should be the final date (i.e., no other signatures should be dated after the completion date). To a contract laboratory this poses problems for getting the portions of the report requiring our sponsor's signatures (i.e., Data Confidentiality Claims, Compliance Statement, and Flagging Statement) signed before finalizing the report. Is it acceptable to have the sponsor's signatures dated after the study director's?

Answer: EPA views the compliance statement as an important document and does not believe that it fulfills its intended function unless it is signed by all parties as specified at 40 CFR 160.12. The regulations also specify at 40 CFR 160.185(b) that the study director must sign and date the study report. While it is clear that the regulations intend that both the compliance statement and the study report be signed by the study director, this could be accomplished by including the compliance statement on the same page of the final report that the study director signs.

Regarding those situations in which the sponsor and applicant are the same person, that person need sign only once, provided that person is clearly identified on the compliance statement as both sponsor and applicant.

In response to your question on whether the individuals signing the compliance statement may sign separate copies, the answer is yes. Where the sponsor, applicant, or study director sign separate copies of the compliance statement, each copy must be identical in content and must be included in the study report with the appropriate signature.

The purpose of the study director's signature is to assure accountability for the contents of the final report. Thus, any amendments to the final report that reflect work that the study director is accountable for require the study director's signature. However, EPA has clarified, at 160.185(c), that reformatting or other modifications to conform with EPA's submission requirements (e.g., to conform with PR Notice 86–5) do not constitute amendments that require study director signature. Insofar as the sponsor-signed items included in the final report do not constitute products intrinsically related to the performance of the study, EPA sees no reason to require that the submission of the report be delayed to acquire the study director's signature. Such contents should be clearly identified as nondata items, and the signature should be clearly identified as the sponsor's signature.

(2) Advisory No. 34

Question: Is a compliance statement required for the analytical report for which a subcontractor with an independent quality assurance unit is performing work and generating a report for a sponsor?

Answer: Under GLPs, there is no requirement to have a compliance statement in each subunit, such as the analytical report, of the overall study. However, assurance of compliance may be needed by the sponsor or study director so that they can truthfully sign the compliance statement. Arrangement for such assurances must be worked out between the subcontractor, the study director and the sponsor.

Computer Validation

(1) Advisory No. 1

Question: How does one perform computer validation of analytical results?

Answer: Design and use of an automated data-collection system falls under GLPs if the system is used in the generation of raw data as defined in § 160.3 and/or in the conduct of the study as addressed in § 160.130(e). In such cases, the equipment must be of "appropriate design and adequate capacity" as specified in § 160.61 and be "adequately tested, calibrated and/or standardized", maintained, and records kept, as specified in § 160.63. Written standard operating procedures (SOPs) are required for this equipment. No distinction is made between the automated data-collection equipment and other analytical instrumentation, so the same standards must be interpreted to apply to both. As long as the data-capture system meets these criteria, there is latitude in its design. Again, the testing facility will be expected to provide documentation in the form of SOPs and other written records to support the validity of the system, however it is designed.

GOOD LABORATORY PRACTICE STANDARDS

Facilities

(1) Advisory No. 21

Question: Clarification is requested concerning a proposed setup where mixed feeds would be stored in the same rooms as the feeding studies, following preparation elsewhere, for 90-day feeding studies.

Answer: Your procedures include storage of the unmixed test material in a separate, locked room, while preparation of the feed mixture is performed in another separate room on a weekly basis. Storage of each mixture is in protected containers (double bagged, tied, and in a separate covered, labeled container) on a nonwooden pallet covered with plastic sheeting in a far corner of the animal room. The animal room, as described in your letter, is designated to that study only.

The GLPs at 40 CFR 160.47(a) require separate areas for test and control substance receipt and storage, mixing, and mixture storage. At 40 CFR 160.47(b), the storage areas for test and control substances and mixtures are required to be separate from areas housing the test systems, as well as adequate to preserve the identity, strength, purity, and stability of the substances and mixtures. There is a need to exercise judgement, since there is no specific definition of separate area or of the time frame intended by the term storage. Consideration must be given to the use of animal testing rooms primarily as testing rooms and not as storage rooms. EPA interprets the rule as allowing working quantities only of a mixture to be kept in the room housing the test system. Working quantities of a feed mix are typically considered to consist of a one- to two-week supply. It would not normally be acceptable to store longer-term supplies of mixtures in a room that is dedicated to housing test systems. It would in any case be necessary to follow procedures that assure that the placing of working quantities of the mixture in the test room does not compromise the integrity of the mixture or otherwise jeopardize the study.

The procedures that you described in your letter are in accordance with our interpretation of 40 CFR 160.47 and would thus be considered by our office to comply with the FIFRA GLP rule.

(2)

Editor's Note: The following question was addressed by the EPA, but there is no corresponding advisory.

Question: Testing facilities may consist of a test substance/analysis laboratory, the test system exposure facility (the field), and the sample analytical laboratory. As you pointed out, the separation of locations and sites is very apparent in the conduct of field residue studies.

Answer: The definition of study reads "any experiment at one or more test sites, in which a test substance is studied in a test system under laboratory conditions or in the environment to determine or help predict its effects, metabolism, product performance (efficacy studies only as required by 40 CFR 158.640), environmental and chemical fate, persistence and residue, or other characteristics in humans, other living organisms, or media. The term 'study' does not include basic exploratory studies carried out to determine whether a test substance or a test method has any potential utility."

A single study may require several experiments to be conducted at several locations. The separation of locations by a physical distance does not in itself justify defining experiments as separate studies and assigning separate study directors to them. According to the language in both the 1983 and 1989 GLP rules, there shall be one overall study director for each reported study. Although there may be distinct locations and/or disciplines involved, there is still a need for each study to have a single scientist or other professional of sufficient knowledge and training to act as the single point of study control.

However, if the experiment is a study as defined in the GLPs, then the requirements of the GLPs need to be complied with. A study director, protocol, quality assurance unit, report, and all other applicable provisions would be required.

(3) Advisory No. 4

Question: When contracting field residue studies, can the sponsor define the contractor's location as a facility?

Answer: The GLPs define the testing facility as "the person who actually conducts the study". This includes a contractor who performs the experimental work involved in the study. Please note that the term "testing facility" will include multiple locations. It will also include more than one physical location if the study is performed at multiple locations. It will also include more than one organizational unit if more than one organizational unit is involved in the study's conduct.

Good Laboratory Practice Defined

(1) Advisory No. 4

Question: Does GLP cover only the quality and integrity of study data generated and not the science of the study?

Answer: GLPs are primarily intended to ensure data quality and integrity. Insofar as scientific methodologies affect this (e.g., instrument calibration and test article characterization), they are subject to the GLPs. Also, a GLP violation would occur if a protocol was not followed. Please note that EPA has regulations and guidelines besides GLPs that determine what studies are required and how such studies are to be conducted.

(2) Advisory No. 18

Question: Clarification is requested on an apparent contradiction between Standard Evaluation Procedures (SEP) for terrestrial field dissipation studies and the GLPs. The SEP in question states that the report should contain (1) a stated goal of the study and (2) sufficient information on the test protocol and the analytical protocol. Since separate protocols are mentioned in the SEP for the test and analysis, these should be allowed to be separate studies under GLPs.

Answer: This approach does not comply with the GLP requirements. The GLPs define a study as an experiment to determine or help predict the effects or characteristics of a test substance. The separate analytical phase of an experiment does not meet this criterion. Further, the SEP that you cited refers to a single study and would itself be contradicted by breaking the experimental effort into two separate studies.

The SEPs may be used to provide guidance for the performance of studies, but they do not supersede the requirements of the GLP regulations. Should there be a terminology difference between the GLPs and an SEP, the requirements of the GLPs take precedence.

To comply with GLPs, each study must have one protocol and one study director. As stated in the GLPs, the protocol must contain but is not limited to the information stated at 40 CFR 160.120(a). The separate test and analytical methodologies must be included or referenced in the single protocol that covers the entire study.

(3) Advisory No. 31

Question: What are EPA's plans concerning making compilations of interpretations and clarifications of the GLPs available to the public?

Answer: EPA is preparing a question-and-answer document to improve communication of interpretations regarding GLPs to the regulated community.

Master Schedules

(1) Advisory No. 2

Question: Master schedule sheets are maintained by the quality assurance units of the testing facilities as a record of all studies conducted at the facilities. The main purpose of these sheets is to facilitate audits by EPA or FDA.

Answer: Master schedule sheets are required by the regulations as described. However, we do not believe that your statement is accurate in describing the main purpose of the master schedule as facilitating audits by the EPA. EPA uses the master schedule to determine the adequacy of the testing facility, that is, whether the facility is of sufficient size and has sufficient personnel to perform the studies that are listed. Finally, the master schedule is presumed to be necessary to the testing facility itself (e.g., for use by the QAU to assist in tracking of studies and scheduling of internal audits).

Method Confirmation and Validation

(1) Advisory No. 14

Question: Should the sponsor have the independent confirmatory method trial performed in compliance with the GLP Standard 40 CFR Part 160 which includes a study director, protocol, full documentation under GLPs, a quality-assurance-conducted audit during the experimental portion, and a final report? Or can this be considered part of method development and basic exploratory work?

Could the Office of Compliance Monitoring give some guidelines for determining at what point basic exploratory work may develop into a study that must be performed under full GLP compliance?

Answer: The scope of the FIFRA GLP requirements covers studies conducted to support tolerance petitions. As such, the confirmatory method trials fall under the scope of GLP requirements.

The FIFRA GLPs are explicit in stating that "basic exploratory studies carried out to determine whether a test substance or test method has any potential utility" do not fall under the definition of "study" and consequently are not under GLPs. Confirmatory method trials are not basic exploratory studies since it is assumed that their potential utility has been established by the time that such trials are performed.

General guidance regarding the point at which an "exploratory" study should be regarded as potentially requiring GLP compliance is provided directly by the GLP regulation. At 40 CFR 160.1 the standards state that GLPs are prescribed for the conduct of studies "intended to support applications for research or marketing permits...." Thus at any time where it is known that study data are intended to be submitted to EPA under the scope and definition given in the regulation, that study must be performed according to GLPs. However, we would advise that at any time that it is known that the data from a study may be submitted to EPA under the scope and definition given in the regulation, that study should also be conducted according to GLPs. Such data, if later reported to EPA, would be required to be accompanied by a valid compliance statement. The data submission may be rejected if the compliance statement indicates GLPs were not followed regardless of whether the data were intended for submission to EPA at the time that the study was performed.

(2) Advisory No. 33

Question: Must a company revalidate a method under GLPs in the case where it had established the analytical methodology and included validation data in an EPA study submitted prior to the effective date of the GLP requirements?

Answer: The GLPs do not specifically require that each method used in a study be revalidated under GLPs. However, there are certain cases where such method validation, or the gathering of data to support method validation, would be required to be performed under GLPs:

1. If the protocol of the study includes method validation, this information must be gathered under GLPs.

2. Regardless of whether a separate validation study is performed, calibration data must be gathered and maintained under GLPs, and any other applicable portions of GLPs must also be adhered to (i.e., maintenance and calibration of equipment and standard operating procedures).

3. If the EPA determines that revalidation of a particular method must be performed as part of a study, or for subsequent use in studies, that would be under GLPs regardless of whether a previous study had been submitted.

If there is a question concerning the acceptability of an analytical method, that is, a question on whether further validation is required, it is appropriate to contact the Office of Pesticide Programs of the EPA.

Non-GLP Studies

(1) Advisory No. 2

Question: Basic exploratory studies are those involved in method development, such as the development and validation of analytical methods.

Answer: The term "exploratory studies" does not have the regulatory meaning as suggested. Certain method development and validation studies could be exploratory, but others may be required in support of registrations. Also, some exploratory studies may have nothing to do with method development. Basically, studies being performed for submission to EPA should be regarded as being subject to the GLPs, while studies performed entirely for internal use would not require compliance.

(2) Advisory No. 2

Question: GLPs do not apply to routine manufacture or to starting materials used.

Answer: This is true where such processes are not performed as studies intended for submission to EPA. Should the submission of such data be required by EPA, it would be subject to the regulations.

(3) Advisory No. 28

Question: Based on our review of 40 CFR 158.640, we have concluded that collection of such efficacy data as horn fly and tick control on cattle, ked control on sheep, lice control on swine, and flea control on dogs during the research stage prior to submission for registration and product approval, is not subject to 40 CFR Chapter I Part 160—Good Laboratory Practice Standards. Further, we believe that if the studies are not subject to GLP Standards at the time they are conducted, that requirement cannot be imposed after the fact if EPA calls in the data at some future date after product registration. Is our interpretation correct?

Answer: The GLPs state at 40 CFR 160.3 under the definition of the term "study" that it applies to efficacy studies as required at 40 CFR 158.640. The table at 40 CFR 158.640 lists only antimicrobial agents, nematicides and fungicides, and vertebrate control agents. It does not currently include herbicides or insecticides, and hence does not cover the testing you mentioned, except when specific requests for such data are made under FIFRA 3(c)(2)(B).

If EPA specifically requires an efficacy study to be submitted, it is considered to be required under 40 CFR 158.640 which states: "The Agency reserves the right to require, on a case-by-case basis, submission of efficacy data for any pesticide product registered or proposed for registration." Even though such data may not be explicitly listed in the data requirement table of that section, and may involve an insecticide or herbicide, it must be accompanied by a statement of compliance or noncompliance.

In cases for which GLPs were not required at the time that a study was performed (e.g., the study was not performed specifically to meet the requirements of a data call in), the EPA does not expect to find such study in total conformity with GLPs after the fact, and will not automatically reject the data submission for failure to comply. However, the EPA reserves the right to determine the adequacy of such data, including whether it can be reconstructed, and to reject data of questionable or unknown integrity. A non-GLP study submitted without a statement (i.e., as specified at 40 CFR 160.12) that states whether or not the study did meet all GLP requirements or that identifies all discrepancies with GLPs may be rejected as not providing sufficient information for the EPA to make an informed decision. Finally, deviations from the GLPs may in fact result in rejection of the study as insufficient to meet the data requirement.

(4) Advisory No. 29

Question: Please clarify the applicability of FIFRA GLP Standards to certain pesticide registrations (regarding certain data submissions required to support pesticide registrations).

Answer: This is in response to your letter in which you requested clarification regarding applicability of Federal Insecticide, Fungicide, and Rodenticide Act (FIFRA) Good Laboratory Practice (GLP) Standards to certain data submissions required to support pesticide registrations held by you. Specifically, you asked: (1) whether an analytical method submitted under 40 CFR 158.120 (Guideline 62–3) or under 40 CFR 158.240

(Guideline 171–4) is considered a study and (2) if so, how GLP protocol and quality assurance unit (QAU) record-keeping and inspection requirements apply when the "data" are developmental in nature.

The GLPs as stated at 40 CFR 160.1 apply to the conduct of all studies which support, or are intended to support, pesticide registrations. When test method verification work is performed per the cited guidelines for submission to EPA it must be conducted according to GLPs. However, as defined under the FIFRA GLPs, the term "study" does not apply to basic exploratory studies performed "to determine whether...a test method has any potential utility." Therefore data which are developmental in nature would not be under GLP protocol and QAU requirements.

Please note that the Federal Register of May 4, 1988 (53 FR 15952) recodified certain data requirements. The enforcement analytical method requirement, previously located at 40 CFR 158.120, is now located at 40 CFR 158.180. The residue chemistry method requirement is still located at 40 CFR 158.240.

Protocol

(1) Advisory No. 9

Question: 40 CFR § 160.3 defines a test substance and includes any degradation products or metabolites of the original test substance in the definition. Per 40 CFR Part 160.120(2), is it necessary to add metabolites and reference substance used to determine metabolites to the protocol? Determining metabolites is often the stated objective of the protocol and including them in the protocol would necessitate many amendments after the fact.

Answer: The term "test substance", as defined in § 160.2, does include any degradation product or metabolite that is used in a study to assist in characterizing the toxicity, metabolism, or other characteristics of a substance that is the subject of an application for a research or marketing permit. However, in the case in which determining metabolites or degradation products is the stated objective of the study, such determination constitutes the characteristic that is being determined. Thus, in such a study metabolites or degradation products do not constitute the test substance. However, any reference substances intended to characterize such metabolites and degradation products should be identified in the protocol, by amendment if necessary.

Quality Assurance Unit and Study Director Duties

(1) Advisory Nos. 15 and 34

Question: Do the GLPs mandate that the quality assurance inspections conducted at the subcontractor's facility be reported to both the study director and the "management"? Should management include management at both the sponsor and the subcontractor facilities?

Clarification is requested regarding the duties of the quality assurance unit and the study director under GLPs in the situation in which contract laboratories are cooperating on studies. A study director employed by a competitor with whom we are cooperating on a GLP study would be in a position to gain access to confidential business information. This would occur through QAU reports, as required by the GLPs, and through inspections by this study director at our facility.

Also, it is difficult to gain timely approvals of procedure changes by off-site study directors, and the involved nature of field studies makes it impossible for one person to be completely responsible for such studies. Could study director oversight responsibilities be transferred to on-site principal investigators?

Answer: The GLPs require at 40 CFR 160.35(b)(4) that status reports be periodically submitted to both the study director and management. "Management" refers to testing facility management (i.e., the person whose responsibilities are stated under 40 CFR 160.31). It is not necessary under GLPs for the quality assurance unit (QAU) to submit multiple management reports (i.e., to both sponsor and subcontractor management) as long as reports are submitted to the correct person to discharge the duties specified at 40 CFR 160.31.

The GLPs require at 40 CFR 160.35(b)(3) that the QAU bring to the attention of the study director and management any problems that are likely to affect the integrity of the study. At 40 CFR 160.35(b)(4), the QAU is further required to submit written status reports to the study director, noting problems and corrective actions. Since these reports need only contain study-performance information, our office does not believe that they involve confidentiality issues, and thus they must be submitted to the study director as required.

The GLPs do require at 40 CFR 160.33 that there be one study director to provide assurance that certain tasks are properly performed. As you pointed out, the study director must authorize deviations in standard operating procedures (SOPs) and must sign protocol revisions. While such approvals should be done as early as possible, conduct of the study is not required to cease before the approval action. Consequently, these requirements are not in conflict with having an off-site study director. Further, the study director has no explicit site-inspection duties under the GLPs. While the study director must maintain overall responsibility, delegation of the practical oversight of technical efforts is not prohibited by the regulation. This allows necessary technical duties to be assigned to on-site individuals (e.g., to principal investigators) and should relieve your concerns regarding the presence of persons who may be security risks. Please note that the study director must sign the compliance statement for the study.

It is our opinion that there is no inherent conflict with GLPs when more than one contracting facility and/or location is involved in a study.

(2) Advisory No. 17

Question: Clarification is requested regarding whether the existence of a client's internal study officer and quality assurance unit (QAU) would obviate the need for the duplication of such personnel at contract facilities.

Answer: The responsibilities of the study director and quality assurance unit (QAU) include assurance that the performance of a study complies with GLP standards. Their duties include on-site inspection and monitoring of study operations. These activities could be adequately conducted by a client-maintained study director and QAU that are located off-site, if provisions are made to ensure that all required on-site duties are performed.

Compliance with GLPs is a responsibility of the testing facility as well as of the registrant. This is true whether the testing facility is contracted for a portion or for the entirety of the technical effort of a GLP study. The contract testing facility must, therefore, ensure adequate oversight of study conduct and quality assurance activities. While adequate standard operating procedures (SOPs) must also be maintained at the contract laboratory, the existence of these would not be sufficient in themselves to ensure GLP compliance.

Raw Data Submission

(1) Advisory No. 4

Question: Will it be necessary to submit copies of individual field trial notebooks as part of a multiple trial data submission, or may summaries of raw data on test conditions for each location be submitted?

Answer: Under GLPs, a summary and analysis of the raw data are required in the study report [Section 160.185(a)(11)], but not a complete copy of raw data. Program-specific guidelines that indicate a need for submission of raw data should be referenced to determine whether a copy of raw data (i.e., field records) is required in the report. In any case, the original raw data would need to be archived at the completion of the study, with the location of the archive(s) listed in the final report.

Soil Analysis and Weather Data

(1) Advisory No. 11

Question: Please provide the current conclusion for weather data and soil analysis regarding field residue studies under 40 CFR Part 160. Do these data require GLP procedures or can they be documented from the nearest weather station? In addition, soils analyses are generally not analyzed in a GLP laboratory; will the current soils laboratories (state university labs) be adequate?

If the state, federal, or independent labs are used for residue studies, what statement would be required in the compliance statement submitted with the residue study to EPA?

Answer: Any study performed in support of a FIFRA research or marketing permit must be in accordance with applicable GLP standards. If soil analyses or weather measurements are made as part of such a study, they must comply with the GLPs. In the case that nonstudy information is referenced (e.g., local weather data), the data generation need not be under GLPs as long as it is clear that a reference to such data is being made. Finally, the statement of compliance or noncompliance is necessary regardless of the affiliation(s) of testing laboratories employed in the study.

Study Director

(1) Advisory Nos. 4, 6, and 8

Question: Editor's Note: There were three inquiries regarding the definition and functions of a study director and how this person would operate under EPA's policy of one study director when the field or biological portion of a study was conducted at a separate site from the analytical portion of the study. The inquirers described their programs in raising their questions. Because EPA's answer was basically the same to all, the answers to the three questions has been combined here.

Answer: This approach does not meet GLP requirements. The term "study", as defined at 40 CFR 160.3, is an experiment "in which a test substance is studied in a test system...to determine or help predict its effects...." The coverage of portions of studies (analytical phases or field application phases) as complete and separate studies under this definition is problematic.

The term "study director" is defined as the individual responsible for the overall conduct of the study. At 40 CFR 160.33, it is further explained that the study director is the single point of study control and is responsible for the interpretation, analysis, documentation, and reporting of results. This clearly indicates that a need is perceived for an individual who has overall responsibilities that would encompass all technical aspects of a study. When a study is submitted to EPA, it is assumed that there was one study director responsible for the overall conduct of the study.

Some technical difficulties arise from the breaking of studies into component parts identified separately as studies. For example, there could be difficulty in assessing who has responsibility during certain critical phases of studies, such as the transfer of sample material from application sites following analysis and the archiving of data. The study directors of each unit of such a subdivided study would also have the authority to account for protocol changes as provided in 40 CFR 160.120(b). This would be expedient, but there would be a loss of assurance that such changes conform to the overall purpose of the study.

Subdividing a study could also increase the overall burden associated with performing the study. It would be necessary to address the entire GLP standard from the viewpoint of each subunit that is described as a study. For example, cooperators involved in application of agricultural chemicals would have to address test substance characterization as required at 40 CFR 160.105 if the application work is a separate study. Each time that a subunit exceeds 4 weeks' duration, the affected contracting testing facility would also be responsible for assuring that a reserve sample of the test substance is retained. Reporting requirements at 40 CFR 160.185 would have to be met for each unit. And since EPA interprets 40 CFR 160.35(b)(3) to require a quality assurance unit (QAU) inspection of each study, each subdivision would require at least one QAU inspection.

It should be possible to accommodate entire field residue or environmental fate experiments as single studies under GLPs. The testing facility that is involved in such studies would encompass all organizational entities involved in conducting the actual work. Testing facility management duties may be predominantly assumed by the sponsor facility or by a lead contracting laboratory, depending on needs and capabilities. Certain overall responsibilities, such as that of the study director to assure that the study is conducted according to GLPs, must be centralized and cannot be delegated. However, the study could be divided into units based on practical considerations, with many technical details and the responsibility of monitoring these details delegated as is seen fit by management. Thus, study project directors or other appropriately identified individuals may be responsible for assuring that day-to-day operations are carried out.

Please note that the testing facility management is responsible for designating the study director and making the appropriate assurances as specified at 40 CFR 160.31 (e.g., that there is a QAU). The regulations do not state that the testing facility management actually performs the duties it is providing assurances for under this section; consequently, there is considerable flexibility for contracted persons to provide their own standard operating procedures (SOPs), QAUs, or other requirements, provided that the overall testing facility management (i.e., at the sponsor or perhaps at the prime contractor) can provide assurance that compliance with GLPs occurred.

(2) Advisory Nos. 20 and 22

Question: Editor's Note: Two letters requested clarification on the single-study-director requirement for studies involving minor-use crops.

Answer A: An approach that uses more than one study director per study would not comply with the GLPs. The requirement at 40 CFR 160.33 states that the study director represents the single point of study control and is responsible for the overall conduct of the study. Dividing a technical effort into multiple studies creates multiple points of control and means that there is no individual with overall responsibility. The accountability provided by a single study director (who plans, oversees, and controls the interpretation, analysis, documentation, and reporting of the results) is one of the most important aspects of the GLP Standards. In addition, the GLPs define a study as a complete experimental effort. The definition does not suggest that separate phases either by location or type of work performed (i.e., analytical versus field) constitute separate studies.

A single study director may take overall responsibility for adequate completion of the study but does not have to be directly involved in performance of each technical effort. The study director could oversee the performance of on-site technical directors who are responsible for the individuals carrying out field and analytical duties.

Answer B: The work on a given project may be performed at several different field or laboratory sites, each with its own management unit. While the unit coordinates all research, it does not have direct technical control of the performance of work at any of the field or laboratory sites. You proposed that we accept a "multistudy director" concept that defines each field trial at each location with a single pesticide or commodity as a study. You would further define the analytical phase of each project as a separate study.

I believe that you could meet the requirements of a study director, and we need to explore this option rather than the multistudy-director system that you have suggested. The latter would require reproposing and amending the GLP regulations and would, in my view, compromise one of the primary tenets of the current regulations, accountability. The requirement at 40 CFR 160.33 states that the study director represents the single point of study control and is responsible for the overall conduct of the study. Dividing a technical effort into multiple studies creates multiple points of control and means that there is no individual with overall responsibility. The accountability provided by a single study director (who plans, oversees, and controls the interpretation, analysis, documentation, and reporting of the results) is one of the most important aspects of the GLP Standards. In addition, the regulations define a study as a complete experimental effort. Separate phases whether by location or type or work performed (i.e., analytical versus field) do not constitute separate studies under the current regulations.

I would recommend to you a single study director who takes overall responsibility for assuring the adequate completion of the entire research project but who need not be directly involved in performance of each technical effort. Thus, for projects coordinated by your group, the study director would oversee the performance of on-site technical directors (or assistant study directors) who are responsible for the individuals carrying out field and analytical duties.

Study Inspections

(1) Advisory No. 3

Question: The preamble for the GLP Standards indicated that "each study, no matter how short, must be inspected at least once while in progress." However, Subpart B, § 792.35 (b)(3) and Subpart B, § 160.35 (b)(3) indicate that studies are to be inspected at intervals adequate to ensure the integrity of the study. Does the preamble comment exclude the possibility of "interval inspections" as allowed by FDA? How does the preamble comment relate to screening, pilot, and range-finding studies that are not routinely inspected and are not considered GLP studies by the FDA? I was told by an EPA staff member that a range-finding teratology study was subject to GLPs and required laboratory inspection. Could you explain why these studies (run to establish dosage levels for a definitive study) would be viewed differently from other range-finding studies? If EPA does not feel that a random inspection program is an appropriate method of evaluating a study (as indicated in a statement in the preamble) what is suggested for appropriate quality assurance (QA) review of short-term repetitive type studies?

Answer: The GLPs state in §§ 160.35(b) and 792.35(b) that the quality assurance unit (QAU) "inspect each study at intervals adequate to ensure the integrity of the study." While this does not specify the number of intervals, we believe that in any case where a study is not inspected at least once, there is a clear GLP violation. Our preamble statement concerning random inspections was directed at the concept of allowing random selection of some studies for inspection, instead of inspecting all studies. For the reasons stated, we do not believe that the regulations allow this.

Finally we do not view range-finding, pilot, or screening studies as requiring GLP compliance unless they are being performed either as a study or a part of a study (i.e., specified in the protocol of a study) that is to be submitted to EPA. There is no specific regulatory requirement that each pilot or screening study performed as part of a larger study be inspected by the QAU as long as each larger study is inspected as required. The QAU may find it necessary to inspect some pilot or screening studies to assure the integrity of data from these portions of the studies.

(2) Advisory No. 13

Question: The way the current GLPs are written, quality assurance (QA) must audit "at intervals adequate to insure the integrity of the study." Would it be possible to get some further guidance from the agency on this? In typical studies, of course, it is no problem to interpret the regulations. The problem arises with protocols covering multiple sites. We realize that each cooperator will have to be visited for a facility audit, but what percentage of cooperators do you feel would be adequate to visit during the in-life phase of a study? Most of these protocols deal with 15 to 25 sites, very narrow crop windows, and cooperators with no QA unit. There seems to be disagreement throughout industry between management or study directors and QA as to what constitutes a reasonable amount of auditing (with QA being on the high side and management on the low side). This is where your input would be extremely beneficial and appreciated, as we would like to be in complete compliance from the start.

Answer: The GLPs state at 40 CFR 160.35(a) that a testing facility shall have a quality assurance unit (QAU) that shall monitor each study to assure management that the facilities, equipment, personnel, methods, practices, records, and controls are in conformance with the GLPs. The GLPs further state at 40 CFR 160.35(b)(3) that the QAU shall inspect each study at intervals adequate to ensure the integrity of the study.

Clearly, the QAU must conduct inspections adequate to provide the assurances required at 40 CFR 160.35(a) and, in the course of so doing, must inspect each study at least once. All parameters must be verified adequate for each site, but it is acceptable to use inspections conducted during other studies to provide necessary assurances. It is also acceptable to use inspections conducted when no study is in progress to assure that methods, personnel, [and other conditions] at a particular site are in conformance with GLPs. However, acceptability of such inspections is contingent on assuring that the facilities, personnel, and methods that are inspected are representative of those used in the study. Please note that it is necessary to reinspect facilities periodically to account for changes in personnel and equipment. Finally, no matter how complete QAU inspection coverage is regarding the sites involved in a study, it is still necessary to conduct at least one inspection of study activities while the study is in progress.

(3) Advisory No. 34

Question: The preamble of the GLPs regulations states at 160.35(b)(3) that each study, no matter how short, must be inspected at least once while in progress. To what extent would the following fulfill this requirement: (1) a protocol audit, (2) a raw data audit, or (3) a draft report review including confirmation that the report reflects the raw data?

Answer: All of these types of audits may be useful. But please note that there must be enough coverage for a given facility during all audits and inspections so that all aspects of testing are covered. As stated at 40 CFR 160.35(a), the coverage must include facilities, equipment, protocols, personnel, methods, practices, records, and controls. It is not necessary to separately address all of these aspects for each study, as long as overall coverage, that is, the sum of all inspections, is balanced to include all aspects. Therefore, protocol audits, raw data audits, and draft report reviews must not be the sole focus of inspections.

Test Chemical Storage Containers

(1) Advisory Nos. 5, 24–27, and 32

Question: Editor's Note: There were several advisories that pertained to requesting a waiver for the requirement of test substance storage containers until the end of a study. All of the replies to these waiver requests were similar and are cited in the following answer.

Answer: EPA believes that the provision for assignment of storage containers for the duration of the study at 40 CFR 160.105(c) is a logical and necessary provision that in most cases provides accountability of test material in a manner that imposes no unusual burden. In this particular case, however, the number of containers and limited storage facilities may pose unusual encumbrance problems.

Our staff has reviewed your request in light of the need to provide complete accountability of test material and the potential burden involved in storing and accounting for the containers in many different locations. It is our opinion that certain record-keeping steps could provide a basis for establishing an acceptable alternate method for the accounting of test substance storage containers in lieu of actual storage of the containers for the duration of this study, and are willing to allow a conditional exception to this requirement.

This exception is applicable to the study that you cited in your letter and is conditional on the following:

1. You shall maintain records fully accounting for each container, and its contents, from receipt of the test substance to the ultimate disposition (i.e., disposal, reclamation, or recycling) of the container. These records shall be maintained as raw data to this study. These records shall include, but not be limited to (1) information on shipments pertaining to each container leaving the storage site (examples of such records are shipping request records, bills of lading, carrier bills, and monthly inventories of warehouse activity; certified copies dated and signed by responsible persons will be considered adequate); (2) test substance receipt records at the testing facility

and/or testing site(s); (3) complete use logs of material taken from containers, including quantitation of amounts; and (4) a record of the disposition of the container, including the place, date, and any appropriate receipts from the recycler or reclaimer (again, dated and signed certified copies will be adequate).

2. A statement certifying that all the conditions outlined in this letter were complied with shall be included with the statement of compliance or noncompliance required at 40 CFR 160.12 [or 40 CFR 792.12(b)].

3. A copy of this letter and a statement certifying that all the conditions outlined in this letter are being complied with shall be maintained as raw data for this study and shall be presented upon any inspection involving this study.

4. You shall maintain complete records of the transfer of any partially used containers to a cooperator or any other person, including identification and amount of the material transferred, the type of container, and the date of the transfer.

5. You shall prepare an inventory of empty containers before disposal, including sufficient information to uniquely identify containers, and shall maintain this inventory in an up-to-date manner recording all shipments and arrivals of empty containers and their disposal. This record shall be maintained as raw data for this study.

6. You shall identify the locations of facilities where test substance is stored; where empty containers are stored prior to disposition; where records of use, shipment, and disposal or recycling of containers are maintained; and where the test substance is used in studies (i.e., testing sites). Within two weeks of receipt of notification of any pending inspection involving this study, you shall report the location of each of these facilities to:

David Dull, Director
Laboratory Data Integrity Assurance Division
Office of Compliance Monitoring (EN–342)
Office of Pesticides and Toxic Substances
U.S. Environmental Protection Agency
401 M Street, SW
Washington, DC 20460

Should these conditions not be fully met, all of the provisions of 40 CFR 160 (GLPs) [or 40 CFR 792.105(c)], including assignment of storage containers for the duration of the study, apply.

In addition, you are reminded that storage, disposal, or recycling of containers must be done in a manner pursuant to all applicable Federal, State, County, or local laws.

(2) Advisory No. 27

Question: May photographs be taken to satisfy the requirements of 40 CFR 160.105(c)?

Answer: Specifically, you asked about the standard at 40 CFR 160.105(c) which requires test substance storage containers to be assigned for the duration of a study. You described this as causing problems during residue field trials where the standard may result in the retention of many empty or near-empty containers for a substantial period of time; this could result in space and safety problems. As a solution, you proposed that a photograph of the container (label intact and punctured or otherwise rendered unusable) be kept in study files throughout the study. You stated that this would satisfy the requirements of 40 CFR 160.105(c) by providing documented evidence that the container was only used for that particular substance for the duration of the study.

Assignment of test substance containers for the duration of a study can best be assured by container retention, which is normally not a great burden. On several occasions, persons conducting or sponsoring studies have written EPA to explain that, in the case of a specific study that they were conducting or planning to conduct, the retention of empty containers would cause encumbrance or safety problems. We have reviewed such cases and made exception on an individual basis, with conditions documented to assure container accountability.

We are presently continuing our policy of individual review of such cases. Your suggested approach of using photographic evidence may be useful as a supplemental condition to assure container accountability, but only in the case that it is in conjunction with and under the conditions provided for in a specific, approved exception. We currently consider 40 CFR 160.105(c) as requiring retention of test substance storage containers unless such an individual exception is made.

Editor's Note: Requests for an exception to the GLPs must be made on an individual basis.

Appendix C: The EPA Enforcement Response Policy for the FIFRA Good Laboratory Practice Standards

On September 30, 1991, the U.S. Environmental Protection Agency issued a supplement to the July 2, 1990 Federal Insecticide, Fungicide, and Rodenticide Act (FIFRA) Enforcement Response Policy (ERP). This supplement describes the appropriate enforcement responses for violations of the FIFRA Good Laboratory Practice (GLP) Standards.

For ease of use by the reader, the text contained in this appendix is a combination of the *entire* GLP ERP and *excerpts* from the FIFRA ERP. Wherever the GLP ERP refers to the FIFRA ERP, the appropriate section from the FIFRA ERP has been added into the GLP ERP. These "added-in" sections are marked so that you will be able to distinguish them from the actual text of the GLP ERP.

Whenever portions of a text are removed from that text and placed elsewhere, there is a risk of topics being taken out of context. Please note that the introduction to the GLP ERP explicitly states that the GLP ERP does not supersede the FIFRA ERP, but that the two documents should be used in conjunction with each other.

For completeness, you should obtain copies of the FIFRA ERP and the GLP ERP. Contact Sharvon Warren or Tracie Cook, OCM, P&GD, PEPB, U.S. Environmental Protection Agency, 41 M. Street, S.W., Washington, DC 20460 (telephone 703–308–8383) to obtain copies of both documents.

Reprinted from *Enforcement Response Policy for the Federal Insecticide, Fungicide, and Rodenticide Act Good Laboratory Practice (GLP) Regulations*, Office of Compliance Monitoring, Office of Pesticides and Toxic Substances, U.S. Environmental Protection Agency, Sept. 30, 1991.

FIFRA inserts are reprinted from *Enforcement Response Policy for the Federal Insecticide, Fungicide, and Rodenticide Act (FIFRA)*, Office of Compliance Monitoring, Office of Pesticides and Toxic Substances, U.S. Environmental Protection Agency, July 2, 1990. This appendix is not subject to U.S. copyright. Published 1992 American Chemical Society.

Enforcement Response Policy for the Federal Insecticide, Fungicide, and Rodenticide Act Good Laboratory Practice (GLP) Regulations

Office of Compliance Monitoring, Office of Pesticides and Toxic Substances, U.S. Environmental Protection Agency

September 30, 1991

Table of Contents

	Page Numbers	
Heading	Original	This Book
Introduction	1	477
Requirements of the FIFRA GLP Standards	2	477–478
Regulated Community	2	478
Liability	2	478
Studies Covered Under the FIFRA GLPs	3	478–479
Violations	3	479
Violations Related to a False Compliance Statement	4	479
FIFRA § § 12(a)(2)(M) and 12(a)(2)(Q)	4	479
Violations of the GLPs Not Related to the GLP Compliance Statement	4	481
FIFRA § 12(a)(2)(B)(i)	5	481
FIFRA § 12(a)(2)(M)	6	481–482
FIFRA § 12(a)(2)(Q)	6	482
FIFRA § 12(a)(2)(R)	7	482
Multiple Violations	7	483
Levels of Action	8	483
Notices of Warning	8	483–485
Civil Administrative Penalties	9	486
Criminal Proceedings	10	496
Referrals	11	496
Office of Pesticide Programs	11	496
Office of Administration	12	497
Appendix GLP–A: FIFRA Charges and Gravity Levels for Civil Penalties Assessed for Violations of the FIFRA GLPs	GLP-A-1	497–498
Appendix GLP–B: Guidance for Determining Whether To Assess a GLP Violation Under FIFRA § 12(a)(2)(Q) as a High-, Middle-, or Low-Level Violation	GLP-B-1	498–500

INTRODUCTION

This policy sets forth the procedures that will be used to determine the appropriate enforcement response for violations of the Federal Insecticide, Fungicide, and Rodenticide Act (FIFRA) Good Laboratory Practice Standards (GLPs) found at 40 CFR Part 160. This policy is a supplement to the July 2, 1990 FIFRA Enforcement Response Policy (ERP) and is to be used in conjunction with the policies and matrices found in that ERP.

The EPA relies on data submitted by registrants as the basis for the Agency's regulatory decisions involving pesticide product registrations, tolerances, experimental use permits, special local needs registrations, emergency exemptions, or any other research or marketing permit for a pesticide (hereafter referred to as "research or marketing permits"). In conjunction with the EPA's data audit program, the FIFRA GLPs are intended to ensure the quality and integrity of this data.

Violations of the FIFRA GLPs may impact: (1) the reliability or scientific merits of test data; (2) the ability of the EPA to validate or reconstruct test results; (3) the ability of the Agency to make sound and timely regulatory decisions regarding a pesticide; and, (4) the EPA's administration of the GLP inspection and enforcement program. Therefore, noncompliance with the FIFRA GLP regulations may result in very serious harm to the EPA's regulatory mission and, ultimately, human health and the environment.

Violations of the FIFRA GLPs may involve violations of FIFRA sections 12(a)(2)(B)(i), 12(a)(2)(M), 12(a)(2)(Q), or 12(a)(2)(R). Appropriate enforcement responses for violations of the FIFRA GLPs include notices of warning, civil penalties of up to $5,000 per offense, and criminal penalties. In addition to these enforcement responses, the EPA may take regulatory action for violations of the GLPs, including: rejection of studies which do not comply with the FIFRA GLPs; cancellation, suspension, or modification of a pesticides research or marketing permit; or denial or disapproval of an application for such a permit. Further, in order to help assure that the Federal Government is dealing with responsible contractors, and for the purposes of the Federal Government's protection, pesticide testing facilities responsible for significant or major GLP violations may also be suspended or debarred from Government contracts or subcontracts. To address these types of actions, this policy includes a section on referrals to other EPA offices.

-2-

REQUIREMENTS OF THE FIFRA GLPS

The FIFRA GLP standards, found at 40 CFR Part 160, prescribe the minimum requirements that a pesticide testing facility (i.e., the laboratory, field site, etc.) and the sponsor must fulfill in the following areas:

1. Organization and personnel.
2. Facilities.
3. Equipment.

4. Testing facilities operation.
5. Test, control, and reference substances.
6. Protocol for and conduct of a study.
7. Records and reports.

Regulated Community

Any person, including a sponsor, pesticide testing facility, or registrant, who conducts, initiates, or supports a study required by the Agency under FIFRA sections 3, 4, 5, 18, or 24(c), or sections 408 or 409 of the Federal Food, Drug, and Cosmetic Act (FFDCA).

Liability

The EPA may pursue enforcement actions for violations of the FIFRA GLPS against any of the persons listed above depending on the specific facts of the case. Generally, the EPA will pursue separate civil administrative enforcement actions against the study sponsor, applicant for the research or marketing permit, and the pesticide testing facility since each of these parties have affirmative obligations to assure a study complies with the GLP regulations. If the sponsor, applicant for the research or marketing permit, or the pesticide testing facility are the same entity, the EPA will generally pursue a single enforcement action against that single entity. The signers of the GLP compliance statement may also be liable as individuals if the compliance statement required by 40 CFR 160.12 is false. However, in most cases, EPA will pursue enforcement actions against the company for which those individuals are employees or Agents.

-3-

Studies Covered Under the FIFRA GLPs

The FIFRA GLPs, as published in the Federal Register on November 29, 1983 (48 FR 53946), apply to all studies performed to determine the toxicity, metabolism, or other effects in humans and domestic animals which were conducted, initiated, or supported on or after May 2, 1984.[1] The FIFRA GLPs, as amended (August 17, 1989; 54 FR 34052), also apply to all studies performed to determine the effects, metabolism, product performance (with the exception of certain efficacy studies), environmental and chemical fate, persistence and residue, or other characteristics in humans, other living organisms, or media, conducted, initiated, or supported on or after October 16, 1989.[1]

As per the scope of the GLP regulations found at 40 CFR 160.1 and the definition of a "research or marketing permit" found in 40 CFR 160.3, the FIFRA GLP standards apply to all studies as defined above which are performed to support: (1) an application

---

[1] The term "supported" includes studies which have been submitted to the EPA after the effective date of the GLP regulations. Therefore, studies which have been conducted or initiated before the effective date, but have been submitted to the EPA in support of a pesticide product research or marketing permit after the effective date, must be submitted with the GLP Compliance Statement required by 40 CFR 160.12.

for registration, amended registration, or re-registration of a pesticide product under FIFRA sections 3, 4, or 24(c); (2) an application for an experimental use permit under FIFRA section 5; (3) an application for an emergency exemption under FIFRA section 18; (4) a petition or other request for establishment or modification of a tolerance, for an exemption for the need for a tolerance, or for other clearance under FFDCA section 408; (5) a petition or other request for establishment or modification of a food additive regulation or other clearance by EPA under FFDCA section 409; (6) a submission of data in response to a notice issued by EPA under FIFRA section 3(c)(2)(B); or (7) any other application, petition, or submission sent to EPA intended to persuade EPA to grant, modify, or leave unmodified a registration or other approval required as a condition of sale or distribution of a pesticide.

Violations

Violations of the FIFRA GLP Standards will be charged as unlawful acts of FIFRA under sections 12(a)(2)(B)(i), 12(a)(2)(M), 12(a)(2)(Q), or 12(a)(2)(R). The determination of the appropriate unlawful act to charge a violator will depend on the specific facts of the case, based on the following guidance.

-4-

Violations of the GLPs Related to a False GLP Compliance Statement - FIFRA §§ 12(a)(2)(M) and 12(a)(2)(Q)

Under 40 CFR 160.12, any person who submits to EPA data from a study which falls under the scope of the GLPs must submit a statement, signed by the applicant of the pesticide product research or marketing permit, the sponsor, and the study director, that: (1) the study complies with the GLP requirements; (2) describes the differences between the practices used in the study and those required by the FIFRA GLPs; or (3) the person was not the sponsor of the study, did not conduct the study, and does not know whether the study complies with the FIFRA GLP requirements. If a study is submitted to EPA with a GLP compliance statement which states that the study complies with the GLP requirements, and GLP violations have occurred, then EPA will consider that compliance statement to be false. Similarly, if a study is submitted to EPA with a GLP compliance statement which incorrectly describes the differences between the practices used in the study and those required by the GLPs, then EPA will also consider the compliance statement to be false.

Submission of a false compliance statement is a violation of FIFRA section 12(a)(2)(M) or FIFRA section 12(a)(2)(Q). If the statement was *knowingly* falsified, EPA may issue a civil penalty for a violation of FIFRA section 12(a)(2)(M) or pursue a criminal action. Otherwise, submission of a false GLP compliance statement will be pursued as a violation of FIFRA section 12(a)(2)(Q), as either a "high level," "middle level," or "low level" GLP violation (see Appendix GLP-A for gravity levels and Appendix GLP-B for guidance for determining whether to assess the violation as a high, middle, or low level violation).

Each independent violation of the GLP regulations which causes the GLP compliance statement to be false may be assessed as a separate violation of either FIFRA section 12(a)(2)(M) or 12(a)(2)(Q), as appropriate. See the "Multiple Violations" section of this ERP for a further discussion. Also see the July 2, 1990 FIFRA ERP, page 25, for a discussion of independently assessable charges.

FIFRA

Independently Assessable Charges

A separate civil penalty, up to the statutory maximum, shall be assessed for each independent violation of the Act. A violation is independent if it results from an act (or failure to act) which is not the result of any other charge for which a civil penalty is to be assessed, or if the elements of proof for the violations are different. Dependent violations may be listed in the complaint, but will not result in separate civil penalties.

Consistent with the above criteria, the Agency considers violations that occur from each shipment of a product (by product registration number, not individual containers), or each sale of a product, or each individual application of a product to be independent offenses of FIFRA.* Each of these independent violations of FIFRA are subject to civil penalties up to the statutory maximum of $5,000 for section 14(a)(1) and $1,000 for section 14(a)(2). For example, when the EPA can document that a registrant has distributed a misbranded product (one single EPA product registration number) in four separate shipments (filling four orders), EPA will charge that registrant with four counts of selling or distributing a misbranded product, and assess the registrant civil penalties of up to $20,000. Similarly, when the EPA can document that a registrant has shipped four separate misbranded products (four separate EPA product registration numbers) in a single shipment, EPA will charge the registrant four counts of selling or distributing a misbranded product, and assess civil penalties of up to $20,000. A commercial applicator that misuses a restricted use product on three occasions (either three distinct applications or three separate sites) will be charged with three counts of misuse, and assessed civil penalties of up to $15,000. A dealer that sells a restricted use pesticide (RUP) to six uncertified persons, other than in accordance with FIFRA section 3(d), will be charged with six violations of FIFRA, and assessed civil penalties of up to $30,000.

On the other hand, a single event or action (or lack of action) which can be considered as two unlawful acts of FIFRA (section 12) cannot result in a civil penalty greater than the statutory limit for one offense of FIFRA. For instance, a person can be assessed a civil penalty of up to $5,000 for selling and distributing a product in violation of a cancellation order. However, while the Agency considers a cancelled product to be no longer registered, that same person should not also be assessed an additional civil penalty of up to $5,000 for sale and distribution of the same unregistered product. In this example the violation of the cancellation order is dependent on the sale and distribution of the unregistered/cancelled product.

Another example of a dependent violation is multiple misbranding on a single product label. If a single product label is misbranded in one way or ten ways, as defined by FIFRA section 2(q), it is still misbranding on a single product label and is considered a single violation of FIFRA section 12(a)(1)(E). As a single violation of FIFRA, the maximum civil penalty that may be assessed is $5,000. However, EPA may assess a count of misbranding each time that a misbranded product is sold or distributed. For example, a registrant who sells or distributes four distinct shipments of a misbranded pesticide product may be assessed a civil penalty of up to $20,000.

---

* Independent violations which can be documented as both per sale and per shipment are to be calculated only as either per sale or per shipment, whichever is more appropriate based on the supporting documentation, and whichever approach yields the highest civil penalty. For example, if Person A has a violation involving 1 sale and 2 shipments, and Person B has a violation involving 2 sales and 1 shipment, both persons would be charged for 2 violations of FIFRA (Person A is charged for 2 shipments and Person B is charged for 2 sales).

Violations of the GLPs Not Related to the GLP Compliance Statement

Certain violations of the GLPs may result in an unlawful act under FIFRA section 12 irrespective and independent of the truthfulness of the GLP compliance statement required by 40 CFR Part 160.12. These unlawful acts include FIFRA sections 12(a)(2)(B)(i); 12(a)(2)(M); 12(a)(2)(Q); and, 12(a)(2)(R).

-5-

FIFRA Section 12(a)(2)(B)(i)

Section 12(a)(2)(B)(i) of FIFRA states that it shall be unlawful for any person to refuse to prepare, maintain, or submit any records required under sections 5, 7, 8, 11, or 19. The FIFRA GLP records which registrants, applicants for registration, and producers are required to maintain are, in part, required under the authority of FIFRA section 8. Therefore, failure by a registrant, applicant for registration, or producer to prepare, maintain, or submit any of the records required by the GLPs, including those required under 40 CFR 169.2(k), may be charged as a violation of FIFRA section 12(a)(2)(B)(i).

While almost all of the requirements under the GLPs provide in part for the production and retention of certain records, violations by a registrant, applicant for registration, or producer of the requirements under 40 CFR Part 160.190 - Storage and retrieval of records and data, and Part 160.195 - Retention of records, are particularly associated with recordkeeping and should be assessed as a violation of FIFRA section 12(a)(2)(B)(i). However, in cases where the raw data or other records are retained, but not according to the requirements in the GLP standards, unlawful acts under FIFRA sections 12(a)(2)(M), 12(a)(2)(Q), or 12(a)(2)(R) should be charged, rather than FIFRA section 12(a)(2)(B)(i).

Because FIFRA section 8 does not currently authorize EPA to require pesticide testing facilities to maintain records, recordkeeping violations by a pesticide testing facility should not be assessed as a violation of FIFRA section 12(a)(2)(B)(i), unless the study is being submitted under FIFRA section 5 for an experimental use permit, or under section 19. Instead, most GLP related recordkeeping or reporting violations by a pesticide testing facility will be charged through enforcement of the truthfulness of the GLP compliance statement [FIFRA sections 12(a)(2)(M) or 12(a)(2)(Q)] or through FIFRA section 12(a)(2)(R).

An unlawful act under FIFRA section 12(a)(2)(M), 12(a)(2)(Q), or 12(a)(2)(R) may be charged in addition to the recordkeeping violation charged under FIFRA section 12(a)(2)(B)(i), if the recordkeeping violation also results in the submission of a false GLP compliance statement [§§12(a)(2)(M) or 12(a)(2)(Q)] or the knowing submission of false data [§12(a)(2)(R)]. The EPA considers these unlawful acts to be independently assessable because it is possible to be in violation of the recordkeeping requirements of the GLPs and still submit a true and correct GLP compliance statement which indicates that the required records have not been maintained. It is also possible to maintain the required records, and therefore comply with section 12(a)(2)(B)(i), but to have maintained false records or to knowingly submit false data to the Agency.

-6-

FIFRA Section 12(a)(2)(M)

Section 12(a)(2)(M) of FIFRA states that it shall be unlawful for any person to *knowingly* falsify all or any part of an application for registration, application for an experimental use permit, any information submitted under section 7, any records required to be maintained by the Act, any report filed under the Act, or any information marked as confidential and submitted to the Administrator under any provision of the Act. Compliance with the GLPs is required as part of an application for registration or an application for an experimental use permit, and GLP compliance entails the maintenance of records (personnel records, Quality Assurance Unit (QAU) records and reports to management, etc.) and filing of reports (final study reports, including the submission of a GLP compliance statement, QAU reports, etc.). *"Knowing* falsification" of the GLP records or reports as related to these provisions constitutes a violation of FIFRA section 12(a)(2)(M).

FIFRA Section 12(a)(2)(Q)

       Section 12(a)(2)(Q) of FIFRA states that it is unlawful for any person to falsify all or part of any information relating to the testing of any pesticide (or any of its ingredients, metabolites, or degradation products) which the person knows will be furnished to the Administrator, or will become a part of any records required to be maintained by this Act.

       Regardless of the truthfulness of the GLP compliance statement, through this unlawful act, EPA may pursue an enforcement action for a violation of any requirement of the GLPs which involves the falsification of testing information which was submitted to the EPA, or for which a testing facility or sponsor knows will eventually be submitted to the Agency, or will be required to be maintained as a record under the GLPs or 40 CFR Part 169. The EPA is not required to assert that the falsification was "knowing," only that the information was "false".

       Additionally, under this unlawful act, EPA may pursue an enforcement action for a GLP violation for an ongoing study for which no final report has been submitted to the Administrator and for which no compliance statement under 40 CFR 160.12 has yet been signed, provided the EPA can document that information which was required to be documented as the study proceeded was false, and the pesticide testing facility knew that the information was being generated with the intention of being submitted to the EPA (note the requirement in 40 CFR 160.10).[2] The language of FIFRA section 12(a)(2)(Q) provides this authority since the unlawful act applies to the falsification of information relating to the testing of a pesticide "... that the person knows will be furnished to the Administrator or will become a part of any records required to be maintained by this Act."

-7-

FIFRA Section 12(a)(2)(R)

       Section 12(a)(2)(R) of FIFRA states that it is a violation of FIFRA to submit data known to be false in support of a registration. Studies required under FIFRA sections 3, 4, and 24(c) are clearly required to support a pesticide registration. Additionally, studies conducted under FIFRA section 5 and 18 may be used to support a pesticide product registration at some time. Therefore, *knowing* submission of false data, including false records/reports required under the FIFRA GLPs will constitute a violation of this provision of FIFRA. Unlike FIFRA section 12(a)(2)(Q), the applicability of this unlawful act is dependent on a finding that the data submitted was *"known"* to be false by the violator.

---

[2]   The appropriate enforcement response for GLP violations in on-going studies will generally be a notice of warning (NOW), unless the violation involves a "knowing" violation. Further, if the violation for which an NOW was issued is not corrected by the time the study is submitted to the Agency, the EPA will pursue a civil or criminal action.

Multiple Violations
_____

A statement, under 40 CFR 160.12, which certifies that a study complies with the GLPs is a statement that all requirements listed in 40 CFR Part 160 have been met. If requirements of the GLPs have not been met, then the GLP compliance statement is false. Each independent requirement of the GLPs which has been violated, but has been represented through the statement as in compliance, may be considered a separate count of FIFRA section 12(a)(2)(M) or 12(a)(2)(Q), as appropriate, and each count assessed a civil penalty up to the statutory maximum (see the July 2, 1990 FIFRA ERP, page 25, for a discussion of independently assessable charges). For example, a sponsor could be assessed a civil penalty for up to $15,000 because that sponsor submitted a study with a GLP compliance statement which failed to truthfully state that the pesticide testing facility: (1) failed to maintain personnel records; (2) failed to designate a study director; and, (3) failed to record raw data.

Unlawful acts under FIFRA sections 12(a)(2)(B)(i); 12(a)(2)(M); 12(a)(2)(Q); and, 12(a)(2)(R) may be assessed in addition to those violations assessed as a false compliance statement, provided that these additional unlawful acts are independent of the counts charged as a falsification of the GLP compliance statement.

Generally, GLP violations will be assessed on a per study basis. Therefore, multiple violations of the same requirement in separate studies will be considered as separate offenses. These violations are independent violations, and therefore, each violation should be assessed a separate civil penalty of up to the statutory maximum. However, as a matter of policy, multiple violations of the same GLP requirement in a *single* study will not be assessed as separate offense each time the specific requirement is violated in that study. Rather, multiple violations of the same requirement in a single study will be considered as a single offense which may be raised from a low level to a middle level, or a middle level to a high level GLP violation, depending on the significance and frequency of the violation in a single study (see Appendix GLP-B-2). The Agency has taken this approach for this ERP because many of the GLP requirements which require repetitious compliance throughout the life of the study (such as failing to initial data entries (40 CFR 160.130(e)), can occur unchecked in a single study for an undefined period of time, and penalties for violations of a single repetitive requirement could accumulate to inappropriate or unrealistic levels.

LEVELS OF ACTION
_____

The levels of enforcement action for violations of the FIFRA GLPs include notices of warning, administrative civil penalties, and criminal proceedings. Additionally, in accordance with the July 2, 1990 FIFRA ERP, press releases should be issued in conjunction with most enforcement responses (except notices of warning).

Press Releases/Advisories, Etc.
_____

Regions may, at their discretion, issue a press release/advisory to notify the public of a person's violation of FIFRA. However, the issuance of press release/advisory must not be an item of negotiation during settlement.

A press release/advisory can be a useful tool to notify the public of a person's noncompliance with FIFRA and to educate the public on the requirements of FIFRA.

GOOD LABORATORY PRACTICE STANDARDS

Notices of Warning (NOW)

Notices of Warning (NOW) are the appropriate enforcement response in the following circumstances:

o   First-time violations by an independent pesticide testing facility which has been contracted by a study sponsor to conduct testing which falls under the scope of the GLPs.[3] Under FIFRA section 14, any person not listed in section 14(a)(1), who is not a "for-hire applicator", may only receive a civil penalty subsequent to receiving a written notice of warning from the Administrator (see page 4 of the July 2, 1990 FIFRA ERP). Pesticide testing facilities are not included under FIFRA section 14(a)(1) and, therefore, fall under the category of persons under FIFRA section 14(a)(2) who must receive an NOW prior to being assessed a civil penalty for violations of FIFRA.

o   First-time "low level" GLP violations assessed as violations of FIFRA section 12(a)(2)(Q) (see Appendix GLP-B).

---

[3]   Pesticide testing facilities which are owned or operated by the registrant may be charged a civil penalty of up to $5,000 per offense for the first violation of the GLPs under FIFRA section 14(a)(1).

Notices of Warning

FIFRA sections 14(a)(2), 14(a)(4), and 9(c)(3) provide EPA with the authority to respond to certain violations of FIFRA with a Notice of Warning to the violator.

Section 14(a)(2) Notices of Warning

Under section 14(a)(2) of FIFRA, a written warning for a violation of FIFRA must be issued to a private applicator or other person not covered by section 14(a)(1) prior to the assessment of a civil penalty. Applicators who apply a registered general use pesticide as a service in controlling pests but who do not deliver any unapplied pesticides ("for hire" applicators), are also included in section 14(a)(2) but are not subject to this limitation. A "for hire" applicator may be assessed a penalty up to $500 for the first offense.

Sections 9(c)(3) and 14(a)(4)

Section 14(a)(4) of FIFRA states that EPA may choose to issue a Notice of Warning in lieu of a civil penalty if EPA determines that the violation occurred despite the exercise of due care or the violation did not cause significant harm to health or the environment. Section 9(c)(3) also permits the EPA to issue a written Notice of Warning in lieu of instituting a proceeding for minor violations of FIFRA if the Administrator believes that the public interest will be adequately served through this course of action.

Generally, a violation will be considered minor, and a section 9(c)(3) notice of warning may be issued in lieu of a civil complaint if the total "gravity adjustment value", as determined from Appendix B of this ERP, is less than three (see the section

484

FIFRA

of this ERP entitled "Gravity of the Violation" and Appendix B, "Gravity Adjustment Criteria"). A Notice of Warning may also be appropriate for certain first-time record keeping violations as listed in Appendix A of this ERP (e.g., late section 7 reports).

-9-

o   GLP violations assessed as violations of FIFRA sections 12(a)(2)(B)(i), or 12(a)(2)(R) which are clerical or technical violations which either separately or collectively have a relatively minor impact on: (1) the reliability or scientific merits of the test data; (2) the Agency's ability to make a regulatory decision regarding a pesticide product's registration or other research or marketing permit; (3) the ability of the Agency to be able to validate the test results or reconstruct the study; and, (4) the EPA's administration of the GLP inspection and enforcement program (i.e., impairment of the Agency's inspection targeting ability or the efficiency of the GLP compliance inspections or data audits, etc.).[4]

o   Falsification of records required to be maintained in an ongoing study for which no final report has been submitted to the Administrator and for which no compliance statement under 40 CFR 160.12 has yet been signed.

Generally, a notice of warning will not be appropriate if the violator has previously violated the GLP regulations (criteria for establishing "compliance history" may be found in the July 2, 1990 FIFRA ERP, Appendix B, page B-3, footnote number 4) or the violator has received a notice of warning for a previous GLP related violation.

---

[4]   An example of a clerical or technical violation which has a relatively minor impact on the criteria listed above is a one-time failure to fulfill one of the GLP repetitive requirements, such as, failure to sign or initial, and date a data entry [160.130(e)]. Another example is a transcription error which can be verified or corrected by other means and which did not result in an erroneous conclusion for the overall study.

FIFRA

APPENDIX B FOOTNOTES

[1]   The gravity adjustment criteria in Appendix B should not be used for recordkeeping and reporting violations. Therefore, first-time civil penalties for recordkeeping or reporting violations should be assessed at the matrix value, while subsequent penalties should be increased by an increment of 30% (up to the statutory maximum).

[2]   For the purposes of this ERP, serious or widespread harm refers to actual or potential harm which does not meet the parameters of minor harm, as described below.

[3]   For the purposes of this ERP, minor harm refers to actual or potential harm which is, or would be of short duration, no lasting effects or permanent damage, effects are easily reversible, and harm does not, or would not result in significant monetary loss.

[4]   The following considerations apply when evaluating compliance history for the purposes of Appendix B:

    (a)   In order to constitute a prior violation, the prior violation must have resulted in: (1) a final order, either as a result of an uncontested complaint, or as a result of a contested complaint which is finally resolved against the violator; (2) a consent order, resolving a contested or uncontested complaint by the execution of a consent agreement; (3) the payment of a civil penalty by the alleged violator in response to the complaint, whether or not the violator admits to the allegations of the complaint; or (4) conviction under the FIFRA's criminal provisions.

A notice of warning (NOW) will not be considered a prior violation for the purposes of the gravity adjustment criteria, since no opportunity has been given to contest the notice. Additionally, a stop sale, use, or removal order (SSURO) issued under FIFRA section 13 will not be considered as compliance history.

(b)   To be considered a compliance history for the purposes of Appendix B, the violation must have occurred within five years of the present violation. This five-year period begins on the date of a final order, consent order, or payment of a civil penalty.

(c)   Generally, companies with multiple establishments are considered as one when determining compliance history. If one establishment of a company commits a FIFRA violation, it counts as history when another establishment of the same company, anywhere in the country, commits another FIFRA violation.

5   EPA enforcement officials are not required to determine culpability at the time the complaint is issued (especially if this information is not readily available). EPA enforcement officials may instead assign a weighting factor of 2 (culpability unknown), at the time of the issuance of the complaint. Culpability adjustments may be reconsidered during settlement negotiations.

6   The Agency may also consider criminal proceedings for "knowing and willful" violations. See the "Criminal Proceedings" section ERP.

Civil Administrative Penalties

Civil penalties assessed for violations of the FIFRA GLPs are to be calculated according to the procedures and matrices provided in the July 2, 1990 FIFRA Enforcement Response Policy. The gravity levels established for each violation on the FIFRA GLPs are listed in Appendix GLP-A of this ERP and in Appendix A of the July 2, 1990 FIFRA ERP.

Civil penalties may be assessed against both the study sponsor and the pesticide testing facility. A registrant (usually the study sponsor), or pesticide testing facility owned by a registrant, will be assessed civil penalties of up to $5,000 per offense under FIFRA section 14(a)(1). An independent pesticide testing facility who was contracted by the study sponsor will be assessed civil penalties of up to $1,000 per offense, subsequent to a written notice of warning, under section 14(a)(2).

-10-

A civil administrative penalty will be issued for all violations of the FIFRA GLPs which do not qualify for a notice of warning and have an impact on: (1) the reliability or scientific merits of the test data; (2) the Agency's ability to make a regulatory decision regarding a pesticide product's registration or other research or marketing permit; (3) the ability of the Agency to be able to validate the test results, i.e., reconstruct study, verify results; or, (4) the EPA's administration of the GLP inspection and enforcement program (i.e., targeting of inspections, delay in an Agency's inspection activities or data audit because standard GLP procedures or formats are not followed, etc.). In addition, a civil administrative penalty will be issued for repeat violations of the GLP regulations.

Most violations of FIFRA GLP Standards are considered as recordkeeping or reporting violations. As noted in the July 2, 1990 FIFRA ERP, the gravity of recordkeeping and reporting violations are already considered in the dollar amounts presented in the FIFRA civil penalty matrices. Further, recordkeeping and reporting violations do not lend themselves to utilizing the gravity adjustments listed in Appendix B of the July 2, 1990 FIFRA ERP. Therefore, first-time civil penalties are to be assessed at the matrix value, while subsequent civil penalties should be increased by an increment of 30% (up to the statutory maximum). Please note, repeat violations of the identical GLP requirement may indicate a "knowing or willful" violation, and therefore, the need to pursue criminal proceedings.

Use of the FIFRA Civil Penalty Matrix

    The gravity of the violation and the size of the business are considered in the FIFRA Civil Penalty Matrices shown in Table 1. Each cell of the matrix represents the Agency's assessment of the appropriate civil penalty, within the statutory maximum, for each gravity level of a violation and for each size of business category. Since FIFRA imposes different statutory ceilings on the maximum civil penalty that may be assessed against persons listed in FIFRA section 14(a)(1) and persons listed in section 14(a)(2), this policy has separate penalty matrices for section 14(a)(1) violators and section 14(a)(2) violators.

    The section 14(a)(2) penalty matrix will only be used by the Agency for persons falling under FIFRA section 14(a)(2) who have previously been issued a notice of warning or civil complaint (FIFRA section 14(a)(2) states that private applicators are only subject to civil penalties subsequent to receiving a Notice of Warning or following a citation for a prior violation, and "for hire" applicators are only subject to a maximum $500 civil penalty for their first offense of FIFRA). The Agency has only included three levels in the section 14(a)(2) Civil Penalty Matrix, rather than the four levels provided in the section 14(a)(1) matrix. This is because the Agency does not believe that the lower base penalty figure that can be obtained from a "level 4" is appropriate for violations of the statute committed after the receipt of a notice of warning or civil complaint.

    When a civil penalty is the appropriate response for a first-time violation by a "for hire applicator" who violates any provision of FIFRA while holding or applying a registered general use pesticide or a registered unclassified pesticide, that civil penalty will be the statutory maximum of $500. Subsequent violations will be assessed using the FIFRA section 14(a)(2) civil penalty matrix below.

TABLE 1

CIVIL PENALTY MATRIX
FOR FIFRA SECTION 14(a)(1)

SIZE OF BUSINESS

LEVEL	I	II	III
level 1	5,000	5,000	5,000
level 2	5,000	4,000	3,000
level 3	4,000	3,000	2,000
level 4	3,000	2,000	1,000

CIVIL PENALTY MATRIX
FOR FIFRA SECTION 14(a)(2) *

SIZE OF BUSINESS

LEVEL	I	II	III
level 1	1,000	1,000	1,000
level 2	1,000	800	600
level 3	800	600	500

This 14(a)(2) matrix is only for use in determining civil penalties issued subsequent to a notice of warning or following a citation for a prior violation, or in the case of a "for hire" applicator using a registered general use pesticide, subsequent to the issuance of a civil penalty of $500.

FIFRA CHARGES AND GRAVITY LEVELS

FIFRA SECTION	FTTS CODE	VIOLATION	LEVEL
12(a)(1)(A)	1AA	Sold or distributed a pesticide NOT REGISTERED under section 3 or was CANCELLED or SUSPENDED, which was not authorized by the Administrator.	2
12(a)(1)(A)	1AB	Registrant, wholesaler, dealer, retailer, or other distributor ADVERTISED or otherwise "offered for sale," in any medium, a pesticide that was NOT REGISTERED under section 3 or was CANCELLED or SUSPENDED, other than in accordance with Agency policy.	2
12(a)(1)(B)	1BA	CLAIMS made for a pesticide as part of sale or distribution differed substantially from those accepted in connection with registration.	2
12(a)(1)(B)	1BB	Registrant, wholesaler, dealer, retailer, or other distributor ADVERTISED, or otherwise "offered for for sale" in any medium, a REGISTERED PESTICIDE product for an UNREGISTERED USE, other than in accordance with Agency policy.	2
12(a)(1)(C)	1CA	Sold or distributed a pesticide whose COMPOSITION DIFFERED from the composition represented in the registration.	2
12(a)(1)(D)	1DA	Sold or distributed a pesticide which has not been COLORED or DISCOLORED pursuant to section 25(c)(5).	2
12(a)(1)(E) 12(a)(1)(F) 2(q)(1)(A)	1EA	Sold or distributed a pesticide or device which is MISBRANDED in that the label has a statement, design, or graphic representation which is false or misleading.	2

FIFRA CHARGES AND GRAVITY LEVELS

FIFRA SECTION	FTTS CODE	VIOLATION	LEVEL
12(a)(1)(E) 12(a)(1)(F) 2(q)(1)(B)	1EB	Sold or distributed a pesticide or device which is MISBRANDED in that the pesticide is not contained in a package or other container or wrapping which conforms to the standards established pursuant to section 25(c)(3) (e.g., not contained in child-resistant packaging or safety containers).	2
12(a)(1)(E) 12(a)(1)(F) 2(q)(1)(C)	1EC	Sold or distributed a pesticide or device which is MISBRANDED in that it is an imitation of, or is offered for sale under the name of, another pesticide.	2
12(a)(1)(E) 12(a)(1)(F) 2(q)(1)(D)	1ED	Sold or distributed a pesticide or device which is MISBRANDED in that the label did not bear the registration number assigned under section 7.	4
12(a)(1)(E) 12(a)(1)(F) 2(q)(1)(E)	1EE	Sold or distributed a pesticide or device which is MISBRANDED in that any words, statements, or other information required by the Act were not prominently placed on the label in such a way as to make it readable or understandable.	3
12(a)(1)(E) 12(a)(1)(F) 2(q)(1)(F)	1EF	Sold or distributed a pesticide or device which is MISBRANDED in that the label did not contain directions for use necessary to make the product effective and to adequately protect health and the environment.	2
12(a)(1)(E) 12(a)(1)(F) 2(q)(1)(G)	1EG	Sold or distributed a pesticide or device which is MISBRANDED in that the label did not contain a warning or caution statement adequate to protect health and the environment.	2
12(a)(1)(E) 2(q)(1)(H)	1EH	Sold or distributed a non-registered pesticide intended for export which is MISBRANDED in that the label did not have a prominently displayed, "Not Registered for Use in the United States of America."	4
12(a)(1)(E) 2(q)(2)(A)	1EI	Sold or distributed a pesticide which is MISBRANDED in that the label did not bear an ingredient statement on the immediate container which is presented or displayed under customary conditions of purchase.	4
12(a)(1)(E) 2(q)(2)(B)	1EJ	Sold or distributed a pesticide which is MISBRANDED in that the labeling does not contain a statement of the use classification for which the product was registered.	3
12(a)(1)(E) 2(q)(2)(C)	1EK	Sold or distributed a pesticide which is MISBRANDED in that there is not a label affixed to the pesticide container, and to the outside wrapper of the retail package if the required information on the immediate container cannot be clearly read, a label bearing all of the following information: (i) the name and address of the producer, registrant, or person for whom produced; (ii) the name, brand, or trademark under which the pesticide is sold; (iii) the net weight or measure of the content; and, when required by	4

FIFRA CHARGES AND GRAVITY LEVELS

FIFRA SECTION	FTTS CODE	VIOLATION	LEVEL
		regulation, (iv) the registration number assigned to the pesticide and the use classification.	
12(a)(1)(E) 2(q)(2)(D)	1EL	Sold or distributed a pesticide which is MISBRANDED in that the pesticide is sold in quantities highly toxic to man and the label failed to bear a skull and crossbones, and the word "poison" prominently in red on a contrasting background color, and/or the label did not bear a statement of practical treatment.	1
12(a)(1)(E) 2(c)(1) - (3)	1EM	Sold or distributed a pesticide which is ADULTERATED in that: (i) the strength or purity falls below the professed standard of quality expressed on the labeling; (2) any substance has been substituted wholly or in part abstracted; or, (3) any valuable constituent of the pesticide has been wholly or in part abstracted.	2
12(a)(2)(A)	2AA	Person DETACHED, ALTERED, DEFACED, or DESTROYED, in whole or in part, any LABELING required under the Act.	2
12(a)(2)(B)(i)	2BA	Person refused to PREPARE, MAINTAIN, or SUBMIT any RECORDS required under sections 5, 7, 8, 11, or 19.	2
12(a)(2)(B)(ii)	2BB	Person refused to SUBMIT any REPORTS required by or under section 5, 6, 7, 8, 11 or 19.	2
12(a)(2)(B)(ii)	2BC	A registrant refused to submit REPORTS under section 6(a)(2) regarding UNREASONABLE ADVERSE EFFECTS of their pesticide.	1
12(a)(2)(B)(iii)	2BD	Person refused to allow entry, INSPECTION, copying of records, or sampling authorized by this Act.	2
12(a)(2)(C)	2CA	Person gave a GUARANTY or undertaking provided for in section 12(b) which was FALSE in any particular.	2
12(a)(2)(D)	2DA	Person used to their personal advantage or revealed to persons other than those authorized by the Act any INFORMATION acquired under the Act which is CONFIDENTIAL.	3
12(a)(2)(E)	2EA	Registrant, wholesaler, dealer, retailer, or other distributor ADVERTISED a RESTRICTED USE PESTICIDE without indicating that the product was restricted.	2
12(a)(2)(F)	2FA	Person DISTRIBUTED, SOLD, MADE AVAILABLE FOR USE, or USED a RESTRICTED USE PESTICIDE for a purpose other than in accordance with section 3(d) or regulations issued.	2

FIFRA CHARGES AND GRAVITY LEVELS

FIFRA SECTION	FTTS CODE	VIOLATION	LEVEL
12(a)(2)(F)	2FB	Person distributed, sold, or made available for use, or used, a RESTRICTED USE PESTICIDE without maintaining the RECORDS required by regulations (A Notice of Warning should be issued for first-time "partial" violations. Violations continuing subsequent to the issuance of a civil complaint are to result in a suspension - see "Denials, Suspensions, Modifications, or Revocations of Applicator Certifications" section of this ERP).	2
12(a)(2)(G)	2GA	Person USED a registered pesticide in a manner inconsistent with its labeling.	2
12(a)(2)(H)	2HA	Person USED a pesticide which was under an EXPERIMENTAL USE PERMIT contrary to the provisions of the permit.	2
12(a)(2)(I)	2IA	Person violated any order issued under section 13 (e.g., STOP SALE, USE OR REMOVAL ORDER or SEIZURE).	1
12(a)(2)(J)	2JA	Person violated any SUSPENSION ORDER issued under section 6.	1
12(a)(2)(J)	2JB	Person violated any SUSPENSION ORDER issued under section 3(c)(2)(B) or 4.	2
12(a)(2)(K)	2KA	Person violated any CANCELLATION ORDER issued under the Act on the grounds of UNREASONABLE ADVERSE EFFECTS.	1
12(a)(2)(K)	2KB	Person violated any CANCELLATION ORDER issued under the Act on grounds OTHER THAN UNREASONABLE ADVERSE EFFECTS.	2
12(a)(2)(K)	2KC	Person failed to submit a SECTION 6(g) NOTICE when required.	2
12(a)(2)(K)	2KD	Person submitted a NOTABLY LATE SECTION 6(g) NOTICE.	3
12(a)(2)(K)	2KE	Person submitted an INCOMPLETE or INCORRECT SECTION 6(g) NOTICE.	3
12(a)(2)(L) 7(a)	2LA	PRODUCED a pesticide or active ingredient subject to the Act in an UNREGISTERED ESTABLISHMENT.	2
12(a)(2)(L) 7(c)(1)	2LB	Producer FAILED TO SUBMIT, or submitted NOTABLY LATE, a REPORT to the Administrator, under SECTION 7, which indicates the types and amounts of pesticides or active ingredients which they are currently producing, which they	2

FIFRA CHARGES AND GRAVITY LEVELS

FIFRA SECTION	FTTS CODE	VIOLATION	LEVEL
		produced during the past year, and which they sold or distributed during the past year.	
12(a)(2)(L) 7(c)(1)	2LC	Producer submitted a LATE REPORT to the Administrator, under SECTION 7, which indicates the types and amounts of pesticides or active ingredients which they are currently producing, which they produced during the past year, and which they sold or distributed during the past year (civil complaint issued only if the producer does not respond to a Notice of Warning or there is a subsequent violation within a three year timeframe from the first violation).	4
12(a)(2)(L) 7(c)(1)	2LD	Producer submitted an INCOMPLETE SECTION 7 REPORT with MINOR OMISSIONS of the required information (civil complaint issued only if the producer does not respond to a Notice of Warning or there is a subsequent violation within a three year timeframe from the first violation).	3
12(a)(2)(L) 7(c)(1)	2LE	Producer submitted an INCOMPLETE or a FALSE SECTION 7 REPORT with MAJOR OMISSIONS or ERRORS of the required information.	2
12(a)(2)(L) 7(c)(2)	2LF	Upon request of the Administrator for the purposes of the issuance of section 13 Stop Sale Orders, a PRODUCER FAILED TO PROVIDE the names and addresses of the recipients of the pesticides produced in any of his registered establishments.	1
12(a)(2)(M)	2MA	Person KNOWINGLY FALSIFIED all or any part of an application for registration, application for an experimental use permit, any information submitted under section 7, any records required to be maintained by the Act, any report filed under the Act, or any information marked as confidential and submitted to the Administrator under any provision of the Act.	1
12(a)(2)(N)	2NA	A registrant, wholesaler, dealer, retailer, or other distributor FAILED TO FILE REPORTS required by the Act.	2
12(a)(2)(O)	2OA	Person ADDED A SUBSTANCE TO, or TOOK a substance from a pesticide in a manner that may defeat the purpose of this Act.	2
12(a)(2)(P)	2PA	Person USED a pesticide in TESTS ON HUMAN BEINGS in violation of the conditions specified by the Act.	1
12(a)(2)(Q)	2QA	Person FALSIFIED INFORMATION RELATING to the TESTING of any pesticide (or any of its ingredients,	1

FIFRA CHARGES AND GRAVITY LEVELS

FIFRA SECTION	FTTS CODE	VIOLATION	LEVEL
		metabolites, or degradation products) for which the person knows will be furnished to the Administrator, or will become a part of any records required to be maintained by this Act.	
12(a)(2)(Q)	2QB	Person falsely represented compliance with the FIFRA Good Laboratory Practice (GLP) regulations as a result of a HIGH LEVEL GLP violation.	2
12(a)(2)(Q)	2QC	Person falsely represented compliance with the FIFRA Good Laboratory Practice (GLP) regulations as a result of a MIDDLE LEVEL GLP violation.	3
12(a)(2)(Q)	2QD	14(a)(1) person falsely represented compliance with the FIFRA Good Laboratory Practice (GLP) regulations as a result of a LOW LEVEL GLP violation.	4
12(a)(2)(Q)	2QE	14(a)(2) person falsely represented compliance with the FIFRA Good Laboratory Practice (GLP) regulations as a result of a LOW LEVEL GLP violation.	3
12(a)(2)(R)	2RA	Person submitted DATA KNOWN TO BE FALSE in support of a registration.	1
12(a)(2)(S)	2SA	Person sold, distributed, or used an UNREGISTERED pesticide in violation of a REGULATION ISSUED UNDER SECTION 3(a).	*
12(a)(2)(S)	2SB	Person violated any REGULATION ISSUED UNDER SECTION 19.	

---

* Gravity levels for these violations will be assigned in subsequent ERPs.

APPENDIX B
GRAVITY ADJUSTMENT CRITERIA [1]

VIOLATION	VALUE	CIRCUMSTANCES
GRAVITY OF HARM		
Pesticide	2	Toxicity - Category I pesticides, Signal Word "Danger", restricted use pesticides (RUPs), pesticides with flammable or explosive characteristics (i.e., signal words "Extremely Flammable" or "Flammable"), or pesticides that are associated with chronic health effects (mutagenicity, oncogenicity, teratogenicity, etc.).
	1	Toxicity - Categories II through IV, signal word "Warning" and "Caution," no known chronic effects.
Harm to Human Health	5	Actual serious or widespread[2] harm to human health.
	3	Potential serious or widespread[2] harm to human health.
	3	Harm to human health is unknown.
	1	Minor[3] potential or actual harm to human health, neither serious nor widespread.
Environmental Harm	5	Actual serious or widespread[2] harm to the environment (e.g., crops, water, livestock, wildlife, wilderness, or other sensitive natural areas).
	3	Potential serious or widespread[2] harm to the environment.
	3	Harm to the environment is unknown.
	1	Minor[3] potential or actual harm to the environment, neither widespread nor substantial.

VIOLATION	VALUE	CIRCUMSTANCES
GRAVITY OF MISCONDUCT		
Compliance[4] History	5	If a violator is a 14(a)(1) person with more than one prior violation of FIFRA, and at least one prior violation was a level 1 violation. If a violator is a 14(a)(2) person with more than two prior FIFRA violations, and at least one prior violation was a level 1 violation.
	4	If a violator is a 14(a)(1) person with more than one prior violation of FIFRA, and no prior level 1 violations. If a violator is a 14(a)(2) person with more than two prior FIFRA violations, and no prior level 1 violations.
	2	If a 14(a)(1) person, one prior violation of FIFRA. If a 14(a)(2) person, two prior FIFRA violations.
	0	No prior FIFRA violations.
Culpability[5]	4	Knowing or willful violation of the statute.[6] Knowledge of the general hazardousness of the action.
	2	Culpability unknown.
	2	Violation resulting from negligence.
	0	Violation was neither knowing nor willful and did not result from negligence. Violator instituted steps to correct the violation immediately after discovery of the violation.

Criminal Proceedings

Criminal proceedings may be initiated against an applicant for registration, or study sponsor who is a registrant, applicant for registration, or pesticide producer, for knowing and willful violations of the FIFRA GLPs, under FIFRA section 14(b)(1)(A) for criminal penalties of up to $50,000 and/or imprisonment for up to one year. Any commercial applicator of a restricted use pesticide, or any person not described in FIFRA section 14(b)(1)(A) who distributes or sells pesticides or devices, who knowingly violates any provision of FIFRA shall be fined up to $25,000 and/or imprisoned for up to one year. The EPA may also pursue criminal proceedings against the pesticide testing facility, or other person for knowing and willful violations of the GLPs under FIFRA section 14(b)(2) for criminal penalties of up to $1,000 and/or imprisonment for up to 30 days. Additionally, criminal proceedings may be pursued against the registrant, pesticide testing facility, or other person for violating Title 18 of the U.S. Code.

-11-

REFERRALS

In addition to the levels of enforcement action listed in the previous section, the EPA may take regulatory action for violations of the GLPs, including: rejection of studies which do not comply with the FIFRA GLPs; cancellation, suspension, or modification of a pesticides research or marketing permit; or denial or disapproval of an application for such a permit. Further, in order to help assure that the Federal Government deals with responsible contractors, and for the purposes of the Federal Government's protection, pesticide testing facilities responsible for significant or major GLP violations may also be suspended or debarred from Government contracts or subcontracts. Until further guidance is issued, the Agency's regulatory response for violations of the GLPs will be addressed on a case-by-case basis and, therefore, will not be addressed in detail in this policy.

Office of Pesticide Programs

If the Agency discovers any significant or major GLP violations or data concerns in the course of a facility inspection or study audit, the Office of Compliance Monitoring will notify the Office of Pesticide Programs so that Office may consider if any regulatory action would be appropriate. These regulatory actions include: (1) rejection of studies which do not comply with the FIFRA GLPs; (2) cancellation, suspension, or modification of a pesticide's research or marketing permit; or, (3) denial or disapproval of an application for such a permit.

Pursuit of an enforcement action by the EPA, such as the issuance of a civil or criminal complaint, does not obligate the Agency to pursue a regulatory response, such as study rejection or cancellation/suspension of a pesticides research or marketing permit. Similarly, a regulatory response by the Agency does not obligate the Agency to pursue an enforcement action. The EPA's decision to pursue an enforcement response and/or regulatory response to a GLP violation will, by administrative necessity, occur on different tracks and will be based on the individual merits of each approach on a case-by-case basis.

-12-

Office of Administration
_____

Pesticide testing facilities responsible for significant or major GLP violations may be suspended or debarred from Government contracts, subcontracts, and assistance loan and benefit programs. This action is not for the punishment of the violator nor is it an enforcement tool, but rather it is for the protection of the Federal Government by assuring that the Government will be dealing with responsible contractors.

The Office of Compliance Monitoring will notify the Compliance Branch of the Grants and Administration Division, Office of Administration of the identity of pesticide testing facilities which are responsible for a significant or major GLP violation and have been assessed a civil penalty through a final order or when there is evidence of a criminal offense for violations of the FIFRA GLPs, so that Office may decide whether they wish to pursue suspension or debarment proceedings in accordance with the Federal Acquisition Regulations (FAR) at 48 CFR Subpart 9.4, and the EPA Suspension and Debarment regulations found at 40 CFR Part 32.

APPENDIX GLP - A

FIFRA CHARGES AND GRAVITY LEVELS
FOR CIVIL PENALTIES ASSESSED FOR VIOLATIONS OF THE
FIFRA GOOD LABORATORY PRACTICE STANDARDS

FIFRA SECTION	FTTS CODE	VIOLATION	LEVEL
12(a)(2)(B)(i)	2BA	Person refused to PREPARE, MAINTAIN, or SUBMIT any RECORDS required under sections 5, 7, 8, 11 or 19.	2
12(a)(2)(M)	2MA	Person KNOWINGLY FALSIFIED all or any part of an application for registration, application for an experimental use permit, any information submitted under section 7, any records required to be maintained by the Act, any report filed under the Act, or any information marked as confidential and submitted to the Administrator under any provision of the Act.	1
12(a)(2)(Q)	2QA	Person FALSIFIED INFORMATION RELATING to the TESTING of any pesticide (or any of its ingredients, metabolites, or degradation products) for which the person knows will be furnished to the Administrator, or will become a part of any records required to be maintained by this Act.	1

FIFRA SECTION	FTTS CODE	VIOLATION	LEVEL
12(a)(2)(Q)	2QB	Person falsely represented compliance with the FIFRA Good Laboratory Practice (GLP) regulations as a result of a HIGH LEVEL GLP* violation.	2
12(a)(2)(Q)	2QC	Person falsely represented compliance with the FIFRA Good Laboratory Practice (GLP) regulations as a result of a MIDDLE LEVEL GLP* violation.	3
12(a)(2)(Q)	2QD	14(a)(1) person falsely represented compliance with the FIFRA Good Laboratory Practice (GLP) regulations as a result of a LOW LEVEL GLP* violation.	4
12(a)(2)(Q)	2QE	14(a)(2) person falsely represented compliance with the FIFRA Good Laboratory Practice (GLP) regulations as a result of a LOW LEVEL GLP* violation.	3**
12(a)(2)(R)	2RA	Person submitted DATA KNOWN TO BE FALSE in support of a registration.	1

---

*   Guidance on the parameters for determining whether a GLP violation assessed as an unlawful act under FIFRA section 12(a)(2)(Q) is a HIGH, MIDDLE, or LOW LEVEL violation is found in Appendix GLP-B.

** A higher level has been assigned for FIFRA section 14(a)(2) persons because a civil penalty which is assessed under this provision represents the second violation of that person. Violators who fall under the category of persons listed in FIFRA section 14(a)(2) must receive a Notice of Warning for the first GLP violation.

APPENDIX GLP - B

GUIDANCE FOR DETERMINING WHETHER TO ASSESS A GLP VIOLATION
UNDER FIFRA SECTION 12(a)(2)(Q) AS A
HIGH, MIDDLE, OR LOW LEVEL VIOLATION

When assessing a civil penalty for violations of FIFRA section 12(a)(2)(Q) for submission of a false compliance statement (FTTS codes 2QB, 2QC, 2QD, and 2QE, as listed in Appendix GLP-A of this ERP and Appendix A of the July 2, 1990 FIFRA ERP), the parameters listed in this appendix will be used to determine whether a GLP violation is a "high", "middle", or "low level" GLP violation. Because of the expertise that is necessary to make an assessment of the impact of a GLP violation, the determination of whether a violation will be considered as a high, middle, or low level will be made at EPA Headquarters by the Office of Compliance Monitoring with input from the Office of Pesticide Programs and the Office of Enforcement based on the criteria listed in this appendix. A brief summary of the rationale for the categorization of the impact of the GLP violation should be included as part of the civil complaint sent to the respondent.

High Level GLP Violations

A "high level" violation of the FIFRA GLPs involves a substantial failure to comply with the regulations. "High level" GLP violations will have a substantial impact on: (1) the reliability or scientific merits of the test data; (2) the Agency's ability to make a regulatory decision regarding a pesticide product's registration or other research or marketing permit; (3) the ability of the Agency to be able to validate the test results, i.e., reconstruct the study, verify results; or (4) the EPA's administration of the GLP inspection and enforcement program (i.e., impairment of the Agency's inspection targeting ability or the efficiency of the GLP compliance inspections or data audits, etc.). Violations which will be considered as "high level" based on the above criteria include, but are not limited to:

1) Failure to notify a person performing under contract of the applicability of GLPs - §160.10.
2) Failure to keep personnel records - §160.29.
3) Falsification of personnel records - §160.29.
4) Failure to designate a study director - §160.31.
5) Failure to assure the existence of a Quality Assurance Unit (QAU) - §160.31.
6) Failure of the QAU to conduct any inspections or maintain any records - §160.31.
7) Failure to maintain Standard Operating Procedures - §160.81.
8) Failure to follow laboratory SOPs without documentation in the raw data and/or written authorization from management - §160.81(a).
9) Failure to isolate all newly received animals from outside sources until their health status has been evaluated - §160.90(b).
10) Failure to characterize the test, control, or reference substances - §160.105.
11) Failure to have a protocol - §160.120.
12) Deviation from the protocol without documentation and/or study director written signoff - §160.120(b).
13) Failure to record raw data - §160.130.
14) Falsification of raw data - §160.130.
15) Failure to retain raw data and specimens - §160.51, §160.190, and §160.195.

Partial compliance with a "high level" GLP violation, such as the examples listed above, may justify considering the violations as "middle level" GLP violations.

Middle Level GLP Violations

All violations of GLPs which are not considered "high level" violations are considered to be "middle level" GLP violations UNLESS the violation is determined to be a "low level" violation (see the subsequent section for the criteria for determining a "low level" violation).

Partial compliance with any of the violations which could be considered as a "high level" GLP violation (such as the violations listed in the "High Level GLP Violation" section above), may qualify for consideration as a "middle level" GLP violation. For example, a pesticide testing facility may be charged a "middle level" GLP violation rather than a "high level" GLP violation for failure to maintain personnel records, provided that the personnel records for that facility are mostly complete.

Low Level GLP Violations

A GLP violation will be considered "low level" in cases where the violative act was purely clerical or technical in nature with relatively minor impacts on: (1) the reliability or scientific merits of the test data; (2) the Agency's ability to make a regulatory decision regarding a pesticide product's registration or other research or marketing permit; (3) the ability of the Agency to be able to validate the test results, i.e., reconstruct the study, verify results; or, (4) the EPA's administration of the GLP inspection and enforcement program (i.e., targeting of inspections, delay in the Agency's inspection or data audit activities because standard GLP procedures or formats are not followed, etc.). "Low level" violations are appropriate where there is a general program in place by a violator to comply with the GLPs, but instances of noncompliance occur anyway due to apparent inadvertent error, equipment failures, or other similar occurrences. Violations which will be considered as "low level" based on the above criteria include, but are not limited to:

1) A facility generally maintained a current summary of training and experience and job description for each individual engaged in the study but failed to do so for one or two individuals - §160.29(b).

2) A facility generally maintained records which documented equipment inspection, maintenance, testing, calibrating, and/or standardizing operations, but failed to document these operations in one instance - §160.63(c).

3) A facility maintained all revisions to the standard operating procedures (SOPs) but failed to note the date of the revision in one or two instances - §160.81(d).

4) A reagent or solution which is documented not to degrade does not have an expiration date on the label - §160.83.

5) One-time failure to fulfill one of the GLPs repetitive requirements, such as failure to sign or initial, and date a data entry - §160.130(e).

6) A clerical error or transposition of numbers in the final study report which can be verified or corrected by other means and which did not result in an erroneous conclusion for the overall study - §160.185 or §160.195.

If the EPA believes that justice may be served, EPA may issue a Notice of Warning under FIFRA section 9(c)(3) for first-time "low level" GLP violations. However, subsequent "low level" violations will result in the assessment of a level 4 civil penalty for a FIFRA section 14(a)(1) violator and a level 3 civil penalty for a FIFRA section 14(a)(2) violator.

Appendix D: Representative Forms Used by Companies for Compliance with GLP Standards

Table of Contents

Title	Page
On-the-job training form for the analytical laboratory (Reproduced with permission from EPL Bio-Analytical Services, Inc.)	504–506
Training form for personnel engaged in terrestrial field studies (Reproduced with permission from Wildlife International, Ltd.)	507–508
Form for documentation of personnel training (Reproduced with permission from Colorado Animal Research Enterprises, Inc.)	509
Format for the master schedule (Reproduced with permission from Pharmacology and Toxicology Research Laboratory)	510
Facility audit checklist form (Reproduced with permission from Environmental Technologies Institute, Inc.)	511–512
Facility inspection checklist form (Reproduced with permission from Quality Associates Incorporated)	513
QAU facility SOP checklist form (Reproduced with permission from Environmental Technologies Institute, Inc.)	514
Protocol evaluation checklist form (Reproduced with permission from Quality Associates Incorporated)	515
Study inspection checklist form (Reproduced with permission from Merck and Company, Inc.)	516
Raw data file/field trial logbook audit checklist form (Reproduced with permission from Environmental Technologies Institute, Inc.)	517–518
Critical phase inspection checklist form (Reproduced with permission from Quality Associates Incorporated)	519
QAU inspection schedule form (Reproduced with permission from Ciba-Geigy Corporation)	520
Analytical laboratory critical phase inspection checklist/report form (Reproduced with permission from McKenzie Laboratories, Inc.)	521
Record of QA activities (Reproduced with permission from Environmental Technologies Institute, Inc.)	522
SOP review form (Reproduced with permission from Ciba-Geigy Corporation)	523

2192–8/92/0501$14.50/0 © 1992 American Chemical Society

Title	Page
SOP retirement form (Reproduced with permission from Ciba-Geigy Corporation)	524
Standard procedure for explaining notebook entry errors (Reproduced with permission from Analytical Development Corporation)	525
Form for crop and pesticide history (Reproduced with permission from Valent U.S.A. Corporation)	526
Form for documenting field and crop characteristics (Reproduced with permission from Valent U.S.A. Corporation)	527
Maintenance log forms for spray booms, spray rigs, and aerial application equipment (Reproduced with permission from Agricultural Chemicals Development Services, Inc.)	528–531
Form for field treatment preparation (Reproduced with permission from California Agricultural Research, Inc.)	532
Form for sprayer calibration (Reproduced with permission from California Agricultural Research, Inc.)	533
Form for sprayer calibration data (Reproduced with permission from Valent U.S.A. Corporation)	534
Residue field sample history sheet (Reproduced with permission from Pan-Agricultural Laboratories, Inc.)	535
Processed residue sample history sheet (Reproduced with permission from Pan-Agricultural Laboratories, Inc.)	536
Sample arrival check sheet (Reproduced with permission from IR–4 National Pesticide Clearance Research Program)	537
Sample receipt form (Reproduced with permission from Merck and Company, Inc.)	538
Sample movement and tracking form (Reproduced with permission from EN–CAS Analytical Laboratories)	539
Freezer storage log for sample storage (Reproduced with permission from California Agricultural Research, Inc.)	540
Chemical check-in form (Reproduced with permission from Landis Associates, Inc.)	541
Receipt record and inventory log for test chemicals and analytical reference standards (Reproduced with permission from Pharmacology and Toxicology Research Laboratory)	542
Receipt record and inventory log of radioactive substances (Reproduced with permission from Pharmacology and Toxicology Research Laboratory)	543
Balance calibration log (Reproduced with permission from Agrisearch Incorporated)	544

Title	Page
Balance calibration record (Reproduced with permission from Colorado Animal Research Enterprises, Inc.)	545
Standardization record for pH meter (Reproduced with permission from FMC Corporation)	546
Record for thermometer calibration (Reproduced with permission from Colorado Animal Research Enterprises, Inc.)	547
Centrifuge temperature check log (Reproduced with permission from Colorado Animal Research Enterprises, Inc.)	548
Form for inspection of the deionized water system (Reproduced with permission from Colorado Animal Research Enterprises, Inc.)	549
Form for the gas chromatograph maintenance log (Reproduced with permission from Analytical Development Corporation)	550
Form for the high performance liquid chromatograph maintenance log (Reproduced with permission from Analytical Development Corporation)	551
Form for documenting telephone calls (Reproduced with permission from Rhone-Poulenc Ag Company)	552
Archive index forms (Reproduced with permission from Blasland, Bouck, and Lee)	553
Form for requesting archived material (Reproduced with permission from BASF Corporation)	554

TECHNICAL DOCUMENTATION

NAME:_____ DATE OF EMPLOYMENT:_____

TECHNICAL FUNCTIONS

I. RADIATION LABORATORY

	DATE	AUTHORIZED BY	COMMENTS
RADIATION SAFETY			
DISPOSAL OF RADIOACTIVE WASTE			
LIQUID SCINTILLATION COUNTER			
COMBUSTION/OXIDIZER(S)			
THIN LAYER CHROM. W/ISOTOPES AND BETA RAM OPERATION			
DECONTAMINATION PROCEDURES			

II. SAMPLE PREPARATION

SAMPLE RECEIPT PROCEDURE			
SAMPLE IDENTIFICATION			
WILEY MILL OPERATION			
SOIL GRINDER OPERATION			
PLANT GRINDER OPERATION			
EXPULSION GRINDER OPERATION			
FREEZE DRYING PROCEDURES			
FOOD PROCESSOR(S) OPERATION			

III. SAMPLE EXTRACTION
         (Extraction and Separation Procedures)

SOXHET PROCEDURES/OPERATION			
ROTOVAP PROCEDURES/OPERATION			
DIGESTION PROCEDURES/OPERATION			
CARBON DIOXIDE INCUBATOR PROCEDURES/OPERATION			
BLENDERS AND HOMOGENIZER OPERATION			
POLYTRON OPERATION			

IV. CHROMATOGRAPHY

	DATE	AUTHORIZED BY	COMMENTS
COLUMN CHROMATOGRAPHY			
THIN LAYER CHROMATOGRAPHY			
GAS CHROMATOGRAPHY			
HIGH PERFORMANCE LIQUID CHROMATOGRAPHY			
GPC; GPC AUTOPREP			
USE OF DATA RECORDS			

V. SPECTROPHOTOMETRY

AA PROCEDURES/OPERATIONS			
UV/VIS PROCEDURES/OPERATIONS			
INFRARED PROCEDURES/OPERATIONS			

VI. DATA HANDLING AND PROCESSING

WEIGHING SYSTEM PROCEDURES			
VG SYSTEM OPERATION			

USE OF COMPUTER SOFTWARE:
  1. MICROCOMPUTER PROGRAMS
  2. MINICOMPUTER PROGRAMS			

ENTERING OF DATA AND FORMULA USED TO CALCULATE RESULTS			

VII. QUALITY ASSURANCE

QUALITY ASSURANCE ORIENTATION			
STANDARD OPERATING PROCEDURES ORIENTATION			
DATA PACKAGE PREPARATION			

VIII. WET CHEMISTRY

VOLUMETRIC GLASSWARE AND PIPETTING (MANUAL & PIPETTORS)			

USE OF LABORATORY EQUIPMENT
  1. SHAKERS
  2. pH METERS
  3. OVENS
  4. CENTRIFUGES/ ULTRACENTRIFUGES
  5. VACUUM PUMPS
  6. HOT PLATES/STIRRERS
  7. WATER BATHS			

HOOD USE			
ENVIRONMENTAL CHAMBER OPERATION			

IX. METHOD DEVELOPMENT/VALIDATION

	DATE	AUTHORIZED BY	COMMENTS
APPROACH AND TECHNIQUE			
REPORT PREPARATION			
REQUIREMENTS - EPA PHILOSOPHY			

X. GENERAL LABORATORY

INTRODUCTION TO COMPANY			
SAFETY TRAINING PROGRAM			
SECURITY/CONFIDENTIALITY ORIENTATION			
WEIGHING/PIPETTING PROCEDURES			
WATER PREPARATION PROCEDURES			
EQUIPMENT MAINTENANCE PROCEDURES			
ORDERING/RECEIVING SUPPLIES			
SOIL/PLANT DESTRUCTION			
FREEZERS - USE AND OPERATION			
DISPOSAL OF LABORATORY WASTE			
REAGENT STORAGE			
WASHING OF LABORATORY GLASSWARE			
FIRE EXTINGUISHER TRAINING			
FREEZER OPERATION TRAINING			
GAS CYLINDER PROCEDURES			

XI. ADMINISTRATIVE PROCEDURES

WORD PROCESSING			
CLIENT COMMUNICATION			
SERVICE ORIENTATION			
PURCHASE/ORDERING SYSTEM			

Reproduced with permission from EPL Bio-Analytical Services, Inc.

WILDLIFE INTERNATIONAL, LTD.

TRAINING LOG FOR PERSONNEL IN TERRESTRIAL FIELD STUDIES

Name of Employee:_____

Page 1 of 2

	TECHNICIAN		APPROVED TO PERFORM		APPROVED TO INSTRUCT	
	Inits.	Date	Super's Inits.	Date	Super's Inits.	Date
QUALITY ASSURANCE Reviewed and discussed QA function						
RECORDING DATA Read SOP # 120						
Record data correctly						
Proper corrections to data						
Use codes for changes to data						
PESTICIDE SAFETY Read SOP # 175 and safety manual						
USE OF PESOLA SCALE Read SOP # 705						
Calibrates properly						
WEATHER MONITORING Read SOP # 706						
RECORD WILDLIFE FIELD OBSERVATIONS Read SOP # 701						
CASUALTY SEARCHING Read SOP # 707						
CARCASS DETECTABILITY TRIAL Read SOPs # 708 and 702						
ESTIMATING NUMBER OF BIRDS Read SOP # 709						
Viewed slide presentation on bird identification						
Reviewed tape recordings of common bird songs						
Practiced distance estimation						
Practiced circular plot technique						
Practiced transect technique						
Practiced scan sampling technique						
Practiced marked bird search technique						
Practiced call count technique						
Practiced roadside count technigue						
LIVE TRAPPING BIRDS Read SOP # 710						
Practiced trapping technique(s)						
Practiced banding techniques						
Practiced radio tagging techniques						

WILDLIFE INTERNATIONAL, LTD.

TRAINING LOG FOR PERSONNEL IN TERRESTRIAL FIELD STUDIES

Name of Employee:_____

Page 2 of 2

	TECHNICIAN		APPROVED TO PERFORM		APPROVED TO INSTRUCT	
	Inits.	Date	Super's Inits.	Date	Super's Inits.	Date
RADIO TELEMETRY TRACKING OF ANIMALS						
Read SOP #711						
Practiced homing techniques						
Practiced triangulation techniques						
NEST MONITORING						
Read SOP # 712						
Practiced using ligatures						
SMALL MAMMAL LIVE TRAPPING						
Read SOP # 713						
Reviewed identification of mammals						
RESIDUE SAMPLE COLLECTION						
Read SOPs # 714 and 702						
Viewed field demonstration technique						
Vegetation						
Soils						
Invertebrates						
Water						
Small mammals						
Reptiles and amphibians						
Fish						
Aquatic invertebrates						
CALIBRATION OF APPLICATION EQUIPMENT						
Read SOP # 716						
Assisted in actual calibration						
DOCUMENTATION OF TEST COMPOUND APPLICATION						
Read SOP # 717						
COLLECTING AND PREPARING EGGS FOR SHIPMENT						
Read SOP # 718						
VEGETATION CHARACTERIZATION						
Read SOP # 719						

MISCELLANEOUS

Reproduced with permission from Wildlife International, Ltd.

DOCUMENTATION OF PERSONNEL TRAINING

Person Trained: _____

Date Training Received	Training was Provided by	New Methods Learned (SOP's may be referenced)	Additional Training or Supervision Needed? Training Yes	No	Supervision Yes	No

Please provide this completed form to:

Beverly George
_____

|___| Consolidated to permanent CARE training file by: _____ _____
Date

Reproduced with permission from Colorado Animal Research Enterprises, Inc.

PTRL MASTER SCHEDULE: November 1, 1989

TEST ARTICLE	PROJECT NUMBER	SPONSOR	REG. AGENCY	NATURE OF STUDY	TEST SYSTEM	STUDY INIT.	STUDY COMPLET.	STUDY DIR.	CURR. STATUS

Reproduced with permission from Pharmacology and Toxicology Research Laboratory.

ETI QUALITY ASSURANCE UNIT
FACILITY AUDIT CHECKLIST

Facilty Inspected:

Location:

Inspector:

Inspection Date:

	GLP/SOP compliance		
GENERAL	YES	NO	COMMENTS
1. Master schedule for facility			
2. Fireproof archive.			
3. Trial RDFN on-site			
4. Field trial logbook(s) in place.			
5. Trial protocol(s) on-site.			
6. Personnel CVs.			
7. Personnel training records			
8. Quality Assurance Unit properly established. .			

TEST SUBSTANCE STORAGE, HANDLING, USAGE

9. Facility chemical log.			
10. Separate storage building.			
11. Storage temperature recorded			
12. Storage building fireproof			
13. Storage building properly labeled.			
14. Proper weighing/measuring/mixing equipment . .			
15. Application equipment stored separately. . . .			
16. Application equipment SOPs			
17. Application equipment calibration records. . .			
18. Appropriate data recording forms			
19. Weighing/mixing equipment logs			

SAMPLE COLLECTION

20. Sample collection SOPs			
21. Proper equipment available			
22. Equipment maintenance log.			
23. Separate sample storage area			
24. Sample weighing equipment calibration log. . .			
25. Freezer storage isolated			
26. Freezer temperature recorded			
27. Freezer maintenance log.			
28. Proper packaging/shipping materials.			
29. Chain of custody forms/SOPs.			

PERSONNEL INTERVIEWED:

_____

_____

_____

COMMENTS

Inspector: _____   Date: _____

Reproduced with permission from Environmental Technologies Institute, Inc.

QUALITY ASSOCIATES INCORPORATED
F A C I L I T Y   I N S P E C T I O N   C H E C K L I S T

Dates of Inspection:_____   QAI Inspector:_____

Name of Facility:_____   Laboratory Management:_____

REQUIREMENTS	YES	NO	REQUIREMENTS	YES	NO
ORGANIZATION AND PERSONNEL			**LABORATORY FACILITIES**		
Personnel trained/experienced			LAB OPERATIONS AREA		
Records of training/experience maintained			Separate work areas are provided for:		
FARM FACILITIES			▢ routine procedures		
ANIMAL CARE			▢ maintenance of equipment & supplies		
Facility has animal rooms/areas to assure:			Test systems appropriately identified		
▢ separation of different species and test systems			Lab areas have SOPs or references for procedures		
▢ isolation of different projects			**EQUIPMENT: MAINT & CALB**		
▢ quarantine of animals			All equipment is/was:		
▢ routine/specialized housing			▢ inspected, cleaned, maintained		
▢ isolation of studies ⁕/ biohazardous, volatile aerosols, radioactive or infectious agents			▢ tested, calibrated/standardized during generation of data for a study		
▢ diagnostic treatment & control of experimental animal diseases			▢ detailed SOPs for all equipment used in data generation		
Facilities exist for collection/disposal/ storage of animal wastes			▢ written records for all equipment		
SOP for animal care and ID			▢ designated responsible person		
ANIMAL SUPPLY			**TEST SUBSTANCE HANDLING**		
Specified separated storage areas			Storage containers labeled		
Protection against infestation/ contamination			Reserve samples of each batch of test/ref./cont. subst. retained		
Refrigeration (if necessary)			Stored as per SOP or Protocol		
LABORATORY FACILITIES			Distributed prevent contamination		
TEST SUBSTANCE HANDLING			Proper ID maintained for duration		
There are separate areas for:			Recpt/Distrib. records maintained.		
▢ receipt and storage			Exp. Date on each container		
▢ mixing substances with a carrier			**SAFETY & HEALTH**		
▢ storage of mixture			Laboratory has/maintains:		
Storage area for test/reference/control substances and mixtures are separate			▢ shower, eyewash, toilets, separate eating areas		
Test/reference/control/mixture storage is separate from test system housing			▢ washing facilities, lockers, provides work clothing		
Reagents/Solutions are labeled for ID, storage conditions, expiration dates			▢ fire extinguishers, central fire suppression or alarm system		
Storage areas sufficient/clean			▢ chemical spill kits		
Storage areas are secured/locked			▢ appropriate equipment		
Radioactive material so labeled			▢ SOPs for above procedures		
Materials disposed of properly			Radiation license compliance		

Reproduced with permission from Quality Associates Inc.

ETI QUALITY ASSURANCE UNIT
FACILITY SOP CHECKLIST

Facilty Inspected:

Location:

Inspector:	Inspection Date:

	SOP on file		
	YES	NO	COMMENTS
1. Master schedule.			
2. Facility archives			
3. Quality Assurance			
4. Test system (crop) preparation			
5. Test system (crop) maintenance/care.			
6. Receipt & storage of test substance.			
7. Handling & use of test substance			
8. Chemical log			
9. Test system (crop) observations.			
10. Harvest & crop destruct.			
11. SOPs			
12. Crop sampling.			
13. Soil sampling.			
14. Sample handling/compositing.			
15. Sample storage			
16. Weighing equipment/calibration			
17. Data recording, storage and retrieval.			
18. Maintenance & calibration logs			
19. Sample labeling/identification			
20. Application equipment calibration.			
21. Test substance mixing.			
22. Test substance application			
23. Additional SOPs (as necessary)			

_____ . .

_____ . .

_____ . .

_____ . .

_____ . .

_____ . .

_____ . .

Reproduced with permission from Environmental Technologies Institute, Inc.

QUALITY ASSOCIATES INCORPORATED
PROTOCOL EVALUATION CHECKLIST

Project Number: _____   Date of Protocol Evaluation: _____

Study Director: _____   QAI Inspector: _____

REFERENCE: EPA GLP; Section 160.120 (FIFRA), 792.185 (TSCA)			
SECTION	YES	NO	Protocol for the above listed study contains the following:
a. (1)			1. Descriptive title and statement of purpose of the study.
a. (2)			2. Identification of the test, reference and control substances by name, chemical abstracts number (CAS) or code number.
a. (3)			3. Name and address of the Sponsor.
a. (4)			4. Name and address of the testing facility/laboratory.
a. (5)			5. Proposed experimental start and termination dates.
a. (6)			6. Justification for selection of the test system.
a. (7)			7. Number, body weight range, sex, source, species, strain, substrain and age of the test system, where applicable.
a. (8)			8. Procedure for identification of the test system.
a. (9)			9. Description of the experimental design, including the methods for the control of bias.
a. (10)			10. Description or identification of the diet and any material used to solubilize or suspend the test, reference or control substance mixed with a carrier.
a. (11)			11. Specifications for acceptable levels of contaminants that are reasonably expected to be present in the dietary material and known to be capable of interfering with the purpose or conduct of the study, or a statement that states there are no such contaminants, where applicable.
a. (12)			12. Route of administration and the reason for its choice.
a. (13)			13. Each dosage level of the test, reference or control substance to be administered, expressed in the appropriate units of measure.
a. (14)			14. Method and frequency of administration of the test, reference or control substance.
a. (15)			15. Type and frequency of tests, analyses and measurements to be made and recorded.
a. (16)			16. The records to be maintained.
a. (17)			17. Dates of approval of the Protocol by the Sponsor.
a. (18)			18. Dated signature of the Study Director.
a. (19)			19. Proposed statistical methods used.

Reproduced with permission from Quality Associates Inc.

STUDY INSPECTION CHECKLIST
MSDRL Protocol Number:
Date of Inspection:
Q.A. Inspector:
             Checklist for Presence/Correctness
             (see discussion/comment page)

	Yes	No	Q.A. Inspector's Initials

PROTOCOL:

1. Descriptive title
2. Statement of objective or purpose
3. Name/address of Merck
4. Names and signatures of responsible individuals
5. Designated study director (Merck)
6. Name/address of testing facility
7. Description of test materials
8. Proposed starting/completion dates
9. Each dosage level designated
10. Description of experimental design
11. Description of spray schedule
12. Route of administration of test material
13. Description of sampling schedule
14. Description of records to be maintained
15. Signature page (protocol approval)

FIELD ACTIVITIES:

1. Protocol at site
2. SOPs covering key activities
3. SOPs at site
4. Adequate number of personnel
5. Plot design
6. Equipment calibration
7. Test material calculations
8. Recording of Merck L number from sample container(s)
9. Confirmation of calibration/calculation
10. Time between mixing and application
11. Application technique
12. Environmental parameters re: application
13. Sampling procedure
14. Sample handling technique
15. Sample storage

For any items checked NO, explain why the item was not inspected.

          Q.A. Inspector's Name_____
                               (Print)

         Q.A. Inspector's Signature_____
880316c:FORMS

ETI QUALITY ASSURANCE UNIT
Raw Data File/Field Trial Logbook Audit Checklist

Chemical:	Crop:

Trial No.:	State:	Type Study:

ETI Study Coordinator:

Field Investigator:

Inspector:	Audit date:

RDFN SECTION COMPLETE	YES	NO	COMMENTS
Protocol			
Logbook			
Signature page			
Curricula vitae			
GLP & QA statements			
FI GLP & QA statements			
Amendments/Deviations			
ETI SOPs			
FI/RC SOPs			
Field Trial Report Form			
Site location map			
Plot plan			
Soil analysis			
Soil temperature data			
Meteorological data			
Met. station location			
On-site precip. data			
Chemical log/MSDS			
Chemical storage data			

Equipment descriptions			
Calibration information			
Application information			
Sample log sheets			
Freezer data			
Bills of lading			
RDF transfer form			

ADDITIONAL COMMENTS

Inspector: _____ _____
                    Signature   Date

QA Officer:_____ _____
                    Signature   Date

Reproduced with permission from Environmental Technologies Institute, Inc.

```
                        SPRAY DRIFT TASK FORCE
                        QUALITY ASSOCIATES, INC.
                     Droplet Size Spectrum Inspection Checklist

   Study Number: _____      Inspector: _____

   Location: _____      Date: _____
```

YES	NO	N/A	DESCRIPTION

DESCRIPTION

1. Test substance is a formulated end-use product of the same formulation category as the end-use product to be registered, i.e., wettable powders, emulsifable concentrates, etc.

2. Label recommended or commonly used nozzles are used. They should similar to the nozzles used in the associated Spray Drift Study.

3. For wind tunnel studies, the product is tested at various temperatures from 10 to 35 C.

4. The air flow (velocity) in the wind tunnel may be adjusted to relate to the type of application equipment to be used:
   a. Fixed wing: 80 to 140 mph [130 to 225 kmph]
   b. Rotary wing: 40 to 70 mph [65 to 110 kmph]
   c. Ground Appl: 3 to 25 mph [5 to 40 kmph]

5. Meteorological conditions should be those most conductive to spray drift, i.e. relatively high temp., low relative humidity, inversion

6. Collection devices should be, at least, one of the following:
   a. Laser particle measuring system (_____)
   b. Collection cards, flat horizontal or vertical surface
   c. Volume air samplers
   d. Other (_____)

7. The following information must be documented in the raw data:
   a. Weather conditions (except air temp./RH% for wind tunnel)
   b. Nozzle type, orifice size, and core identification
   c. Nozzle pressure and flow rate
   d. Nozzle discharge orientation to the airstream
   e. Air velocity past the nozzle
   f. Description of measuring techniques and instrumentation
   g. Product formulation, diluents, mixtures, adjuvants, and their physical properties (surface tension, viscosity, density, etc)

8. Instrumentation and data collection equipment have proper calibration maintenance and repair documentation. (logbooks, records, forms)

9. Facility has management approved Standard Operating Procedures for all procedures performed and equipment used for this study.

10. Measuring devices automatically calculate, or a separate program is used to determine Particle size distribution vs. cumulative percent volume, and PSD vs. droplet number

11. Training records and job descriptions available for all employees

12. All test substance, reagent and waste containers as well as the test system are properly labeled. Sufficient tracking records exist for test substance.

Ciba-Geigy Corporation
Agricultural Division
Quality Assurance Unit   QAU INSPECTION SCHEDULE

METABOLISM STUDY - PLANT

STUDY TITLE:_____

PROTOCOL NUMBER:_____ STUDY DIRECTOR:_____

PROPOSED EXPERIMENTAL START DATE:_____ PROPOSED EXPERIMENTAL TERMINATION DATE:_____

CRITICAL PHASE	DATE SCHEDULED	DATE INSPECTED	DATE REPORTED	DATE RESPONSE DUE	DATE RESPONDED	AUDITOR
PROTOCOL	N/A			N/A	N/A	
PLANTING						
APPLICATION PREPARATION						
APPLICATION						
SAMPLE COLLECTION						
BIOLOGY REPORT (If applicable)						
SAMPLE PREPARATION						
BALANCE COMBUSTION/COUNTING						
CHARACTERIZATION						
IDENTIFICATION						
FINAL REPORT						

Reproduced with permission from Ciba–Geigy Corporation.

McKenzie Laboratories, Inc.
Inspection Report Form

Project/Study ID:	Project Director:

Title:

PHASES

Test System

☐ Receipt	_____	☐ Column Clean-up	_____
☐ Preparation	_____	☐ Concentration	_____
☐ Weighing	_____	☐ G/L Chromatography	_____
☐ Fortification	_____	☐ Raw Data	_____
☐ Extraction	_____	☐ Report	_____
☐ Partitioning	_____	☐ _____	_____

Analytical Reference Standard

☐ Receipt	_____	☐ Preparation	_____

Findings/Actions Recommended	Comment and/or Action Taken	Init

Inspector/Date:	Project Dir./Date:
Response Due:	Management/Date:

---MKL-F006 07/90---   Page _____ of _____

Reproduced with permission from McKenzie Laboratories, Inc.

ETI QUALITY ASSURANCE UNIT
RECORD OF QA ACTIVITIES

Chemical:		Crop:

Trial No.:	State:	Type Study:

ETI Study Coordinator:

Field Investigator:

Date	Phases Inspected	Inspector	Reported to D. of Res.	Reported to ETI Mgmt.	Certified

This document accurately reflects the QA activities for this trial:

QA Officer   :_____   _____
                          Signature   Date

Director of Research:_____   _____
                          Signature   Date

ETI Management   :_____   _____
                          Signature   Date

Reproduced with permission from Environmental Technologies Institute, Inc.

```
CIBA-GEIGY
AGRICULTURAL DIVISION
METABOLISM/RESIDUE CHEMISTRY DEPARTMENT
```

<u>SOP REVIEW FORM</u>

SOP No.: _____

Title: _____

_____

_____

Effective Date (Revision 0): _____

Reviewed By	Completion Date	NO	REVISION
			YES - Revision No. & Effective Date

Reproduced with permission from Ciba–Geigy Corporation.

SOP RETIREMENT FORM

Date: _____

To: _____

From: _____

The following SOP is being retired for the following reason(s):

SOP Number: _____

SOP Title: _____

                _____

                _____

Retirement Date: _____

Reason(s):   _____

                _____

                _____

                _____

                _____

                _____

                _____

Management Approval: _____

Approval Date: _____

Reproduced with permission from Ciba–Geigy Corporation.

ANALYTICAL DEVELOPMENT CORPORATION

STANDARD PROCEDURE FOR EXPLAINING
NOTEBOOK ENTRY ERRORS

When a notebook entry error is made it is necessary to
explain the error. In order to speed-up the process, conserve
notebook space and add some consistency throughout the organization to these explanations, the following numerical listing has
been developed:

1. Misspelled
2. Mathematical error
3. Wrong entry (date, sample no., word, etc.)
4. Transposition error
5. Transcription error
6. Procedural change
7. Wrong conclusion
8. Illegible entry
9. Unnecessary entry
10. Footnoted explanation
11. Additional information

An error is to be crossed out with one line only. Each
time an error is made it will be initialed, dated, and one of
the above numbers will be placed next to the initials and
circled. A copy of this list of errors will be taped in the
front of each notebook for reference.

SOP-33.1
3/1/88
MJG

Reproduced with permission from Analytical Development Corporation.

VALENT TRIAL NUMBER: _____   8

E.2 CROP AND PESTICIDE HISTORY

CROP HISTORY

YEAR: _____ CROP(S): _____

YEAR: _____ CROP(S): _____

YEAR: _____ CROP(S): _____

PESTICIDE HISTORY

CHEMICAL - FORMULATION	RATE LBS ai/A	PURPOSE	DATE APPL.

COMMENTS: _____

COMPLETED BY (SIG): _____ DATE: _____

VALENT U.S.A. CORPORATION ♦ ♦ ♦ FIELD RESIDUE DATA BOOK

Reproduced with permission from Valent U.S.A. Corporation.

VALENT TRIAL NUMBER: _____ 7

E.1 FIELD AND CROP CHARACTERISTICS

PLOT DESCRIPTION AND EXPERIMENTAL DESIGN:

_____

_____

_____

_____

_____

CROP CHARACTERISTICS

CROP: _____   VARIETY: _____

PLANTING DATE: _____   TYPE OF PLANTER: _____

PLANTING DEPTH: _____   PLANTING RATE: _____

ROW WIDTH: _____   PLANT SPACING: _____

SOIL AND GROUND WATER CHARACTERISTICS

SOIL SERIES NAME: _____
                  (eg: Hanford Sandy Loam)

SOIL TEXTURE: (From %Sand, Silt, & Clay)_____

SOIL ANALYSIS REPORT DATE: _____
                  (Report must be included with FRDB)

IS THERE MANMADE SUBSURFACE DRAINAGE?: _____

DEPTH OF WATER TABLE: _____

DESCRIBE SEASONAL VARIATION IN WATER TABLE DEPTH: _____

_____

DESCRIBE METHOD OF MEASURING AND/OR SOURCE OF DATA: _____

_____

COMPLETED BY (SIG): _____ DATE: _____

VALENT U.S.A. CORPORATION ♦ ♦ ♦ FIELD RESIDUE DATA BOOK

Reproduced with permission from Valent U.S.A. Corporation.

ACDS. Inc. Maintenance Logs

SPRAY BOOM

Type of Equipment: _____   ACDS ID No.: _____
Manufacturer: _____   Model & Serial No.: _____

          DATE-   ____   ____   ____   ____   ____   ____   ____   ____

Nozzle(s)
Replaced   ____   ____   ____   ____   ____   ____   ____   ____

Screen(s)
Replaced   ____   ____   ____   ____   ____   ____   ____   ____

Nozzle Size
Changed   ____   ____   ____   ____   ____   ____   ____   ____

Fittings Tightened   ____   ____   ____   ____   ____   ____   ____   ____

Boom Sections
Repaired, Replaced   ____   ____   ____   ____   ____   ____   ____   ____

Valves,Gates,Gauges
Repaired, Replaced   ____   ____   ____   ____   ____   ____   ____   ____

Equipment Drained
Winter Storage   ____   ____   ____   ____   ____   ____   ____   ____

Equipment Cleaned Yrly.
Per SOP/E/06/R1   ____   ____   ____   ____   ____   ____   ____   ____

      or

Cleaned By Use
of: _____   ____   ____   ____   ____   ____   ____   ____   ____

ACDS, Inc. Maintenance Logs

SPRAY RIG COMPLETE

Type Of Equipment: _____   ACDS ID No.: _____
Manufacturer: _____   Model & Serial No.: _____

DATE-   ____ ____ ____ ____ ____ ____ ____ ____

Nozzle(s)
Replaced   ____ ____ ____ ____ ____ ____ ____ ____

Screen(s)
Replaced   ____ ____ ____ ____ ____ ____ ____ ____

Nozzle Size
Changed   ____ ____ ____ ____ ____ ____ ____ ____

Fittings Tightened   ____ ____ ____ ____ ____ ____ ____ ____

Valves,Gates,Gauges
Repaired, Replaced   ____ ____ ____ ____ ____ ____ ____ ____

Spray Pump
Repaired, Replaced   ____ ____ ____ ____ ____ ____ ____ ____

Boom Sections
Repaired, Replaced   ____ ____ ____ ____ ____ ____ ____ ____

Framework
Repaired, Replaced   ____ ____ ____ ____ ____ ____ ____ ____

In-Line Filters
Cleaned, Replaced   ____ ____ ____ ____ ____ ____ ____ ____

Elec. Parts
Repaired, Replaced   ____ ____ ____ ____ ____ ____ ____ ____

Oil Changed
Or Lubrication   ____ ____ ____ ____ ____ ____ ____ ____

Safety Equip.
Checked   ____ ____ ____ ____ ____ ____ ____ ____

Equip. Drained
Winter Storage   ____ ____ ____ ____ ____ ____ ____ ____

Equip. Cleaned Yrly.
Per SOP/E/06/R1   ____ ____ ____ ____ ____ ____ ____ ____
     Or
Cleaned By Use
Of:_____   ____ ____ ____ ____ ____ ____ ____ ____

ACDS. Inc. Maintenance Logs

SPRAY OR DRY APPLICATOR RIG

Type of Equipment: _____   ACDS ID No.: _____

Manufacturer: _____   Model & Serial No.: _____

DATE - ____ ____ ____ ____ ____ ____ ____ ____

Valves, Gates, Gauges
Repaired, Replaced   ____ ____ ____ ____ ____ ____ ____ ____

Spray Pump
Repaired, Replaced   ____ ____ ____ ____ ____ ____ ____ ____

Fittings Tightened   ____ ____ ____ ____ ____ ____ ____ ____

In-Line Filters
Cleaned, Changed   ____ ____ ____ ____ ____ ____ ____ ____

Electrical Parts
Repaired, Replaced   ____ ____ ____ ____ ____ ____ ____ ____

Oil Changed or
Lubrication   ____ ____ ____ ____ ____ ____ ____ ____

Tire Pressure
Checked   ____ ____ ____ ____ ____ ____ ____ ____

Safety Equipment
Checked   ____ ____ ____ ____ ____ ____ ____ ____

Framework
Repaired, Replaced   ____ ____ ____ ____ ____ ____ ____ ____

Equipment Drained
Winter Storage   ____ ____ ____ ____ ____ ____ ____ ____

Equipment Cleaned Yrly.
Per SOP/E/06/R1   ____ ____ ____ ____ ____ ____ ____ ____
     or
Cleaned By Use
of: _____   ____ ____ ____ ____ ____ ____ ____ ____

ACDS, Inc. Maintenance Logs

AERIAL APPLICATION EQUIPMENT

Type Of Equipment: _____   ACDS ID No.: _____
Manufacturer: _____   Model & Serial No.: _____

DATE-   ___ ___ ___ ___ ___ ___ ___ ___

Nozzle(s)
Replaced   ___ ___ ___ ___ ___ ___ ___ ___

Screen(s)
Replaced   ___ ___ ___ ___ ___ ___ ___ ___

Nozzle Size
Changed   ___ ___ ___ ___ ___ ___ ___ ___

Fittings Tightened   ___ ___ ___ ___ ___ ___ ___ ___

Valves,Gates,Gauges
Repaired, Replaced   ___ ___ ___ ___ ___ ___ ___ ___

Spray Pump
Repaired, Replaced   ___ ___ ___ ___ ___ ___ ___ ___

Boom Sections
Repaired, Replaced   ___ ___ ___ ___ ___ ___ ___ ___

Elec. Parts
Repaired, Replaced   ___ ___ ___ ___ ___ ___ ___ ___

Framework
Repaired, Replaced   ___ ___ ___ ___ ___ ___ ___ ___

Oil Changed
Or Lubrication   ___ ___ ___ ___ ___ ___ ___ ___

In-Line Filters
Cleaned, Replaced   ___ ___ ___ ___ ___ ___ ___ ___

Safety Equip.
Checked   ___ ___ ___ ___ ___ ___ ___ ___

Equip. Drained
Winter Storage   ___ ___ ___ ___ ___ ___ ___ ___

Equip. Cleaned Yrly.
Per SOP/E/07/R0   ___ ___ ___ ___ ___ ___ ___ ___
    Or
Cleaned By Use
Of:_____   ___ ___ ___ ___ ___ ___ ___ ___

Reproduced with permission from Agricultural Chemicals Development Services, Inc.

California Agricultural Research, Inc.

STANDARD OPERATING PROCEDURES

Form SOP TREAT-1   TREATMENT PREPARATION

Study No. _____   Sponsor _____

Application Equipment _____

SPRAY MIX VOLUME (Total Gallons)

Plot Size: _____ ft. X _____ ft. = _____ sq. ft.

Acreage Basis: _____ sq. ft./43,560 = _____ Acres

                    _____ Acres X _____ No. of Reps. = _____ A-Plot

Allowance: Percentage Factor (At least 1.00) _____

           Acreage = A-Plot X Factor = _____ A-Treated

Carrier Volume Measured into Sprayer = GPA _____ X A-Treated = _____

Special Instructions: _____

_____

_____

TREATMENT DOSAGE, SPRAY MIX AMOUNTS, AND MEASUREMENT RECORDS

Trt. No.	Test Substance	Form	Lot No.	Appl. Rate AI/A	Appl. Rate Form/A	Desired* Amount Form/Mix Volume	Final** Volume of Water

*Desired Amount Form./Mix Volume = Appl. Rate Form./A X A-Treated

**Final Volume = Carrier Volume - Desired Amount Form/Mix Volume

Initials of Person Recording Data _____ Date _____ Time _____

Initials of Person Confirming Data _____ Date _____ Time _____

Reproduced with permission from California Agricultural Research, Inc.

California Agricultural Research Inc.

STANDARD OPERATION PROCEDURES

Form SOP WKS1-E2 UNMOUNTED SPRAYER CALIBRATION

Volume Output Calibration (Liquid Spray Applicator)

Study No. _____

Sprayer _____ Boom type _____

No. of Nozzles _____ Nozzle Type and Size _____

Screen Size _____ mesh Boom: Nozzle Spacing _____ in.

Height From Target _____ in. Spray Swath _____ ft. Pressure _____ psi

VOLUME (ml) PER NOZZLE PER 10 SECONDS:

Trial	No. 1	No. 2	No. 3	No. 4	No. 5	No. 6	No. 7	No. 8	Mean
1	___	___	___	___	___	___	___	___	___
2	___	___	___	___	___	___	___	___	___
3	___	___	___	___	___	___	___	___	___
Mean	___	___	___	___	___	___	___	___	

Grand
Mean

Total Sprayer Output per Second = Grand Mean X Number of Nozzles ÷
10 seconds = _____ mls.

(1.) Plot size _____ acres (2.) Desired GPA _____ gallons

(3.) Volume to be sprayed per plot = (1) X (2) X 3,785 mls/gal = _____mls/plot.

Spray time per plot = (3) ÷ Total Sprayer Output per second = _____ seconds*.

No. of Spray swaths/plot = _____.

Spray time per swath = Spray time per plot ÷ number of spray swaths/plot =
_____ seconds.

Initials of person recording data _____ Date _____ Time _____

Initials of person confirming data _____ Date _____ Time _____

*For foliar treatments in orchards or vineyards, the spray is directed evenly
over the canopy.

Reproduced with permission from California Agricultural Research, Inc.

VALENT TRIAL NUMBER: _____   11

F.2 SPRAYER CALIBRATION DATA

APPLICATION NO: _____.__ TREATMENT NUMBER(S): _____

SPRAYER OUTPUT (ml/30 sec.)
NOZZLE NUMBER ALONG BOOM

CAL#	1	2	3	4	5	6	7	8	9	10	11	12	TOT	AVG
1														
2														
3														
4														
5														

AVERAGE OUTPUT: _____

SOP NUMBER FOLLOWED: _____

COMPLETED BY (SIG): _____ DATE: _____

APPLICATION NO: _____ TREATMENT NUMBER(S): _____

SPRAYER OUTPUT (ml/30 sec.)
NOZZLE NUMBER ALONG BOOM

CAL#	1	2	3	4	5	6	7	8	9	10	11	12	TOT	AVG
1														
2														
3														
4														
5														

AVERAGE OUTPUT: _____

SOP NUMBER FOLLOWED: _____

COMPLETED BY (SIG): _____ DATE: _____

VALENT U.S.A. CORPORATION ♦ ♦ ♦ FIELD RESIDUE DATA BOOK

Reproduced with permission from **Valent U.S.A. Corporation.**

PESTICIDE RESIDUE SAMPLE HISTORY SHEET - FIELD SAMPLES (MGMT)

Chemical: _____   Protocol and Site No: _____

Formulation(s): _____

Crop & variety: _____   Trialist: _____

Date of planting: ___/___/___   Trialist location: _____

Row spacing (Field/Veg. Crops): _____   Location of test plot (Town): _____

(County): _____ (State): _____

Amount(s) of active ingredient per spray volume applied: _____

Length & type of storage after harvest: _____

Sample ID	Rate (lbs ai/A)	Formu- lation	Date(s) of Application	Sampling Date	Sample Description

YELLOW   PINK

ATTENTION FIELD TRIALIST: White and Pink - Enclose with the shipment to the lab. Yellow - To Pan-Ag Project Management at time of shipment. Goldenrod - For the field trialist (to be included with logbook).

TRANSFER OF SAMPLES:

Shipped from: _____   Shipped to: _____

(Trialist) _____   (Lab) _____

Date shipped: _____   Date received: _____

Carrier: _____   Sample condition upon receipt: _____

Signature of shipper: Signature of recipient:

Name (Print): _____   Name (Print): _____

ATTENTION LABORATORY: To comply with GLP Requirements, the "Transfer of Samples" section MUST be completed, signed by your designated recipient, and the WHITE copy of this form returned IMMEDIATELY to Pan-Ag Project Management Group. Retain the pink copy for your records.

PAN-AGRICULTURAL LABORATORIES, INC.
32380 Avenue 10, Madera, California 93638, Tel: (209) 675-0889

Page_____of_____

Rev. 10/89 9-PRSHS/MGMT.PM(FIELD)

PESTICIDE RESIDUE SAMPLE HISTORY SHEET - PROCESSING SAMPLES

Chemical: _____   Name and location of field cooperator: _____

Project number: _____   _____

Crop: _____   _____

Date of shipment by cooperator: _____   _____

Date samples received: _____   Condition upon receipt: _____

Courier: _____   _____

Sample ID	Treatment Rate	Processed Fraction	Processing Date	Days Elapsed (receipt to processing)

ATTENTION LABORATORY: To comply with GLP Requirements, The WHITE copy of this transfer of samples section MUST be completed, signed by your designated recipient, and returned IMMEDIATELY to Pan-Ag.

White - By mail to lab; signed, dated and returned to Pan-Ag   Yellow - To Pan-Ag (at time of shipment)

Pink - For your records (to be included with logbook)   Goldenrod - Enclosed with shipment

Shipped from: _____   Shipped to: _____

(Processor) _____   (Lab) _____

    _____

Date shipped: _____   Date received: _____

Signature of shipper: Signature of recipient:

Name (Print) _____   Name (Print): _____

PAN-AGRICULTURAL LABORATORIES, INC.
32380 Avenue 10, Madera, California 93638 Tel: (209) 675-0889

Page_____ of _____

Reproduced with permission from **Pan-Agricultural** Laboratories, Inc.

IR-4 NATIONAL PESTICIDE CLEARANCE RESEARCH PROGRAM

RESIDUE DATA REPORTING FORM (Part 6)
SAMPLE ARRIVAL CHECK SHEET

INSTRUCTIONS: Complete all blanks and boxes that apply. Sign (or initial) and date sections as completed adding comments as necessary. Keep this original sheet in notebook of IR-4 Sample Arrival Check Sheets and place copy and any accompanying shipping documentation, such as Federal Express receipts, cooperator's residue sample shipping and identification forms, etc., in the notebook for this project.

PR No.: _____ Trial ID No.: _____ LAB ID No.:_____

CHEMICAL: _____COMMODITY: _____

COOPERATOR: _____

SHIPPER: [ ] ACDS   [ ] FEDERAL EXPRESS   [ ] _____

SHIPPING REFERENCE NO.: _____ No. OF BOXES: _____

DATE RECEIVED: _____ BY: _____

A. CONDITION OF SAMPLES:

      [ ] FROZEN   [ ] DRY ICE PRESENT
      [ ] THAWED   [ ] FRESH, NEVER FROZEN
      [ ] RESIDUE SAMPLE SHIPPING & IDENTIFICATION FORM ENCLOSED
      [ ] ALL SAMPLES PRESENT IN AGREEMENT WITH SHIPPING FORM
      [ ] SAMPLE BAGS INTACT
      [ ] SAMPLE BAGS BROKEN OR OPEN AND CONTENTS MIXED

B. FORM OF SAMPLES AS RECEIVED:

      [ ] WHOLE   [ ] QUARTERED   [ ] SLICED
      [ ] OTHER ___ Total # samples - _____

C. NOTIFICATION OF ARRIVAL: (send copies of completed forms to the cooperator, Regional Coordinator, and IR-4 Headquarters)

      [ ] COOPERATOR, DATE: _____.
      [ ] REGIONAL COORDINATOR, DATE: _____.
      [ ] IR-4 HEADQUARTERS, DATE: _____

D.   SAMPLE LOG:

      [ ] LAB NUMBERS ASSIGNED: _____
      [ ] PROJECT LISTED ON MASTER SCHEDULE

E.   COMMENTS: _____

IR-4 Form PC 12 (Part 6)   10/89

Reproduced with permission from IR-4 National Pesticide Clearance Research Program.

SAMPLE RECEIPT FORM   N° 131

MERCK and Co., Inc.   Shipped Via:
Hillsborough Rd.   [_] Fed Exp
Three Bridges, NJ 08887   [_] Trammel
201 363-3015   [_] Other (specify)
                                       _____

SAMPLE INFORMATION

study no.	crop type	treatment	no. of boxes	condition of sample **	COMMENTS

** Condition of Shipment:
[1] completely thawed   [2] partially thawed
[3] frozen center/thawed edges [4] completely frozen

SHIPPING INFORMATION

Shipped From:_____   Drivers signature
             _____   _____
             _____   Date_____
Time of Arrival_____
Received by:_____(print or type)_____
Date_____

Reproduced with permission from Merck and Company, Inc.

```
DATE NEEDED _____           |Copies to:
                                             |----------------------
         SAMPLE MOVEMENT AND TRACKING FORM   |White  -
                                             |Yellow -
Samples needed for Processing                |Pink   -
                                             |----------------------

|      |       |           |Chop/      |Weigh   |Moisture|        |      |
| Pull | Cut   |Composite  |Homogenize |Sub-    |Determi-|Skin    |      |
|Samples| Cores|Samples    |Samples    |Samples |nation  |Samples | Other|

Processing Method Ref.: Attached _____ SOP # _____ Other _____

REQUESTER: _____ DATE_____

PROJ. NO. _____ PROJ. NAME _____ JOB # _____

SAMPLE E # RANGE _____ Set # (if applicable) _____

1. _____  5. _____  9. _____  13. _____  17. _____

2. _____  6. _____ 10. _____  14. _____  18. _____

3. _____  7. _____ 11. _____  15. _____  19. _____

4. _____  8. _____ 12. _____  16. _____  20. _____

     SAMPLE MOVEMENT TRACKING
     -----------------------                       Start  End
                                     Cond.* Date   Time   Time   Initials  Re-

1. a. Samples removed from Freezer   ____  _____  ____  ____  ____  __
   b. _____ ____  _____  ____  ____  ____  __
2. a. Samples Received for Processing ____ _____  ____  ____  ____  __
   b. _____ ____  _____  ____  ____  ____  __
3. a. Samples Processed              ____  _____  ____  ____  ____  __
   b. _____ ____  _____  ____  ____  ____  __
4. a. Subsamples Weighed             ____  _____  ____  ____  ____  __
   b. _____ ____  _____  ____  ____  ____  __
5. a. Moistures Determined           ____  _____  ____  ____  ____  __
   b. _____ ____  _____  ____  ____  ____  __
6. a. Subsamples Returned to Freezer ____  _____  ____  ____  ____  __
   b. _____ ____  _____  ____  ____  ____  __
7. a. Samples Returned to Freezer    ____  _____  ____  ____  ____  __
   b. _____ ____  _____  ____  ____  ____  __
8. a. Subsamples Received for Analysis ____ _____ ____  ____  ____  __
   b. _____ ____  _____  ____  ____  ____  __

Comments: _____

Condition Codes - F = frozen, C - chilled, TC = thawed but cold, RT = room temp.

If samples are removed twice for processing, use second line.

ENP16
```

Reproduced with permission from EN–CAS Analytical Laboratories.

California Agricultural Research, Inc.
STANDARD OPERATING PROCEDURES

Form SOP FZR1 FREEZER STORAGE LOG

Freezer No._____

Study No.	Sample No. Grouping Start-End	Test Substance	Sample Matrix	In			Out		
				Date	Time	Int.	Date	Time	Int.

5/18/88

Reproduced with permission from California Agricultural Research, Inc.

LANDIS ASSOCIATES, INC.
1989 CHEMICAL LOG BOOK

Chemical Check-In Form

Chemical: _____   Formulation: _____

Pesticide Class: _____   Active Ingredient: _____

Courier: _____   Date Received: _____

Units Received: _____   Quantity/Unit: _____

Condition: _____

Received From: _____   Lot No.: _____

Manufacturer: _____   "LX" No.: _____

Storage Location: _____   Logged By: _____

COOPERATOR CHEMICAL ALLOCATION:

Name & Location: _____

_____

Protocol No.: _____

Courier: _____   Date Shipped: _____

Quantity Shipped: _____   Shipped By: _____

COOPERATOR CHEMICAL RETURN:

Date Chemical Received: _____   Courier: _____

Quantity Received: _____   Received By: _____

Condition: _____

Page 1

Reproduced with permission from Landis Associates, Inc.

ptrl-East

PTRL RECEIPT RECORD AND INVENTORY LOG OF INCOMING TEST,
REFERENCE AND/OR CONTROL ARTICLES/SYSTEMS

Receipt Date: _____   PTRL Log Number: _____

Name on Container Label: _____   Formulation: _____

Active Ingredient: _____   Purity: _____

Batch/Lot No.: _____   Expiration/Re-analysis Date: _____

Other Label Info: _____

Shipper (Name, Address): _____

No. of Containers: _____   Gross Weight Received: _____

Received By: _____   Date: _____

Condition on Arrival: _____

Physical Description: _____/ Solid _____/ Liquid _____/ Solution _____/ Other

Storage Location: _____   Storage Temp: _____

Comments: _____

Shipper: _____Way Bill #: _____

INVENTORY LOG

Project #	Purpose	Gross Wt. ( )	Amount Dispensed ( )	Adjusted Gross Wt. Remaining ( )	Initials	Date	Balance I.D. No.

Disposed: _____   Returned To: _____   Date: _____ __

Initials: _____: Date: _____   Return Approved By: _____

DEV 4/90

Reproduced with permission from **Pharmacology** and **Toxicology Research Laboratory.**

ptrl·East

PTRL RECEIPT RECORD AND INVENTORY LOG OF RADIOACTIVE SUBSTANCES

PTRL No.: _____

Receipt Date: _____ Date: _____

Received By: _____

Test Substance: _____

Specific Activity: _____ Total Activity: _____ Purity: _____ Molecular Weight: _____

Batch/Lot No.: _____ Expiration/Re-Analysis Date: _____

Other Label Info: _____

Physical Description: _____ / Solid _____ / Liquid _____ / Solution _____ / Other _____ No./Size of Containers: _____

Amount Received (mg or ml): _____ Solution Concentration: _____

Condition on Arrival: _____ Storage Temperature: _____ Storage Location: _____

Shipper: _____ Way Bill #: _____

Comments: _____

INVENTORY LOG

Project #	Purpose	Removed (Wt. or Vol.)	Removed (µCi or mCi)	Remaining (Wt. or Vol.)	Remaining (µCi or mCi)	Initials	Date	Balance ID No.	Authorized By

Disposed: _____ Returned To: _____ Date: _____

Initials: _____ Date: _____ Return/Disposal Approved By: _____

Reproduced with permission from Pharmacology and Toxicology Research Laboratory.

AGRISEARCH INCORPORATED
Balance Calibration Log

Balance ID _____

Date	Theoretical Weight (g)	Actual Weight (g)	Maintenance	Init.

AGI-014

Reprinted with permission from Agrisearch Inc.

BALANCE CALIBRATION RECORD
METTLER ANALYTICAL BALANCE

YEAR:_____

Performed and Recorded		Method of Calibration*						Adjusted and OK	Unac-cept.#	Comment
Date	Initials	0.1 gm	1.0 gm	10 gm	100 gm	Other	OK			

CALIBRATE BALANCE DAILY PRIOR TO FIRST USE
* = Standard Weight
# = When unacceptable, please complete an equipment failure report
and notify laboratory director.

Reprinted with permission from Colorado Animal Research Enterprises, Inc.

FMC Corporation
Agricultural Chemical Group
Formulations Department

STANDARDIZATION RECORD FOR pH METER

Manufacturer_____ Model_____

Serial No._____   Electrode Serial No._____

Study Number	Temperature Setting (°C)	Reference Standard 1		Reference Standard 2		Initials	Date
		Type	Meter Reading	Type	Meter Reading		

Reprinted with permission from FMC Corporation.

YEAR _____

THERMOMETER CALIBRATION

Thermometer			Temp 1 NBS=	Temp 2 NBS=	Temp 3 NBS=	Within limits?*	Overall % Accuracy
ID	Range	Location					

Initials (Person Checking and Recording) _____ Date of Test: _____

NBS Thermometer ID No. _____

*± 0.5 degrees of intended

Reprinted with permission from Colorado Animal Research Enterprises, Inc.

Centrifuge Temperature Check Log

Temperature Setting

Year	20°C		10°C		4°C		Temp Accept-able[b]	Ther-mom-meter ID#	Check Motor Brushes[a] (-length)	Test performed and recorded by: Init.	on: Date.
	Centrifuge gauge Reading	Thermometer Reading	Centrifuge gauge Reading	Thermometer Reading	Centrifuge gauge Reading	Thermometer Reading					
Jan											
Feb											
March											
April											
May											
June											
July											
Aug											
Sept											
Oct											
Nov											
Dec											

[a]Replace both brushes when one brush is <¼" long
[b]The temperature on the thermometer should be ±2°C of the reading on the centrifuge gauge

Reprinted with permission from Colorado Animal Research Enterprises, Inc.

YEAR_____   3 Month Inspection of Deionized Water System

(Serial No.: 87-10-055)

Performed and Recorded					
Date	Initials	Megaohm Reading[a]	Water Leaks Found?	System Acceptable for Use?[b]	Comments

[a]Megaohm reading should be > 17.0
[b]If unacceptable, fill out an equipment defects data form to report problem

Reprinted with permission from Colorado Animal Research Enterprises, Inc.

Gases

Date	Init	Proj #	Column Description	Carrier mL/min	Split Vent	Sept Purge	Column Type	Head PSI ISO	Prog	Make-up Type	Total Flow	Detector H₂ mL/min	PSI	Air mL/min	PSI	O₂ mL/min	PSI	Detector Type	Oven Temp	Flows Measured at what

Purge System Time On(min)	Time Off(min)	Signal	Range/Attn	Temperature °C Detector	Inlet ISO	Prog	Oven ISO	Prog	Run Time Prog	Insert Type	Injection/Sample Information Type	#	Vol	Auto	Solvent	Maintenance & Comments Septum, Glasswool, Detector Clean., Etc.

Reprinted with permission from Analytical Development Corporation.

Column

Date	Name	Proj	Make	Packing	Length	Temp

Precolumn

Packing	Date Packed

Injector

Make & Model	Loop Size

Sample Information

Type	#	Inject Volume Au

Recorder

Span	Chart Speed

Detector

Make and Model	Wave-length	Sensitivity Setting

Mobile Phase

Mix	Flow Rate (ml/min)	Pump Pressure psi

Reprinted with permission from Analytical Development Corporation.

BUSINESS CONFIDENTIAL

CHRONOLOGICAL LOG

INSTRUCTIONS: Record,in black ink, values, observations, notes, calculations, phone calls and events that can effect the quality or integrity of the results. Record the trial number on each page. Date and sign or initial all entries. Every action or activity that is performed in regard to this trial should be entered in this log.

PROJECT NUMBER:_____

TRIAL NUMBER:_____

Date	Initials	Activities / Comments

QAU-88

Reprinted with permission from Rhone–Poulenc Ag Company.

BLASLAND, BOUCK & LEE

ARCHIVE INDEX

INDEX NO.	STUDY NO.	TYPE OF STUDY	TEST SYSTEM	TEST SUBSTANCE	STUDY DATES*	RECEIPT DATE

* Study dates - Study Initiaton Date through completion date

CLIENT/STUDY NO. CROSS-REFERENCE

STUDY NUMBER	ARCHIVE INDEX NUMBER	STUDY TITLE	TEST SUBSTANCE

Reprinted with permission from Brasland, Bouck, and Lee.

Archived Material
Request Form

BASF Corporation

The following request for removal of material from the archive is being made with the understanding that the removed material shall not be removed from the premises and is intended for the use by the individual(s) listed on this form.

Document(s) removed

Temporary storage location

Individuals using document(s)

Person removing document(s)

_____   _____
Signature   Date

Returned on: _____
Date

F 4532   **Distribution:** White—QA   Canary—Requester

Reproduced with permission from BASF Corporation.

Indexes

Author Index

Barge, Maureen S., 13, 43
Buckler, Phillip M., 203
Burnett, Gene, 53
Burton, Jesse, 155
Coody, Peter N., 343
Cooper, Sandra C., 333
DeMartinis, James M., 333
Dillon, Frances A., 163
Dull, David L., 375
Ewing, Duane D., 241
Flaherty, Phyllis E., 387
Fuller, S. Rand, 113
Garner, Willa Y., 113, 127, 445
Gerngross, Malcolm F., 249
Giddings, Jeffrey M., 297
Harris, Jeffery L., 163
Hill, Dean F., 361
Hochman, Judy H., 127, 257
Hornshuh, M. Jean, 27
Howie, Stephen J., 387
Huntsinger, Del W., 175

Hyndman, Harry L., 181
Jaber, Mark, 309
Jensen, Markus M., 85
Johnson, David, 155
Johnson, Mary E., 309
Liem, Francisca E., 375
Marco, Gino J., 257
McCann, John A., 317
Merricks, D. Larry, 279
Merricks, Donna H., 279
Novak, Roger A., 257
Platt, James L., Jr., 399
Pollock, Robert J., 95
Royal, Patricia D., 143
Snyder, Frederick G., 419
Spare, William C., 279
Townsend, Joseph B., 3
Ussary, James P., 21, 235
Watson, J. Drew, 217
Wright, John F., 227

Affiliation Index

ABC Laboratories, Inc., 203
Agrisearch Inc., 279
Analytical Development Corporation, 95
BASF Corporation, 175
Bio/dynamics Inc., 3
Blasland & Bouck Engineers, 333
Ciba–Geigy Corporation, 53
Compliance Reviews, 257
E. I. du Pont de Nemours and
    Company, 27
Dynamic Corporation, 399
EPL Bio-Analytical Services, Inc., 217
FMC Corporation, 13, 43, 227
Garndal Associates, 113, 127, 445
Glaxo, Inc., 419
Hazleton Corporation, 419
ICI Americas, Inc., 21, 235
International Chemical Consultants, Inc., 399

Jensen Agricultural Consultants, Inc., 85
Marco-Tech, 257
Monsanto Agricultural Company, 181
NPC, Inc., 257
PTRL East, Inc., 343
Pan-Agricultural Laboratories, Inc., 241
Quality Associates Inc., 113, 127
Spectralytix, Inc., 163
Springborn Laboratories, Inc., 143, 297
Stewart Agricultural Research
    Services, Inc., 155
Stewart Pesticide Registration
    Associates, Inc., 163
Texas A&M University System, 249
U.S. Environmental Protection Agency,
    317, 361, 375, 387
Ussary Scientific Services, 21, 235
Wildlife International Ltd., 309

Subject Index

A

Accelerated-storage data,
product-chemistry package, 232
Accreditation, laboratories generating
environmental and human health data,
381–384
Activity logs, data audit, 178
Adverse effects, pesticide testing, 298
Agricultural chemical products,
registration requirements, 235–240
Agricultural field studies, worker-exposure
testing, 279–295
Agricultural product study, experimental
phase, 21–22
Agrochemical company experience, GLP
Standards, 412
Agrochemical groundwater studies, 333–342
Agrochemical industry, full acceptance of
GLPs, 6
Alterations, protocol, 64
Amended final report, QA statement, 131
Amendments, protocol, 47,64
Analytical balance log, 100,102f,106
Analytical data, EPA audit, 369–370
Analytical facilities
GLP review, 369
receipt of field samples, 97–100
Analytical method trials
data audit, 185–186
description, 285
Analytical phase
inspection, 163–173
major concerns, 370
Analytical reference standards
calibration and preparation, 286
characterization and accountability, 113–126
data audit, 186–187
logbooks, 99–100
Animal studies, dose level, 259
Animal test systems, procedural SOPs, 37–38
Antimicrobial pesticides, efficacy
testing, 381
Aquatic field studies, pesticides, 297–307
Aquatic laboratory study, contrast with
mesocosm study, 303
Aquatic risk assessment, overall hazard,
346–347

Archives
alternate locations, 394
commercial facilities, 153
computer data, 215
data package, 188–189
data retention, 151–152
designated archivist, 153
documentation, 151
ecotoxicological field studies, 328–329
EPA inspection, 371
facility, 150–151,328–329
GLP Standards, 150–153
organization, 158
procedural SOPs, 40
quality assurance unit, 150–154
schedule for complex studies, 393
types of copies, 152
Archivist
conflict of interest, 153
GLP Standards, 153
Assays, test substances and reference
standards, 119
Audits, raw data, records, and other
documents, 361–374

B

Bias, control in field studies, 241–248
Bilateral agreements, data in support of
product registration, 424–425
Bill of lading, 100,103
Bioaccumulation
nonradioactive animal feeding
studies, 271
pesticide residues, 267–270
Biodegradability, testing of industrial
chemicals in Japan, 429
Biological development, mesocosm study,
299–300
Biological interactions, mesocosms,
306–307
Bird surveys, terrestrial field
studies, 312
Black box
performance audit, 220–221
software validation, 220f
Breakthrough trials, description, 285

C

Calculation equation, documentation, 185
Calculations and tabular or graphic
  information in study report, 136–137
Carrier
  analytical documentation, 122
  mixed with test substances, 121–123
Casualty searching, terrestrial field
  studies, 311–312
Chain of custody
  data audit, 177,187
  documentation, 95–112,177
  field residue studies, 85–93
  key points, 92–93
  runoff studies, 348
  test substance, 315
Chain-of-custody form, 294f,349f
Characterization
  data audit, 187
  definition, 265
  physical and chemical studies, 123
  test substances and analytical reference
    standards, 116–121
Checklists
  inspections, 169
  protocol audit, 47
Chemical receipt and use log, 117f
Chemical substances
  example, 70–72
  protocol, 60
Chemistry, experimental design, 61
Color-coding, expiration date, 125
Commitment of managers to GLP
  program, 16
Communication between field and laboratory
  in protocol development, 340
Completion date, definition, 150
Complex field trials, SOP with specific
  instructions, 22
Compliance
  cost of noncompliance, 17
  FIFRA GLP regulations, 334
  inability to verify, 380–381
  personnel proficiency, 7,8
  problems from force-feeding of GLP
    programs, 14
  programs, 5
  protocol, 51,58
  technical problems, 318

Compliance monitoring
  data base information exchange, 422
  field studies, 318
  OECD working group, 421
  publications, 421–422
Compliance statement
  documentation, 328
  GLP, 135
Computer acquisition system
  diagram, 210f
  operation, 209
  validation, 209–213
Computer-calculated area, 211
Computer operations, procedural SOPs,
  39–40
Computer software
  data security, 219
  development cycle, 218f
  hardware resources, 219
  validation for laboratory use, 217–223
Computer system
  environmental safety, 215
  errors, 213
  quality-control samples, 213
  validation and verification, 203–216
  verification of storage capabilities, 213
Computerized master schedule, 149–150
Computerized system, data audit, 185
Confidentiality, processing studies, 253
Conflict of interest
  archivist, 153
  quality assurance unit, 327
Consistency
  international testing guidelines, 420–421
  regulatory language, 392
  validation program, 205
Contract field laboratory survey
  cost increases with GLP
    implementation, 405f
  distribution of respondents, 402f
  GLP training sources, 410f
  improvements with GLP
    implementation, 407f
  number of employees at facilities, 402f
  number of studies conducted
    annually, 403f
  staff time consumption by activities, 408f
  staff time spent on GLP issues, 406f
  studies regulated by GLP Standards,
    403f,404f

Contractor, duties of study director, 320
Control runoff, runoff studies, 348
Control substance, definition, 114
Core samples, representativeness, 242–243
Crop field studies, need for nonradioactive
   animal feeding studies, 271
Cross-contamination, test samples, 289
Custodian, SOP system, 31–32

D

Data acquisition system
   electronic signal transmitted to
      computer, 107
   security codes, 214
   security of hardware, 214–215
Data audit
   areas of concern, 177
   BASIC programs for subsample selection,
      185,192–195
   benefits, 179–180
   field studies, 175–180
   laboratory practice, 181–189
   older studies, 180
   performance, 179
   preparation, 176–178
   procedure, 182–189
   purpose, 182
   quality assurance unit, 146–147
   review of data, 183–187
   study report format, 195–199
Data collection, QAU, 282
Data collection and verification, runoff
   studies, 353–355
Data collection forms
   nonroutine study, 239
   terrestrial field studies, 313–314
Data collection system, scintillation
   counter, 206f
Data confidentiality claims, 134,135f
Data deficiencies, regulatory decision
   making, 388–390
Data evaluation, experimental design, 62
Data integrity
   laboratory vs. field, 156
   study conditions, 145
Data reporting guidelines, 137
Data requirements
   product-chemistry studies, 228–230
   registration, 55–56

Data retention
   archives, 151–152
   FIFRA requirements, 152
   security, 158
Data trail, test substance, 86
Data transfer, preparation for audit, 182–183
Database system, advantage over
   spreadsheets, 206
Dates and schedules
   example, 68–69
   protocol, 59–60
Definitive study, definition, 311
Delegation of responsibility, study
   director and QAU, 328
Detection limit, crop residue analysis, 263
Deviations, protocol, 47,64,321–323
Dietary toxicity, wildlife, 312
Disinfectants, efficacy testing, 381
Dislodgeable residue studies, sampling
   strategy, 246
Disposal, empty test-substance
   containers, 330
Distribution
   protocol, 51
   SOP system, 31–32
Documentation
   analytical procedure, 185–186
   analytical reference standards, 186–187
   archives, 151
   calculation equation, 185
   chain of custody, 95–112,187
   data audit, 183–184
   data-recording forms, 356
   field sample source and receipt, 96–97
   field test data, 322–323,325–326
   GLP compliance, 329
   identity of substances, 119,330
   legibility of field study data record, 240
   method of synthesis, 120
   nonroutine study, 238–239
   notebook, 286–287
   protocol deviations, 64
   QAU inspection, 172–173
   reagents and solutions, 124–125
   runoff studies, 353–355
   sampling, 96
   soil characteristics, 353
   stability, 120
   staff qualifications, 323–324
   staff time, 406

Documentation—*Continued*
 storage
  conditions and duration, 97
  containers, 120
 study directors and sponsors, 320
 test substance, 89,315,364
 training of field personnel, 337–338
Dose level
 animal studies, 259
 choice of high-level dose, 261–263
 feeding trial, 271*t*
 pesticide in the diet, 272
 relationship to tissue residue, 162*t*

E

Ecological monitoring, mesocosm study,
  300–301
Ecosystem metabolism, ecological
  monitoring, 301
Ecotoxicological field studies
 problems, 317–331
 protocol changes, 321
Education, key to change in manager-driven
  GLP program, 17
Enforcement
 low output, 379–380
 penalties provided under FIFRA, 396
 policy, 395–397
 problems, 380
 sanctions, GLP program, 378,396
Environmental assessment, pesticide,
  297–307
Environmental conditions, study
  protocol, 323
Environmental Protection Agency (EPA)
 decisions, integrity of data submitted,
  388–390
 fines to enforce GLP compliance, 7
 forethought about impact of
  regulations, 6–7
 GLP regulations, 3–4
 inspections, measuring device, 10
 laboratory inspection and data audit program
  history, 376–377
  scope, 377–378
 monitoring advisory, compliance
  statement, 135
 pesticide assessment guidelines, 136–137
 responsibility to educate industry, 7

Environmental records, data audit, 178
Environmental studies, 163–173
EPA, *See* Environmental Protection Agency
Equipment
 failures, mesocosm study, 307
 GLP problems, 372
 maintenance record, residue studies, 159
 specification, calibration, and
  maintenance, 314,348–350
Equipment SOPs
 categories, 33
 functional listing, 40–41
 information required, 33
 volume-reduction techniques, 33–34
Erroneous results, computer system, 213
European Communities (EC), series of
  directives, 423
European Communities–OECD initiatives,
  summation, 424
Experimental design
 example, 76–79
 protocol, 61–62
Expert group meetings, international
  testing guidelines, 420
Expiration date
 color-coding, 125
 mixture, 122
 reagents and solutions, 124–125
Exposure assessment, runoff studies, 345–347
Extraction trials, description, 285–286

F

Facilities
 data retention, 152
 field sites, 157–162
 GLP problems, 372
Facility and personnel management,
  procedural SOPs, 38–39
Facility inspection, QA perspective for
  field studies, 155–162
Facility requirements, test substances and
  analytical reference standards, 114–115
Facility responsibilities, processing
  studies, 252–254
Federal Insecticide, Fungicide, and
  Rodenticide Act (FIFRA)
 advisories, 445–474
 changes, 5

Federal Insecticide, Fungicide, and
   Rodenticide Act (FIFRA)—*Continued*
   GLP Standards, 28
   compliance, representative forms, 500–554
   enforcement response policy, 475–500
   text, 445–474
   pesticide registration, 96
   reactions of agrochemical community, 6
   regulations, penalties for violations, 380
Feeding studies
   facilities, 272
   metabolism, 271–272
   sampling strategy, 242–246
Field data, data audit, 187
Field facilities
   equipment, 159–160
   field sites, 161–162
   office, 157–158
   operation areas, 161
   storage facilities, 160–161
Field notebook, items recorded, 286–287
Field phase
   data requirements, 368
   EPA audit, 366
Field sample(s), shipping to analytical
   laboratory, 97–106
Field sample transmittal form, receipt at
   laboratory, 103
Field sampling
   nonbiased, 241–248
   tractor-mounted soil probe, 243–246
Field scientists, developing a study
   protocol, 336–337
Field sites, requirements, 161–162
Field SOPs, 23–27
Field spikes, runoff studies, 348
Field staff, qualifications, 340
Field studies
   chain of custody, 85–93
   data audits, 175–180
   five phases, 363
   GLP compliance, 235–240
   GLP problems, 371–373
   government perspective on audits,
      361–374
   handling of samples, 90
   management responsibility, 24
   personnel requirements, 378
   protocol development, 48–51
   sampling strategy, 246–248

Field studies—*Continued*
   sources of problems, 318
   test substance documentation, 89
Field supervisor, mesocosm study, 304–305
Field team, members and responsibilities,
   287–294
Field training, hands-on experience, 338
FIFRA, *See* Federal Insecticide, Fungicide,
   and Rodenticide Act
Final report, 131
Fish, ecological monitoring, 301
Flexibility
   choice of words in protocols, 64
   cooperator requirements, 341
   protocol study design, 44
   protocol time frames, 340–341
   standard operating procedure, 35,338–340
Flow meter, calibration, 350–351
Flow meter calibration form, 352*f*
Format, SOP, 30–31
Formulation chemist, role in product
   development, 231–232
Formulation label, 280–281
Fractionation scheme
   metabolites, 267
   plant and animal material, 269*f*
Freezer logbook, storage of samples,
   103,105*f*,106
Freezer trucks, residue samples, 91
Freezers, maximum–minimum thermometer
   readings, 160

G

General Accounting Office, neutral-scheme
   inspection program, 381
Generic phases of studies, 147,148*t*
GLP, *See entries under* Good Laboratory
   Practice
Good Laboratory Practice documentation,
   standard interpretation, 412
Good Laboratory Practice inspections
   OECD guidelines, 421
   training course, 422
Good Laboratory Practice program,
   compliance
   issues, 367
   OECD guidelines, 421
   staff time, 407–408

Good Laboratory Practice program,
  OECD and EC part in implementation
  and development, 425
  problems, 378–385
  review committee, 384–385
Good Laboratory Practice regulations
  adaptation to field studies, 155
  field contractor survey to assess
    impact, 400–411
  guidance on compliance, 408–409
  inability to cite testing facilities, 380
  Japanese, 427
  necessity, 388–390
  purpose, 156
Good Laboratory Practice Standards
  adaptation to field environment, 334–335
  additional resources needed, 409
  agrochemical company experience, 412
  agrochemical research, 3–10
  archives, 150–153
  benefits, 8,17
  changes brought about, 8
  compliance, 4,6,8–10
  compliance requirements, 123,235–240
  cost, 410
  definition, 27
  discovery of violations, 395
  economic impact, 399–417
  efficacy studies, 410
  environmental activism, 4,5
  general requirements, 390
  history of programs, 14
  improvements in studies, 412
  inspection and verification, 423–424
  Japanese, 428t
  legislation, 389–390
  mock inspections, 10
  policies and interpretations, 387–398
  processing facilities, 251
  program at a crossroads, 375–385
  promulgation pattern, 4
  public pressure to apply restrictions, 4,5
  questions to be clarified, 390
  range of studies, 389
  reagents and solutions, 124–125
  refinement of process, 17
  regulations, 376
  regulatory interface between laboratory
    and EPA, 10
  revisions to ensure consistency, 390

Good Laboratory Practice Standards—
  Continued
  single generic regulation, 425
  staff training, 409–410
  subcontracted studies, 409
  terrestrial field studies, 313
  transition stages, 14–15
Grid design, study protocol, 339
Groundwater field studies
  agrochemical, 333–342
  small-scale prospective studies and
    retrospective studies, 335–336
Guideline documents, adherence, 138

H

Handling requirements, reference
  standards and test substances, 115–116
Harmonization, laws, regulations, and
  administrative procedures, 423
Hydrogeologically vulnerable site, EPA
  guidelines, 339
Hydrolysis, points of inspection, 166

I

Identification
  definition, 265
  samples in field studies, 89–91
  test substances and reference
    standards, 116
Impurities, test substances and reference
  standards, 119
In-life studies
  protocol, 264
  quality assurance unit, 264–265
Industry survey of contact field
  laboratories, 399–417
Informed consent form, 288f
Initiation date, definition, 150
Input from scientists, development of GLP
  program, 16
Inspection
  announced vs. unannounced, 168
  facility controls, 172
  field studies, 239–240
  generic phases, 147
  GLP requirements, 165

Inspection—*Continued*
laboratory operations, 171–172
noncompliance, 169
phases and frequency, 165–166
procedures, 168–169
protocol, 170
qualifications of study personnel, 169–170
quality assurance unit, 146–147,163
sample receipt and storage, 171
scheduling, 146–147
test substances and analytical reference
    standards, 170
Inspection and audit program, scope, 377–378
Instrument log, 107–110f
International cooperation, legislation
    relative to OECD and EC, 424
International GLP implementation, 425–430
International GLP programs
    European Communities, 423–424
    list of approved facilities, 384
    monitoring systems, 377
    Organization of Economic Cooperation and
        Development, 420–422
Invertebrates, ecological monitoring, 301

J

Job descriptions, field study staff, 324

L

Label format, reagents and solutions, 124f
Laboratory accreditation
    advantages and disadvantages, 383–384
    Environmental Management Monitoring
        Council, 382
    EPA task force, 381
    fundamental issues, 382–383
    list of approved facilities, 384
    Toxicology Laboratory Accreditation
        Board, 381–382
Laboratory input, SOP development, 29–30
Laboratory inspection program, managerial
    support, 164
Laboratory maintenance, equipment SOPs,
    33–34
Laboratory operations, procedural SOPs, 38
Laboratory phase, EPA audit, 366–370

Laboratory protocols, practical, 53–83
Livestock, definition, 258
Livestock diet, raw agricultural commodities
    and feeds, 260t
Livestock studies
    components, 259
    in-life phase, 272–277
    metabolism of pesticides, 257–278
    outline, 274f–275f
Livetrapping small mammals, terrestrial
    field studies, 312
Log sheet, substance distribution, 116

M

Management
    responsibility and leadership, 13–19
    SOP for the laboratory, 30–36
Manager-driven GLP program, benefits, 16
Marketing permit, data retention, 152
Master log
    condition and transport of samples, 103
    field samples, 97–99
Master schedule
    audit trail of changes, 149
    computerized, 149–150
    ecotoxicological field studies, 326–327
    inspections, 147
    maintenance, 150
    purpose and actual use, 149
    quality assurance unit, 147–150
Material Safety Data Sheet
    processing studies, 252
    test substances and analytical reference
        standards, 116
Maximum–minimum thermometers, storage
    areas, 161
Memoranda of understanding, international
    agreements, 424
Mesocosm
    biological effects from specific levels of
        chemical exposure, 346
    definition, 298
Mesocosm study
    biological development, 299–300
    boundaries, 305
    contrast with laboratory study, 303
    development, 298
    documentation, 301

Mesocosm study—*Continued*
ecological monitoring, 300–301
GLP issues relating to test substance, 303
interpretation, 304
nonroutine aspects, 306–307
overview, 298–299
pesticides, 297–307
physical design, 299
program management, 304–306
purpose, 303–304
statistical analysis and interpretation, 303–304
treatment with test material, 302–303
Metabolism studies, 259–270
animal subjects, 258
low residues in crop field studies, 271
pesticides, 257–278
residues in milk, eggs, or meat, 272
Metabolites, 265–270
decision-making scheme for identification, 268*f*
found in edible tissues, 263
modification of existing residue method, 270
separation into components, 267
Ministry of Agriculture, Forestry, and Fisheries
agricultural chemical product registration, 429
GLP principles different from those of EPA, 429
petition for confirmation procedure, 427–429
Mixture
analytical documentation, 122
expiration date, 122
substances with carriers, 121–123
Modifications, statement in protocol, 50
Monitoring, exposure to agrochemicals, 279–295
Monitoring points, variety in mesocosm study, 301
Monitoring wells, clay in unsaturated zone, 339

N

Necropsy data, documentation review, 270
Nesting birds, terrestrial field studies, 312–313

Neutral-scheme inspection, quality of data, 381
Non-health-effects studies, extension of GLP jurisdiction, 362
Noncompliance, QAU action, 169
Nontarget aquatic organisms, exposure, 346
Notebook
documentation, 286–287
specific studies, 106

O

Operation areas, 161
Optional information
example, 81–82
protocol, 63
Organization, protocol, 58–63
Organization for Economic Cooperation and Development (OECD)
GLP inspection and study audits, 421
international guidelines for GLPs, 420–421
primary purposes, 420
SOPs, 28

P

Packaging and labeling, residue samples, 91
Paper trail, chemicals, 160–161
Peak generator, chromatogram, 212*f*
Peak generator method, validation process, 210–213
Penalty determinations, FIFRA violations, 396
Personnel
example, 68
field studies, 238
GLP problems, 371–372
listing format, 129*f*
records
field-site GLP inspection or data audit, 323–324
storage, 158
Pesticide
aquatic field studies, 297–307
data requirements for registration, 228
dislodgeable residue studies, 246
fate of crop residues during processing, 249–255

Pesticide—Continued
  field residue trials, 246–248
  mesocosm studies, 297–307
  terrestrial dissipation studies, 242–246
Pesticide Assessment Guidelines
  Data Requirements for Registration,
    56–57t
  specific data requirements, 229
Pesticide registration
  notice 86–5
  instructions on formatting of final
    reports, 230
  standard format, 131–136
  TSCA GLP Standards, 96
Pesticide runoff
  aqueous and sediment-bound
    fractions, 344
  factors, 344–345
  field studies, 343–357
  pesticide formulation and method of
    application, 344–345
Pesticide transformational processes, 344
Pharmacological effects, residue
  level, 261
Photolysis, points of inspection, 166
Physical and chemical properties, test
  method SOPs, 34–35
Physical property tests, guidelines, 35
Plants, ecological monitoring, 300
Policy decisions, key factors, 392
Policy dissemination
  GLP inspection and audit program, 392
  regulated community, 392–393
Policy issues
  examples, 393–395
  GLP Standards, 390
  sources, 391
Policy-making process within EPA, 391
Policy violations, case-by-case basis, 391
Preconditioning, nonradioactive test
  material, 261
Preprotocol meeting, planning stages,
  45–46
Primary device for flow measurement,
  runoff studies, 350
Procedural SOPs, categories,
  32–33,37–40
Processing data, data audit, 187
Processing facilities, GLP requirements,
  254–255

Processing studies, 249–255
  commercial methods, 253
  confidentiality, 253
  definition and purpose, 250
  GLP compliance, 252–253
  history and growth, 250–251
  reproducibility, 253
  research and information services, 254
Product chemistry
  formulation perspective, 227–234
  purpose, 229
Product-chemistry package
  changing emphasis, 229
  general categories, 228
Product-chemistry studies
  data requirements, 228–230
  testing facilities, 34
  using GLPs, 230–232
Product development
  individual roles, 231–232
  products registered with EPA, 231
Program management, mesocosm study,
  304–306
Project master file, documentation, 103
Prospective studies, definition, 335
Protection, worker exposure, 279–295
Protective equipment, formulation
  label, 281
Protocol
  alterations, 64,83
  amendments, 236–237,321
  approval, 283
  case-specific, 355
  checklist, 276f
  complex nonroutine study, 236–237
  definition, 55–56
  design for field studies, 43–52
  details, 63
  development, 48–51,364
  deviations, 188,263,322–323,339–340
  ecotoxicological field studies, 321
  flexibility, 337
  GLP problems, 373
  groundwater-monitoring field study,
    336–337
  organization, 58–63
  practical, 53–83
  processing study, 251–252
  purpose, 44–45,54–55
  requirements, 56–58

Protocol—*Continued*
  resources available, 45
  review by auditor or inspector, 46,183,364
  revisions, 47
  runoff studies, 347–348
  sampling, 242
  SOPs and checklists, 46–47
  unique identifier, 48
Protocol-evaluation form, QAU, 276*f*
Public health, FIFRA efficacy testing, 362
Pump calibration form, 290*f*
Pump sampler, calibration process, 351
Purpose, statement in protocol, 48

Q

QAU, *See* Quality assurance unit
Qualifications, field study staff, 323–324
Quality assurance (QA) inspections
  auditor responsibilities, 168
  minimum requirements, 394
Quality assurance programs vs. Good
  Laboratory Practices programs, 15
Quality assurance report, data audit, 188
Quality assurance statement, format, 130*f*
Quality assurance unit (QAU), 143–154
  activities at mixing site or application
    site, 330
  animal studies, 265
  archives, 150–154
  audits
    independence of auditor, 164
    QAU data audit, 146–147,175–189
    runoff study, 355
    SOP revision, 30
  checklist, 266*f*
  conflict of interest, 327
  contributions to GLP compliance
    process, 18
  documentation of nonroutine study, 238–239
  ecotoxicological field studies, 321,327–328
  field studies required for registration
    of chemical products, 236–240
  field test preparation, 289
  GLP programs and requirements, 14,408
  historical development, 144
  inspection
    field studies, 239–240
    testing facility, 156–157

Quality assurance unit (QAU)—*Continued*
  international development, 144
  master schedule, 147–150
  monitoring of studies, 156
  multisite locations, 145
  personnel training, 238
  procedural SOPs, 39
  product-chemistry study, 232
  protocol amendments, 236–237
  protocol deviations, 237
  purpose, 144–145
  responsibility in field study, 295
  role, 111
  scientific staff meetings, 17
  SOP approval, 30
  study inspection, 164
  supervision and functions, 327
  terrestrial field studies, 314–315
  test substance documentation, 284–285
  validation procedures, 213–214
Question-and-answer document, GLP
  regulations, 393

R

Radioactive labeling, metabolism
  studies, 259
Radioradiotelemetry, terrestrial field
  studies, 312
Rainfall simulator, uniformity tests, 351–353
Random sampling, field residue trials,
  246–248
Randomization, sampling bias, 241
Raw agricultural commodities
  processing studies, 250
  supplied for processing, 252
Raw data
  archives, 158
  calculations audit, 185
  computer security, 214
  definition, 176
  storage, 158,328–329
  study verification, 389
  traceability from reported values,
    184–185
  transfer statement and checklist, 190–191
Reagents and solutions
  GLP Standards, 124–125
  label formats, 124*f*

Receipt and distribution, test substances
    and reference standards, 116
Receiving body of water, runoff study, 346
Records and reports
    availability to EPA and Food and Drug
        Administration representatives, 173
    GLP problems, 373
    maintenance
        data audit, 183–184
        example, 79–80
        field study, 325–326
        protocol, 50,62
Reference standards
    definition, 114
    handling requirements, 114–116
References or attachments
    example, 80–81
    protocol, 63
Regulatory changes in agrochemical
    research, 3–10
Regulatory decision making, data
    deficiencies, 388–390
Regulatory history, 4–7
Replicated pond studies, biological
    effects from specific levels of
    chemical exposure, 346
Report preparation, EPA inspection, 371
Reporting study results, 127–139
Representativeness, core samples, 242–243
Reproducibility, processing studies, 253
Research, data retention, 152
Research projects, use of GLPs, 232
Reserve samples
    storage, 121
    test substances and reference standards, 118
Residue(s)
    fate during processing, 249–255
    toxicologically significant, 258
Residue analysis, points of inspection, 167
Residue chemistry studies, 163–173
Residue methods, 270–272
Responsibility, statement in protocol, 48
Retrospective studies, definition, 335
Review and revision
    protocol, 47
    SOP, 31
Runoff studies, 343–357
    containment and measurement of runoff
        water, 347–348
    GLP-related overhead, 356

Runoff studies—*Continued*
    GLP requirements, 347–355
    mesocosm study, 302–303
    natural and simulated rainfall, 345
    published guides, 345–346
    regulatory arena, 345–347
Runoff water sampler, 351

S

Saltwater marsh, inadequate planning, 323
Sample amounts, animal study, 264t
Sample analysis, experimental design, 62
Sample collection
    documentation, 86,301
    monitored by QA officer, 267–270
Sample generation
    experimental design, 61–62
    field studies, 89–90
Sample identification, procedure, 91
Sample media, preparation, 286
Sample or specimen transfer form, 293f
Sampling bias, control in field studies,
    241–248
Sampling design, dislodgeable residue
    study, 247f
Sampling locations, grid system, 353,354f
Sampling methods, data audit, 178
Sampling pump, recalibration, 289
Sanctions, testing facility in violation, 377
Scintillation counter
    data collection system, 206f
    development of computer system, 205–207
    sample data, 208f
    security, 214
    verification printer, 209f
Screening study, definition, 311
Security
    software development, 219
    test substance, 330
    validation and verification of computer
        data, 214–215
Sensitivity issues, field studies, 341
Shipment and documentation
    return to analytical laboratory, 91–92
    runoff studies, 355
    statement in protocol, 50
    test substances, 86–88
    verification of sample arrival, 92

Signatures
  example, 66
  protocol, 51,59,64
Simulation
  rainfall, runoff studies, 348
  spray-drift and runoff, 202–203
Site characterization, EPA guidelines, 339
Site inspection, quality assurance unit, 315
Site selection, agrochemical groundwater
    studies, 339
Software, 218–221
Soil characterization
  data audit, 178
  location of borings, 339
  test site, 353
Soil leaching, points of inspection, 167
Soil sampling, runoff study, 353
Solubility, test substances and reference
    standards, 118
SOP, See Standard operating procedure
Specific activity
  metabolism studies, 259
  problems, 259–261
Specific instrument sheet, equipment
    SOPs, 34
Specimen chain of custody, field studies,
    85–93
Sponsor responsibilities, processing
    facilities, 251–252
Spray drift, mesocosm study, 302–303
Sprayer specification sheet, 282f
Stability
  documentation, 120
  test substances and reference standards,
    118,119
Standard Evaluation Procedures,
    138,229–230
Standard operating procedure (SOP)
  authorship, 29
  custodian's job responsibilities, 35
  definition, purpose, and value, 28–29
  distribution, 31–32
  field studies, 21–26,237–238
  flexibility, 338–340
  format, 30–31
  functional categories, 32–35,37–41
  groundwater monitoring field study,
    336–337
  guidelines for writing, 27–31
  inherent challenges, 35

Standard operating procedure (SOP)—
    Continued
  input from technical personnel, 29–30
  management approval, 30
  managing an effective system, 31–36
  need for SOP, 29
  negative aspects, 29
  nonroutine data management, 239–240
  program requirements, 36
  review and revision, 31
  software development, 218–219
  studies done at multiple sites by
    various scientists, 22–23
  terrestrial field studies, 313–314
  validation program, 205
  volume reduction techniques, 33–34
Standard storage log, 100,101f
Statistical analysis
  mesocosm study, 303–304
  source of disagreement, 303
Storage
  conditions, test substances and analytical
    reference standards, 115–116
  empty containers, 120–121,285,315,330
  quality of test substance, 330
  raw data, 328–329
  stability in laboratory, 232,286
Storage containers
  documentation, 120
  test substances and reference standards, 118
Storage facilities
  archiving, 158
  field sites, 160
Storage stability data guideline,
    product-chemistry package, 232
Study dates, statement in protocol, 48–49
Study design, statement in protocol, 49–50
Study director
  documentation, 320
  ecotoxicological field study, 319–321
  mesocosm study, 305
  more than one per study, 319–320
  procedural SOPs, 39
  responsibilities, 319
Study identification number, protocol, 59
Study notebook, field functions, 286–287
Study personnel, protocol, 59
Study protocol
  planning and communication tool, 43,54
  purpose and objectives, 60

Study protocol—*Continued*
  quality assurance unit, 315
  reconstruction of the study, 54
Study report, contents, 136
Study size, metabolism identification,
  263–265
Study termination, runoff study, 355
Study title, 58,66
Submittal package, studies in support of
  single regulatory action, 132–136
Surface runoff, mesocosm study, 302–303
Systematic sampling
  bias, 241–242
  field residue trials, 246–248

T

Table of contents, recommended, 137*t*
Tank mixing
  analysis, 122–123
  data audit, 178
Technical Guidance Document, 138
Terrestrial dissipation study
  randomized sampling design, 243*f*,244*f*
  sampling strategy, 242–246
  tractor-mounted soil probes, 245*f*
Terrestrial field studies, 309–316
  components, 311–313
  determination of need, 311
  objectives and techniques, 310
  screening studies and definitive studies, 311
Test chemicals, field studies, 238
Test, control, and reference substances,
  statement in protocol, 49
Test equipment, in study report, 136
Test facility manager, responsibilities, 321
Test material, treatment in mesocosm
  study, 302–303
Test methods
  general, 41
  laboratory SOP, 34
  product chemistry, 34–35,41–42
Test sites
  current agricultural practices, 284
  location and size, 283–284
Test subjects, informed consent, 287
Test substance
  chain of custody, 85–93
  characterization and accountability, 113–126
  definition, 113

Test substance—*Continued*
  documentation, 284–285,330
  example, 70
  GLP problems, 372–373
  handling, 92,114–116
  homogeneity, 284
  identification, data audit, 187
  mixed with carriers, 121–123
  parameters, 114
  proprietary information, 365
  provision, 364–365
  questions asked by EPA auditors, 365
  reporting format, 129*f*
  responsibilities, 114,315
  shipment problems and documentation,
    86–88
  storage and handling, 315
  use and tracking, 88–89
Test system
  definition, 231
  examples, 72–74
  justification, 60–61,75
  procedural SOPs, 37
  protocol, 60
Test workers, agricultural practices, 289
Testing facilities
  equipment SOPs, 33–34
  fraudulent submissions, 389
  GLP problems, 372
  prestudy evaluations, 156
  sufficiency, 157
Third-party vendors, software
  validation, 221
Timeliness, inspection system, 379
Tissue storage stability studies, residue
  method, 271
Title page, sample, 133*f*
Toxic Substances Control Act (TSCA), GLP
  Standards, 28
Toxicological research, Food and Drug
  Administration and GLP regulations, 6
Toxicological testing
  industrial chemicals in Japan, 429
  laboratories under GLP Standards, 155
Toxicology Laboratory Accreditation
  Board, 381–382
Traceability of raw data, data audit, 184–185
Tractor, soil sampling, 243–246
Training, field personnel, 324,337–338
Training course, GLP inspections, 422

Treatment program, statement in protocol, 49–50
"Two customers" concept, study protocol, 54–55

U

Units of measurement, in study report, 136

V

Validation
computer systems, 203–216
consistency, 205
data stored on diskette after archiving, 215
definition, 218
development, 204
in-house data base system, 205–207
laboratory computer software, 217–223
link between scintillation counter and data-collection device, 207
planning process, 210
process, 218–221
quality assurance unit, 213–214
SOP for software, 220–221
standard operating procedure, 205

Verification, data base system, 207
Violations
case-by-case basis, 391
notices of warning, 396
penalties, 396

W

Water chemistry, ecological monitoring, 300
Weather and schedule, protocol flexibility, 323,340–341
Weather data, field test, 288
Wildlife food and water sources, residue analysis, 312
Wildlife mortality, terrestrial field studies, 311–312
Worker-exposure studies
agricultural field studies, 279–295
QA inspection checklist, 291f–292f
study design, 280–285
Working archive
data audit, 183
study data package, 188–189

Copy editing: Paula M. Bérard and Julie Poudrier Skinner
Production: Margaret J. Brown and Donna Lucas
Indexing: Paula M. Bérard and Colleen P. Stamm
Cover design: Jack Ballestero
Acquisition: Cheryl Shanks

Books printed and bound by United Book Press, Inc., Baltimore, MD

The paper used in this publication meets the minimum requirements of American National Standard for Information Sciences—Permanence of Paper for Printed Library Materials, ANSI Z39.48–1984. ∞

Bestsellers from ACS Books

The ACS Style Guide: A Manual for Authors and Editors
Edited by Janet S. Dodd
264 pp; clothbound, ISBN 0–8412–0917–0; paperback, ISBN 0–8412–0943–X

Chemical Activities and Chemical Activities: Teacher Edition
By Christie L. Borgford and Lee R. Summerlin
330 pp; spiralbound, ISBN 0–8412–1417–4; teacher ed. ISBN 0–8412–1416–6

Chemical Demonstrations: A Sourcebook for Teachers,
Volumes 1 and 2, Second Edition
Volume 1 by Lee R. Summerlin and James L. Ealy, Jr.;
Vol. 1, 198 pp; spiralbound, ISBN 0–8412–1481–6;
Volume 2 by Lee R. Summerlin, Christie L. Borgford, and Julie B. Ealy
Vol. 2, 234 pp; spiralbound, ISBN 0–8412–1535–9

Writing the Laboratory Notebook
By Howard M. Kanare
145 pp; clothbound, ISBN 0–8412–0906–5; paperback, ISBN 0–8412–0933–2

Developing a Chemical Hygiene Plan
By Jay A. Young, Warren K. Kingsley, and George H. Wahl, Jr.
paperback, ISBN 0–8412–1876–5

Introduction to Microwave Sample Preparation: Theory and Practice
Edited by H. M. Kingston and Lois B. Jassie
263 pp; clothbound, ISBN 0–8412–1450–6

Principles of Environmental Sampling
Edited by Lawrence H. Keith
ACS Professional Reference Book; 458 pp;
clothbound; ISBN 0–8412–1173–6; paperback, ISBN 0–8412–1437–9

Biotechnology and Materials Science: Chemistry for the Future
Edited by Mary L. Good (Jacqueline K. Barton, Associate Editor)
135 pp; clothbound, ISBN 0–8412–1472–7; paperback, ISBN 0–8412–1473–5

Personal Computers for Scientists: A Byte at a Time
By Glenn I. Ouchi
276 pp; clothbound, ISBN 0–8412–1000–4; paperback, ISBN 0–8412–1001–2

Polymers in Aqueous Media: Performance Through Association
Edited by J. Edward Glass
Advances in Chemistry Series 223; 575 pp;
clothbound, ISBN 0–8412–1548–0

For further information and a free catalog of ACS books, contact:
American Chemical Society
Distribution Office, Department 225
1155 16th Street, NW, Washington, DC 20036
Telephone 800–227–5558